Semiconductor Lasers

Past, Present, and Future

AIP Series in
Theoretical and Applied Optics

Semiconductor Lasers

Past, Present, and Future

Editor

Govind P. Agrawal

The Institute of Optics
University of Rochester
Rochester, New York

American Institute of Physics   **Woodbury, New York**

AIP Press
American Institute of Physics
500 Sunnyside Boulevard
Woodbury, NY 11797-2999

Library of Congress Cataloging-in-Publication Data
Agrawal, Govind P.
    Semiconductor lasers : past, present, and future /
    Govind P. Agrawal, editor.
        p. cm.
    Includes bibliographical references and index.
    ISBN 1-56396-211-X
    1. Semiconductor lasers. I. Agrawal, G. P. (Govind P.), 1951–.
TA1700.S46 1994   94-39171
621.36'6 --dc20   CIP

10 9 8 7 6 5 4 3 2 1

Contents

List of Contributors

AGRAWAL, GOVIND P., The Institute of Optics, University of Rochester, Rochester, New York 14627

BOWERS, JOHN E., Department of Electrical and Computer Engineering, University of California, Santa Barbara, California 93106

CHANG-HASNAIN, CONNIE J., Ginzton Laboratory, Electrical Engineering Department, Stanford University, Stanford, California 94305

CHINONE, NAOKI, Optoelectronics Research Center, Hitachi Central Research Laboratory, Kokobunji, Tokyo 185, Japan

COLEMAN, JAMES J., Department of Electrical Engineering, University of Illinois, Urbana, Illinois 61802

DUAN, GUNG-HUA, Communication Department, Ecole Nationale Superieure des Telecommunications, 46 rue Barrault, 75013 Paris Cedex, France

GRAY, GEORGE R., Electrical Engineering Department, University of Utah, Salt Lake City, Utah 84112

GUNSHOR, ROBERT L., School of Electrical Engineering, Purdue University, West Lafayette, Indiana 47907

HATAKOSHI, GEN-ICHI, Research and Development Center, Toshiba Corporation 1, Komukai Toshiba-Cho, Saiwai-ku, Kawasaki 210, Japan

HELKEY, ROGER J., Massachusetts Institute of Technology Lincoln Library, Lexington, Massachusetts 02173

KOUROGI, M., Tokyo Institute of Technology, 4259 Nagatsuta, Midori-Ku, Yokohama 227, Japan

NURMIKKO, ARTO V., Division of Engineering, Brown University, Providence, Rhode Island 02912

OHTSU, MOTOICHI, Tokyo Institute of Technology, 4259 Nagatsuta, Midori-Ku, Yokohama 227, Japan

OKAI, MAKOTO, Optoelectronics Research Center, Hitachi Central Research Laboratory, Kokobunji, Tokyo 185, Japan

Series Preface

Vasco Ronchi, in his historical survey *The Nature of Light*, notes that the opening verses of *Genesis* in the Old Testament give to light "an absolute precedence of Creation" and, thereby, an implicit theory that "attributes to light an existence of its own, independent of its source and of its receiver."* Even in pre-civilization times, men and women almost certainly pondered the nature of light as well as its relationship to source and receiver. Indeed, optics, the science of light, is considered by many to be the first of the sciences.

As both a science and as a technology, optics has been characterized by many periods of intense activity brought on by new developments in both theory and application. The major developments of the corpuscular, wave, and quantum theories of light with which most students of optics are familiar have influenced science as a whole as well as refined our understanding of the nature, control, and uses of light. The many applications of optics, ranging from the kaleidoscope to eye glasses to scientific instruments like the microscope, from photography to the laser and fiber optic communication system components, have all had tremendous impact on our day-to-day living and well-being. Optics is a rich and varied subject, and each decade sees exciting new developments. The latter half of the 20th century alone has witnessed the development of the laser, holography, optical disk storage, and fiber optics; of optically-nonlinear polymers, the concept of squeezed-state light, and ultra-short optical pulses and their applications; of computer-generated diffractive optical elements, liquid crystal displays, and a host of other enormously important inventions that impact our lives in a myriad of ways.

The *AIP Series in Theoretical and Applied Optics* is intended to provide the scientist, the engineer, and, at least occasionally, the interested layman with contemporary and authoritative monographs, graduate-level textbooks, and reference books on optics. The Series will address changing reference and textbook needs in classical areas of optics like physical and quantum optics, sources and detectors, and the optical properties of materials, but it will also provide up-to-date monographs in emerging areas of importance like ultra-fast optics, nonlinear optical interactions and applications, and the computer-aided fabrication and application of diffractive structures. Series books are intended to meet the high expectations and standards of

*Vasco Ronchi, *The Nature of Light* (Harvard University Press, Cambridge, 1970; originally published in Italian as *Storia della Luce,* 1939), p. 2.

American Institute of Physics member Societies. In short, we expect Series books to be both good and widely distributed, to the great benefit of the optics community.

William T. Rhodes
Series Editor-in-Chief

Preface

During the past decade semiconductor lasers have had tremendous impact on society, often in ways invisible to the average person. For example, long-distance telephone conversations are now almost always transmitted to their destination over optical fibers by means of light emitted from semiconductor lasers. In another example, wide-scale usage of compact-disk technology has guaranteed the presence of a semiconductor laser in most households in the interior of CD players capable of playing high-quality music. Similarly, most offices contain semiconductor lasers inside laser printers or inside computers containing CD-ROM (read only memory) disk drives. Such large-scale commercialization of semiconductor lasers would not have been possible without significant advances, brought about by extensive research and development, in our understanding of these devices. Indeed, the field of semiconductor lasers has grown tremendously in many different ways. Examples of areas of major development over the past five to ten years include strained quantum-well lasers, narrow-linewidth semiconductor lasers, multiwatt semiconductor laser arrays, semiconductor lasers operating in the visible region, surface-emitting semiconductor lasers, and semiconductor lasers capable of emitting subpicosecond optical pulses.

Because of rapid expansion of the field of semiconductor lasers, most recent research results are not yet available in a book form. Although several monographs and edited volumes exist, they are either significantly outdated or they cover only narrow aspects of semiconductor lasers. As a result, researchers and graduate students are often forced to consult widely scattered research material even when their objective is simply to get acquainted with a specific topic.

This book is intended to fill a major gap by providing in a single volume ten up-to-date review articles written in a pedagogical manner by well-known experts. The topics were selected to cover the entire range of current activity in the field of semiconductor lasers. The last two chapters of the book are devoted to applications of semiconductor lasers and are intended to provide the reader with a perspective on how the research advances described in earlier chapters eventually translate into commercial products.

A multiauthored book is a joint effort that requires considerable cooperation among its contributors. I am most thankful for the efforts of the contributors to this book for their initial acceptance of their responsibilities and for their producing well-balanced, up-to-date chapters in spite of their busy schedules. The efforts of the editor and the contributors will prove successful if this book is found useful by its intended readers.

Govind P. Agrawal
Rochester, New York

Introduction

This book provides an overview of the current research activities in the field of semiconductor lasers. Although semiconductor lasers developed slowly compared with other lasers, they assumed paramount importance in the 1980s through their increasing presence in compact-disk (CD) players, laser printers, and fiber-optic transmitters used for audio and video communications. This introductory chapter discusses briefly the past history, the present status, and the future prospects of these singularly important devices.

HISTORICAL PERSPECTIVE

The history of semiconductor lasers spans three decades. Although the feasibility of stimulated emission in semiconductors was considered earlier,[1-3] it was only in 1962 that the observation of lasing action in semiconductors was made almost simultaneously by four research groups.[4-7] The new laser device consisted of a GaAs p-n junction whose facets, perpendicular to the junction, were polished to form a laser cavity. These first GaAs lasers lased at 0.88 μm. Soon p-n junctions of other semiconductor materials, such as GaAsP, GaInAs, InAs, and InP were used to provide semiconductor lasers that could lase at different wavelengths. The practical utility of these earlier devices was limited since a high threshold current density prevented their continuous-wave (cw) operation at room temperature.

As early as 1963 it was suggested[8,9] that semiconductor lasers would be improved considerably by a heterostructure design in which the active layer of one semiconductor material is sandwiched between two cladding layers of another semiconductor that has a wider band gap. The concept could not be demonstrated until after the technique of liquid-phase epitaxy[10] (LPE) was perfected and used to grow such heterostructure devices. It was only in 1969 that the successful room-temperature operation of a pulsed semiconductor laser was demonstrated using the LPE technique.[11-13] Further work led, in 1970, to the cw operation of semiconductor lasers at room temperature.[14,15] At about the same time, low-loss silica fibers became available.[16] The development of GaAs lasers was fueled by the potential application of such lasers in fiber-optic communication systems. Indeed, GaAs lasers were used in the first generation of lightwave telecommunication systems starting in 1978.[17] Two books on semiconductor lasers published during the late 1970s attest to the rapid progress in this field.[18,19]

By 1979 the loss of silica fibers had been reduced to 0.2 dB/km.[20] However, silica fibers exhibit low dispersion and low loss in the wavelength regions located near 1.3 and 1.55 μm. Thus, during the 1980s the attention shifted toward the development of InP-based semiconductor lasers, since such lasers can operate over the wavelength range 1.1–1.6 μm if the composition of the quaternary InGaAsP material used for the active layer is properly chosen. Such lasers were used in fiber-optic communication systems throughout the 1980s. The same application fueled the development of distributed feedback (DFB) semiconductor lasers, which became commercially available by 1989. Several books and review articles published in the late 1980s cover the rapid progress made in the field of semiconductor lasers.[21–25]

The advent of quantum-well semiconductor lasers led to further improvement in device performance. Although quantum-well lasers were demonstrated[26] as early as 1978, their use did not become common until the advanced epitaxial growth techniques such as metal-organic chemical vapor deposition (MOCVD) and molecular beam epitaxy (MBE) became available in the late 1980s. The active region of such lasers consists of a single or multiple layers, each so thin (~ 10 nm or less) that quantum-confinement effects become important. Such thin layers are referred to as quantum wells. The quantum-confinement effects modify the density of states substantially and improve device performance. Further improvements occurred when, in the late 1980s and the early 1990s, the idea of strained quantum wells was put to use in the design of semiconductor lasers.[27–33] The quantum-well structure is now routinely used in the design of semiconductor lasers.[31] Strained-layer quantum wells have been used to generate laser output at new wavelengths. For example, semiconductor lasers operating at 0.98 μm and 1.48 μm wavelengths have been developed to meet the pumping requirements of erbium-doped fiber amplifiers. Such lasers became available commercially by 1991.

PRESENT STATUS

To meet the demands of different applications, developmental efforts have focused on several fronts during the last five years or so. An indication of the current activities in the field of semiconductor lasers can be obtained by consulting the special June issue of the IEEE Journal of Quantum Electronics, which is published every two years following the International Conference on Semiconductor Lasers. The special issue published in June 1993 features multiple papers on: (i) strained-layer quantum-well lasers, (ii) laser modulation and short-pulse generation, (iii) single-frequency (DFB) lasers, (iv) tunable lasers, (v) visible-wavelength lasers, (vi) high-power lasers, (vii) vertical-cavity lasers, and (viii) laser amplifiers. It is not surprising that the contents of this book largely mirror these topics. Progress in the design and manufacture of quantum-well and DFB semiconductor lasers is being made continuously. Surface-emitting lasers have attracted attention because of their potential applications in opto-electronic integrated circuits. Visible semiconductor lasers are being developed for applications related to optical data recording and laser printing. Red diode lasers are already available commercially and are replacing He–Ne lasers in many applications. The development of blue semiconductor lasers is progressing rapidly. High-power semiconductor lasers are needed for applications such as pumping solid-state lasers, and new power output records are being set regularly. Other research directions include making tunable narrow-linewidth lasers and designing semiconductor lasers capable of producing ultrashort optical pulses. This book attempts to cover as many topics as possible within realistic space limitations. The following is an overview of the coverage offered by the book.

Chapter 1 discusses the operation of quantum-well lasers by considering the fundamental as well as practical aspects. The physics of quantum-well lasers is first presented through the solution to the Schrödinger equation for a single quantum well. Fundamental concepts such as the density of states, the optical gain, and the threshold current density are then introduced. Particular attention is paid to strained-layer quantum wells because of their current importance. The issue of critical thickness of the quantum well is addressed, as are the polarization properties of the emitted light.

Chapter 2 is devoted to DFB semiconductor lasers. It considers theory, fabrication, and performance of different kinds of DFB lasers (such as quarter-wave-shifted and gain-coupled). The performance of DFB lasers as they relate to fiber-optic communication systems is discussed through consideration of spectral and dynamic characteristics such as the laser linewidth and the modulation-induced frequency chirp. Recent progress in the field of DFB lasers is addressed in a separate section on high-speed, narrow-linewidth, wavelength tunable, surface-emitting, and integrated DFB lasers.

Some applications of semiconductor lasers require extreme control of phase noise and linewidth. Several phase-control techniques are described in Chapter 3. These techniques can reduce low-frequency phase noise to such an extent that the spectral linewidth of the laser becomes less than 100 Hz. The chapter also discusses the possibility of a semiconductor laser serving as an accurate, widely tunable, narrow-linewidth source that can be used as an optical frequency synthesizer similar to those available in the microwave domain.

Chapter 4 deals with semiconductor lasers capable of generating ultrashort optical pulses. Both active and passive mode-locking techniques are discussed. Particular emphasis is placed on the role of the nonlinear phenomenon of self-phase modulation, which can limit the pulse width by imposing a frequency chirp. Timing jitter and the phase-noise issue are addressed in a separate section because of their importance for mode-locked semiconductor lasers.

Surface-emitting semiconductor lasers are covered in Chapter 5. These lasers have been developed to meet the requirements of specific applications, such as optical interconnects and optical computing, which require one- or two-dimensional arrays of low-threshold lasers. Although in development throughout the 1980s, the technology has matured only recently to the point where such lasers are now available commercially. Chapter 5 addresses both fundamental and applied issues associated with surface-emitting semiconductor lasers.

Visible light semiconductor lasers are covered in Chapter 6. There has been a continuing trend to shift the wavelength of semiconductor lasers from the infrared to the visible region. There are many potential applications of visible light lasers. For example, semiconductor lasers operating near 630 nm can replace He–Ne lasers in most applications where He–Ne lasers are commonly used. The capacity of optical recording systems increases as λ^{-2} simply because the spot size of a focused optical beam is ultimately limited by its wavelength. Many techniques have been used to reduce the output wavelength for semiconductor lasers based on III–V materials. This chapter reviews these techniques and describes the performance levels achieved so far.

Chapter 7 is devoted to a new technology that uses II–VI materials to fabricate semiconductor lasers that operate in the blue-green region of the optical spectrum. The field is still in its infancy, and new innovations are continuously being made. The chapter attempts to review the progress made up to the fall of 1994 in this rapidly evolving area. Both basic and applied aspects are covered. Specifically, optical and transport phenomena in quantum-well heterostructures made with II–VI semiconductor materials is discussed. The performance of the current state-of-the-art light-emitting diodes and semiconductor lasers operating in the blue-green region is also described.

Chapter 8 considers semiconductor lasers whose design is modified to suppress optical feedback from the cleaved facets so that they can act as optical amplifiers. Such optical amplifiers have a large number of potential applications and were extensively studied during the 1980s. The chapter first discusses the basic concepts common to all amplifiers and then focuses on the design and the performance of semiconductor laser amplifiers. Particular attention is paid to the case in which ultrashort optical pulses are amplified, since several interesting nonlinear effects can affect both the temporal and spectral properties of such pulses.

The applications of semiconductor lasers are covered in Chapters 9 and 10. Chapter 9 focuses on applications in the fields of optical communications and optical data storage. It also covers other applications such as laser printing, medicine, and high-resolution spectroscopy. Applications of semiconductor lasers and amplifiers in the field of optical switching are covered separately in Chapter 10. Even though such applications are not as well developed as those in the fields of optical communications and optical data storage, they are attracting increasing attention because of their relevance to the important information-age technologies of 1990s.

FUTURE PROSPECTS

The continuous development of semiconductor-laser technology has led to a wide variety of applications in such diverse areas as fiber-optic communications, consumer electronics, and optical data recording. The wide-scale commercial success of the compact-disk and laser-printer technologies has guaranteed the presence of a semiconductor laser in most households and offices. With the presence of semiconductor lasers already so common, one may wonder what impact semiconductor lasers will have in the future. Interestingly enough, there remain ample applications of semiconductor lasers yet to be perfected. For example, advances in local-area-network (LAN) technology may eventually bring a semiconductor laser to each household (or at least to the curb side), where it will be used to select among a vast range of audio and video services. The further development of the Internet will only accelerate this particular application of semiconductor lasers. The development of visible semiconductor lasers, operating in the blue-green part of the optical spectrum, should prove a boom to the optical data recording industry since such lasers would eventually permit storage of several tens of billions of bits on a single compact disk.

Surface-emitting semiconductor lasers, by far, are likely to make the most impact in the future. These devices are just being perfected. They can be mass produced on a large scale since each wafer provides millions of tiny semiconductor lasers. Arrays of individually addressable surface-emitting lasers have a host of applications that have remained largely unexplored.[34] A few among these are fiber-optic networks, digital printing and scanning, optical data storage, projection laser display, optical interconnects, and photonic switching. The development of smart-pixel devices is another area where surface-emitting lasers should find applications. In general, optoelectronic integrated circuits, in which optical components such as semiconductor lasers and amplifiers are integrated with electronic components, are likely to provide new devices whose functionality at this stage can only be imagined. It is impossible to predict precisely what the future holds, but certainly semiconductor lasers will be an important part of the information revolution currently underway.

REFERENCES

1. N. G. Basov, O. N. Krokhin, and Yu. M. Popov, Sov. Phys. JETP **13**, 1320 (1961).
2. M. G. A. Bernard and G. Duraffourg, Phys. Status Solidi **1**, 699 (1961).
3. W. P. Dumke, Phys. Rev. **127**, 1559 (1962).
4. R. N. Hall, G. E. Fenner, J. D. Kingsley, T. J. Soltys, and R. O. Carlson, Phys. Rev. Lett. **9**, 366 (1962).
5. M. I. Nathan, W. P. Dumke, G. Burns, F. H. Dill, Jr., and G. Lasher, Appl. Phys. Lett. **1**, 62 (1962).
6. T. M. Quist, R. H. Rediker, R. J. Keyes, W. E. Krag, B. Lax, A. L. McWhorter, and H. J. Zeiger, Appl. Phys. Lett. **1**, 91 (1962).
7. N. Holonyak, Jr., and S. F. Bevacqua, Appl. Phys. Lett. **1**, 82 (1962).
8. H. Kroemer, Proc. IEEE **51**, 1782 (1963).
9. Zh. I. Alferov and R. F. Kazarinov, Authors Certificate 181737 (USSR), 1963.
10. H. Nelson, RCA Rev. **24**, 603 (1963).
11. H. Kressel and H. Nelson, RCA Rev. **30**, 106 (1969).
12. I. Hayashi, M. B. Panish, and P. W. Foy, IEEE J. Quantum Electron. **QE-5**, 211 (1969).
13. Zh. I. Alferov, V. M. Andreev, E. L. Portnoi, and M. K. Trukan, Fiz. Tekh. Poluprovodn. **3**, 1328 (1969) [Sov. Phys. Semicond., 1107 (1970)].
14. I. Hayashi, M. B. Panish, P. W. Foy, and S. Sumuski, Appl. Phys. Lett. **17**, 109 (1970).
15. Zh. I. Alferov, V. M. Andreev, D. Z. Garbuzov, Yu. V. Zhilyaev, E. P. Morozov, E. L. Portnoi, and V. G. Trofim, Fiz. Tekh. Poluprovodn. **4**, 1826 (1970) [Sov. Phys. Semicond. **4**, 1573 (1971)].
16. F. P. Kapron, D. B. Keck, and R. D. Maurer, Appl. Phys. Lett. **17**, 423 (1970).
17. T. C. Canon, D. L. Pope, and D. D. Sell, IEEE Trans. Commun. **COM-26**, 1045 (1978).

18. H. Kressel and J. K. Butler, *Semiconductor Lasers and Heterojunction LEDs* (Academic Press, New York, 1977).
19. H. C. Casey, Jr. and M. B. Panish, *Heterostructure Lasers*, Parts A and B (Academic Press, New York, 1978).
20. T. Miya, Y. Terunuma, T. Osaka, and T. Miyoshita, Electron. Lett. **15**, 106 (1979).
21. G. P. Agrawal and N. K. Dutta, *Long-Wavelength Semiconductor Lasers* (Van Nostrand Reinhold, New York, 1986).
22. Special issue on semiconductor lasers, IEEE J. Quantum Electron. **QE-23**, June 1987.
23. S. E. Miller and I. P. Kaminow, Eds., *Optical Fiber Telecommunications II* (Academic Press, San Diego, 1988).
24. G. P. Agrawal, Chap. 3 in *Progress in Optics*, vol. 26, ed. by E. Wolf (North-Holland, Amsterdam, 1989).
25. Special issue on semiconductor lasers, IEEE J. Quantum Electron. **25**, June 1989.
26. R. D. Dupuis, P. D. Dapkus, N. Holonyak, Jr., E. A. Rezek, and R. Chin, Appl. Phys. Lett. **32**, 295 (1978).
27. E. Yablonovitch and E. O. Kane, J. Lightwave Technol. **LT-4**, 504 (1986); **6**, 1292 (1988).
28. D. P. Bour, D. B. Gilbert, L. Elbaum, and M. G. Harvey, Appl. Phys. Lett. **53**, 2371 (1988).
29. P. J. A. Thijs and T. van Dongen, Electron. Lett. **25**, 1735 (1989).
30. Special issue on semiconductor lasers, IEEE J. Quantum Electron. **27**, June 1991.
31. P. S. Zory, Ed., Quantum-Well Lasers (Academic Press, San Diego, 1993).
32. Special issue on semiconductor lasers, IEEE J. Quantum Electron. **29**, June 1993.
33. G. P. Agrawal and N. K. Dutta, *Semiconductor Lasers*, 2nd ed. (Van Nostrand Reinhold, New York, 1993).
34. J. Jewell and G. Olbright, Opt. Photon. News (March issue) **5**, 9 (1994).

CHAPTER 1

Quantum-Well Heterostructure Lasers

James J. Coleman

*Microelectronics Laboratory and Materials Research Laboratory,
University of Illinois, Urbana, Illinois*

1.1. INTRODUCTION

The semiconductor quantum-well heterostructure laser is as interesting for its fundamental physical properties as it is important for a wide range of commercial applications such as in fiber optic telecommunications networks and compact-disk players. The advantages associated with the quantum-size effect in a semiconductor laser are such that the quantum-well heterostructure laser has superseded the conventional double heterostructure laser in virtually every application. The semiconductor injection laser has a venerable history of more than 30 years, a history which has been punctuated by a number of key advances in the materials of choice, the methods for preparing these materials, and the structures that can be made from these materials. The early history of the injection laser, from the initial reports of GaAs homojunction laser operation through the demonstration of room temperature cw operation in $Al_xGa_{1-x}As$–GaAs double heterostructure lasers, is described nicely in the introductory chapter of *Heterostructure Lasers*[1] by Casey and Panish. Additional detail is provided by the historical retrospectives published in a special issue[2] of the *Journal of Quantum Electronics* on semiconductor lasers and selected reprints[3–5] of many of the key published papers have been assembled.

The quantum-well heterostructure laser traces its origin[6] to original work by researchers at IBM and Bell Laboratories. In 1970, Esaki and Tsu reported[7] the first of a series of studies on the energy band structure and transport properties of superlattices comprised of layers thin enough (100 Å) to be of the order of an electron wavelength. In 1974, Dingle *et al.* reported[8] on the optical absorption properties of thin layer superlattices. These data were remarkable in that they clearly showed[8] the step-like density of states associated with the quantum-size effect, including the splitting associated with the separation of the light-hole and heavy-hole valence bands. The incorporation of thin quantum wells into semiconductor injection lasers required advances in the art of growing epitaxial layers of III–V compound semiconductors. A suitable growth method must certainly provide high optical quality material but also controlled growth of very thin layers and abrupt heterostructure interfaces. In the 1970s two very different epitaxial growth methods, molecular beam epitaxy (MBE)[9,10] and metalorganic chemical vapor deposition (MOCVD),[11] were developed to the extent that high-quality semiconductor injection lasers could be grown and in 1978 the first demonstration of a room temperature quantum-well heterostructure injection laser was reported.[12]

Since that time the growth processes for quantum-well heterostructure lasers have steadily improved, the range of materials and wavelengths available for quantum-well heterostructure lasers has been extended, more complex single quantum-well and multiple quantum-well structures have been developed, and new concepts—such as the elastic accommodation of strain in thin layers—have been incorporated into quantum-well heterostructure lasers. In this chapter we introduce the basic concept of the quantum-size effect and describe the two-dimensional density of states in the frame of reference of bulk semiconductor materials. We describe the relationship between the two-dimensional density of states and the large reduction in the transparency current-density with decreasing well thickness in quantum-well lasers. We will compare optical gain in bulk and quantum-well heterostructure lasers and show the relationship of increased gain and reduced transparency current-density to the threshold current-density in quantum-well lasers. We outline the effects of elastically accommodated strain on the valence-band structure of strained layer quantum-well heterostructure lasers, including the metallurgical effects of critical thickness, and we describe the effects of strain on laser threshold current-density, optical gain, and polarization in strained-layer quantum-well heterostructure lasers. Finally, we take a brief look at the future for quantum-well heterostructure injection lasers.

1.2. THE QUANTUM SIZE EFFECT

The quantum-well heterostructure laser can be considered, to a very good approximation at least in the conduction band, as a physical manifestation of the venerable quantum mechanics[13] problem of a charged particle in a finite potential well. The general solution to the time-independent Schrödinger equation inside the quantum well $(-L_z/2 < z < L_z/2)$ is a superposition of standing waves given by

$$\Psi(z) = A \sin k_1 z + B \cos k_1 z, \tag{1.1}$$

where k_1 is the magnitude of the momentum vector in the well. Only the z-direction need be considered since there is no quantization in the x- and y-directions. The energy E for electrons is referenced to the conduction-band edge (or the valence-band edge for holes) and approximately[14] related to k_1 by

$$(k_1)^2 = \frac{2m_e E}{\hbar^2}, \tag{1.2}$$

where $\hbar$ is Planck's constant and m_e is the effective mass of the electron. Outside the well the wave functions must vanish at $\pm\infty$ so the solutions are given by

$$\Psi(z) = C \exp(k_2 z) \quad \text{for } z < -L_z/2, \tag{1.3}$$

$$\Psi(z) = D \exp(-k_2 z) \quad \text{for } z > L_z/2, \tag{1.4}$$

where

$$(k_2)^2 = \frac{2m_e(\Delta E_c - E)}{\hbar^2}. \tag{1.5}$$

In Eq. (1.5), the term ΔE_c is the height of the potential barrier and is the heterostructure discontinuity for quantum-well heterostructure lasers. The difference between the band-gap energies of the quantum well and the barrier layers is apportioned between the valence-band (ΔE_v) and the conduction-band (ΔE_c) heterostructure discontinuities. For $Al_x Ga_{1-x}As$–GaAs heterostructures, ΔE_v is $\sim 40\%$[15] of the difference between the direct (Γ) band edges for any $Al_x Ga_{1-x}As$ composition, i.e.

$$\Delta E_v = 0.40(E_\Gamma^{AlGaAs} - E_\Gamma^{GaAs}). \tag{1.6}$$

The conduction-band discontinuity ΔE_c is given by

$$\Delta E_c = (E_g^{AlGaAs} - E_\Gamma^{GaAs}) - \Delta E_v, \tag{1.7}$$

where E_g^{AlGaAs} is either the Γ band edge energy for direct gap $Al_x Ga_{1-x}As$ barriers or the X band edge energy for indirect gap $Al_x Ga_{1-x}As$ barriers. After applying the usual boundary conditions that the eigenfunctions and their derivatives must be continuous at $\pm L_z/2$, we arrive at two possible solutions:

$$k_1 \tan \frac{k_1 L_z}{2} = k_2 \quad \text{(even solutions)}, \tag{1.8}$$

$$k_1 \cot \frac{k_1 L_z}{2} = -k_2 \quad \text{(odd solutions)}. \tag{1.9}$$

The eigenfunctions and eigenenergies for the first two conduction-band electron quantum states are shown schematically in Fig. 1.1. By convention the lowest energy quantum state is the $n = 1$ or n_1 state.

The discussion above assumes a simple parabolic conduction band with a constant effective mass [Eq. (1.02)]. This is a reasonable enough assumption for the conduction band in conventional III-V compound semiconductor quantum-well heterostructure lasers in the normal range of operation. The valence band in GaAs[16] and other III-V heterostructure materials[17,18] is much more complicated. The degen-

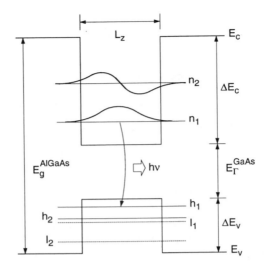

FIGURE 1.1. The eigenfunctions and eigenenergies for the first two conduction-band electron quantum states in an Al_xGa_{1-x}As-GaAs quantum-well heterostructure. By convention the lowest energy quantum state is the $n = 1$ or n_1 state. Also shown are the first two light-hole and heavy-hole valence-band eigenenergies.

eracy of the heavy-hole and light-hole valence bands and the resulting valence-band mixing, and perhaps also the presence of the split-off valence band, must be taken into account in calculating the eigenstates in the valence band and, as we discuss later, optical gain in quantum well heterostructures. The details of such calculations are beyond the scope of this chapter. Suffice it to say that they result in two sets of quantum states for holes—those corresponding to the light-hole effective mass (l_1, l_2, etc.) and those corresponding to the heavy-hole effective mass (h_1, h_2, etc.). Figure 1.2 shows an example of the calculated confined state energy E, which is the energy separation between the quantum state and the bulk band edge, as a function of quantum well width (L_z) for a realistic quantum-well heterostructure having a GaAs quantum well and $Al_{0.20}Ga_{0.80}$As barrier layers. For simplicity, the bulk GaAs band-gap energy is not shown. Positive energy corresponds to the electron energy above the bulk GaAs conduction-band edge and negative energy corresponds to hole energy below the bulk GaAs valence-band edge. From Fig. 1.2 it is apparent that well widths of less than ~ 150 Å are necessary in order to observe the quantum-size effect at room temperature ($kT = 0.026$ eV at 300 K) in these structures.

The normal k-selection rules are generally assumed to apply to recombination in quantum-well heterostructure lasers. This is not strictly true because of valence-band mixing effects and the finite barrier heights in these structures. However, it is apparent from Fig. 1.2 that, based on energy arguments alone, the dominant transition will be the allowed $n_1 \rightarrow h_1$ transition. Later, when we also consider the

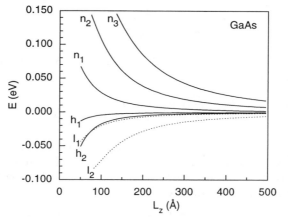

FIGURE 1.2. The confined state energy E as a function of quantum well width (L_z) for a realistic quantum-well heterostructure having a GaAs quantum well and $Al_{0.20}Ga_{0.80}As$ barrier layers. Positive energy corresponds to the electron energy above the bulk GaAs conduction-band edge and negative energy corresponds to hole energy below the bulk GaAs valence-band edge.

density of states and the quasi-Fermi distributions, we shall see that although all of the confined electron and hole states can play a considerable role in determining optical gain and transparency current density, emission wavelength is still generally defined by the lowest energy ($n_1 \rightarrow h_1$) transition.

1.3. TWO-DIMENSIONAL DENSITY OF STATES

Quantization has an enormous effect on the density of states in a quantum-well heterostructure laser. Calculation of the density of states in a quantum-well hetero-structure is most easily described[19] in comparison with the same calculation for a bulk semiconductor. In a bulk semiconductor, the number of states N_k in a spherical volume of k-space from k to $k + dk$ is given by the differential volume of the sphere divided by the k-space volume per state, which is $8\pi^3/(L_xL_yL_z)$. So

$$N_k dk = \frac{L_xL_yL_z}{(2\pi)^3} 4\pi k^2 dk. \tag{1.10}$$

The density of states in k-space per unit volume $\rho(k)dk$ is given by Eq. (1.10) divided by the real space volume $L_xL_yL_z$ and multiplied by a factor of 2 to account for spin degeneracy. Thus

$$\rho(k)dk = \frac{k^2}{\pi^2} dk. \tag{1.11}$$

From the relationship between k and E given by Eq. (1.2) we can calculate

$$dk = \frac{1}{2}\left(\frac{2m_e}{\hbar^2}\right)^{1/2}\frac{dE}{E^{1/2}}, \tag{1.12}$$

and substitution of Eqs. (1.12) and (1.2) into Eq. (1.11) yields the density of states as a function of energy:

$$\rho(E)dE = \frac{1}{2\pi^2}\left(\frac{2m_e}{\hbar^2}\right)^{3/2}E^{1/2}dE. \tag{1.13}$$

We can repeat the calculations of Eqs. (1.10)–(1.13) for a quantum-well hetero-structure by recognizing that, because of quantization in the z-direction, the volume in k-space is given by $2\pi k dk$ and the volume per state is given by $4\pi^2/(L_xL_y)$. For the quantum well, then, $\rho(k)dk$ becomes

$$\rho(k)dk = \frac{1}{L_z}\frac{k}{\pi}dk \tag{1.14}$$

and $\rho(E)dE$ becomes

$$\rho(E)dE = \frac{1}{2\pi}\left(\frac{2m_e}{\hbar^2}\right)\frac{1}{L_z}\sum_{n=1}^{\infty}H(E-E_n)dE. \tag{1.15}$$

Note that $\rho(E)dE$ is a step-like function of energy and, not unexpectedly, contains a term involving L_z. The term $H(E-E_n)$ is the Heaviside function for the quantum state energy E_n. The summation correctly accounts for the contribution of all confined states to the total density of states at any energy. Equations similar to Eqs. (1.13) and (1.15) can be developed for holes in the valence band.

Figure 1.3 shows the conduction-band density of states versus energy from Eq. (1.15) for GaAs quantum-well widths of (a) $L_z = 500$, (b) $L_z = 200$, and (c) $L_z = 100$ Å. These data are for heterostructures having $Al_{0.20}Ga_{0.80}As$ barrier layers, as in Fig. 1.2. Shown for reference as a dashed line in each of the figures is the bulk GaAs density of states from Eq. (1.13). Note that even for the 500 Å quantum well [Fig. 1.3(a)], which is sufficiently thick to be considered bulk-like, there is clearly a stepped density of states, although the steps are small relative to kT at room temperature. For the 200 Å quantum well [Fig. 1.3(b)], the steps are fewer and approach or exceed kT in energy. For the 100 Å quantum well [Fig. 1.3(c)], the departure from the bulk density of states is substantial. Note that only two states are confined and the energy steps are large compared to kT.

1.4. TRANSPARENCY CURRENT DENSITY IN QUANTUM-WELL LASERS

The density of states functions developed above [Eqs. (1.13) and (1.15)] can be used to find the total concentration of electrons in the conduction band or holes in the valence band. This is done by multiplying the appropriate density of states

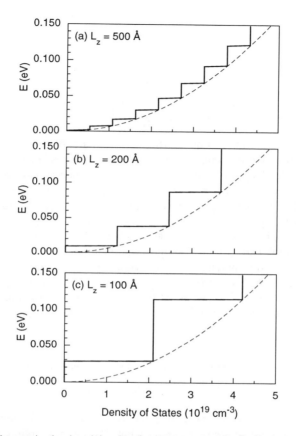

FIGURE 1.3. The conduction-band density of states vs energy for GaAs quantum well widths of (a) $L_z = 500$ Å, (b) $L_z = 200$ Å, and (c) $L_z = 100$ Å. Shown for reference as a dashed line in each of the figures is the bulk GaAs density of states.

function at each energy by the temperature dependent probability that the state is occupied and then summing over all energy and for both light and heavy holes. For electrons in the conduction band and using the appropriate units, this gives

$$n = \int_{E_c}^{\infty} \rho_n(E) f_n(E) dE \quad (\text{cm}^{-3}), \tag{1.16}$$

where $f(E)$ is the Fermi–Dirac distribution function

$$f_n(E) = \frac{1}{\exp[(E - E_f)/kT] + 1} \tag{1.17}$$

and for holes in the valence band, we get

$$p = \int_{-\infty}^{E_v} \rho_p(E) F_p(E) dE \quad (\text{cm}^{-3}) \tag{1.18}$$

and

$$f_p(E) = 1 - \frac{1}{\exp[(E - E_f)/kT] + 1}. \tag{1.19}$$

The limits to the integration in Eqs. (1.16) and (1.18) are, of course, defined by the band-edge energy or the first confined state energy in the case of the quantum-well heterostructure. The density of states in Eq. (1.18) must necessarily include the sum of both heavy holes and light holes. The equilibrium Fermi level degenerates into separate quasi-Fermi levels for electrons and holes under excitation of any sort, although electrical injection is the most interesting to us, and the concentration of injected electrons and holes must be equal. In a laser, this concentration is generally much larger than the background doping in the materials.

The condition for transparency in a semiconductor laser material, defined[20] by Bernard and Duraffourg, is that the energy separation of the n- and p-quasi-Fermi levels (E_{Fn}, E_{Fp}) must be at least equal to the transition energy. For a bulk semiconductor, this condition is given by

$$E_{Fn} - E_{Fp} \geq \hbar\omega \geq E_g \tag{1.20}$$

and for the $(n_1 \rightarrow h_1)$ transition in a quantum-well heterostructure it is given by

$$E_{Fn} - E_{Fp} \geq \hbar\omega \geq E_g + E_{n1} + E_{h1}. \tag{1.21}$$

It is a straightforward exercise to calculate the transparency carrier density in a bulk semiconductor or quantum-well heterostructure by simply finding the electron concentration that satisfies simultaneously the Bernard–Duraffourg condition and the requirement for charge neutrality.

The density of states versus energy at the point of transparency is shown graphically in Fig. 1.4. Shown as solid lines in Fig. 1.4 are the electron and hole density of states versus energy for a GaAs quantum well heterostructure. The dashed lines shown for reference in Fig. 1.4 are the bulk density of states functions. Although the scales have been removed from Fig. 1.4 for clarity, this is not a schematic diagram. Only the bulk band-gap energy has been reduced. The dotted line in Fig. 1.4 is the product of the quantum well density of states function and the Fermi–Dirac distribution function for the quasi-Fermi levels shown. The area under the dotted curve in the conduction band corresponds to the electron concentration [Eq. (1.16)] and the area under the dotted curve in the valence band corresponds to the hole concentration [Eq. (1.18)]. Because the heavy-hole effective mass in GaAs is roughly ten times greater than the electron effective mass, the density of states for the conduction and valence bands also differ greatly. As a result, charge neutrality, which corresponds to equal areas under the dotted curves, and the Bernard–Duraffourg condition are satisfied when the quasi-Fermi level for electrons is well above the

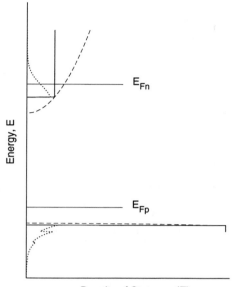

FIGURE 1.4. The density of states vs energy at the point of transparency. Shown as solid lines are the electron and hole density of states vs energy for a GaAs quantum-well heterostructure. The dashed lines shown for reference are the bulk density of states functions. The dotted line is the product of the quantum-well density of states function and the Fermi–Dirac distribution function for the quasi-Fermi levels shown.

band edge (or the first quantum state) and the quasi-Fermi level for holes remains in the gap. The value of n (or p) that satisfies these conditions is the transparency carrier density. For GaAs double heterostructure and quantum well heterostructure lasers, the electron concentration at transparency is $\sim 1.5 \times 10^{18}$ cm^{-3} and essentially independent of quantum well width.

In the diode laser structure, we are less concerned with the total carrier density in the volume than we are with the injected carrier density per unit area, which of course is related to the current density. The injected carrier density is given simply by

$$nL_z = \int_{E_c}^{\infty} \rho_n(E) f_n(E) L_z dE \quad (\text{cm}^{-2}). \qquad (1.22)$$

Figure 1.5 shows the integrand of Eq. (1.22) at transparency as a function of energy in the conduction band for GaAs quantum well widths of $L_z = 100, 200,$ and 500 Å. Shown for reference as a dashed line is the same data for a bulk GaAs laser. Actually, the "bulk" curve in Fig. 1.5 corresponds to a conventional double heterostructure with $L_z = 1000$ Å. The injected carrier density at transparency is given by the area under these curves. A striking characteristic of quantum well heterostruc-

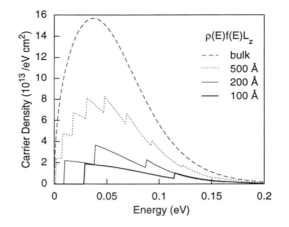

FIGURE 1.5. The integrand of Eq. (1.22) at transparency as a function of energy in the conduction band for GaAs quantum-well widths of L_z = 100, 200, and 500 Å. Shown for reference as dashed line is the same data for a bulk GaAs laser.

ture lasers, then, is that the injected carrier density is inversely proportional to the quantum well width. The transparency current density J_0 is defined at equilibrium by

$$J_0 = \frac{qnL_z}{\tau_{\text{spon}}} \quad (\text{A/cm}^2), \tag{1.23}$$

where q is the fundamental charge and τ_{spon} is the spontaneous recombination lifetime. The lifetime τ_{spon} is a function[21-23] of various material parameters as well as the electron concentration and, as a result, does not vary greatly with quantum well width. Thus, transparency in the active layer and the onset of gain can take place at much smaller injected current densities for a thin quantum well than for a conventional double heterostructure laser. The transparency current density can be expected to play an important, although not necessarily dominant, role in the laser threshold current density described later.

1.5. GAIN IN QUANTUM-WELL LASERS

A calculation of gain in a semiconductor[24] always proceeds from Fermi's golden rule which can be expressed as

$$W = \frac{2\pi}{\hbar} |\langle i|H'|j\rangle|^2 \rho(E_f) \delta(E - E_i - E_f), \tag{1.24}$$

where W is a transition rate (s^{-1}) between initial and final states at energies E_i and E_f, respectively, $\rho(E_f)$ is the density of final states, and $\langle i|H'|j\rangle$ is the perturbation matrix element that arises from the interaction of the vector potential of the elec-

TABLE 1.1. The anisotropy factor A_{ij}

Polarization	Heavy-hole	Light-hole
TE	3/2	1/2
TM	0	2

tromagnetic field with the electron in the initial state. Assuming a sinusoidal electric field and neglecting intraband scattering, we can derive a material gain coefficient for a quantum well heterostructure which is, using the notation of Chinn et al.,[25] given by

$$g(E) = \frac{q^2|M|^2}{\epsilon_0 m_0^2 c \hbar \bar{n}} \frac{1}{EL_z} \sum_{ij} m_r C_{ij} A_{ij}\{f_n(E) - [1 - f_h(E)]\} H(E - E_{ij}),$$

(1.25)

where ϵ_0 is the free space permittivity, m_0 is the free electron mass, c is the speed of light, $\bar{n}$ is the effective refractive index, and $H(E - E_{ij})$ is the Heaviside step function for a transition energy between steps i and j. C_{ij} is a spatial overlap factor which has values of either near-zero or near-unity and is essentially the k-selection rule. The reduced effective mass m_r is given by

$$\frac{1}{m_r} = \frac{1}{m_i} + \frac{1}{m_j}$$

(1.26)

and the bulk momentum matrix element $|M|^2$ is given by

$$|M|^2 = \frac{m_0^2 E_\Gamma (E_\Gamma + \Delta)}{6 m_e (E_\Gamma + 2\Delta/3)},$$

(1.27)

where Δ is the split-off valence-band energy separation. A_{ij} is an anisotropy factor that accounts for polarization of the electromagnetic field. The appropriate values for A_{ij} are given in Table 1.1.

For unstrained quantum wells, heavy-hole transitions dominate because of the much larger effective mass and the smaller size quantization energy shift. As a result transverse electric (TE) polarization is favored. This is not the case for certain strained layer quantum well lasers, as we will discuss below.

The material gain coefficient of Eq. (1.25) is a function of both energy and, perhaps not obviously, injected carrier density, since the injected carrier density defines the quasi-Fermi levels. Calculation of the gain coefficient of Eq. (1.25) is complicated enough when considering real energy band structure and valence-band mixing effects. In addition, other physical phenomena must be included, such as the line broadening that results from many-body effects. In terms of the laser diode, we are interested in the peak gain γ as a function of injected carrier (or current) density above the transparency value, rather than the spectral distribution of the gain $g(E)$ at a single carrier density. Furthermore, the practical gain in a semiconductor laser

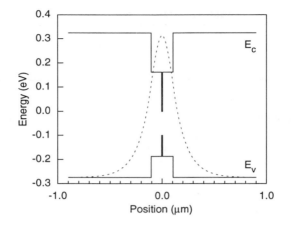

FIGURE 1.6. Energy bands of a typical five-layer Al$_x$Ga$_x$As-GaAs separate confinement hetero-structure. The resulting optical mode intensity profile for this waveguide is shown as a dashed line.

diode is the modal gain, so the structural properties of the laser must be considered. In any semiconductor laser, the structure consists of a relatively large waveguide region (of the order of many wavelengths in the material) which contains a relatively small gain region (smaller than one wavelength), particularly for quantum well heterostructure lasers. The energy bands of a typical five-layer Al$_x$Ga$_x$As-GaAs separate confinement[26] heterostructure are shown in Fig. 1.6. Everything in Fig. 1.6 is to scale, except for the bulk GaAs band-gap energy which has been reduced for clarity. At the center of the structure is the thin (~ 0.01 μm) quantum well. On both sides of the quantum well are thicker (~ 0.1 μm) wider gap, lower refractive index inner barrier layers. Outside the inner barrier layers are the outer confinement (or cladding) layers which are still thicker (~ 1 μm) wider gap, and lower refractive index.

In terms of the propagation of an electromagnetic wave in the laser, the quantum well is so thin as to be negligible and the waveguide properties[27] of the structure are essentially defined by a three-layer slab dielectric in which the inner barrier layers together comprise the core of the waveguide. The resulting optical mode intensity profile for this waveguide is shown as a dashed line in Fig. 1.6. The half-width of the optical intensity profile (~ 0.3 μm) is much larger than the quantum well width, and little change the optical intensity profile results from variation of the quantum well width from 50 to 1000 Å. Thus, only a small portion of the optical field overlaps with the gain in the thin active layer and the bulk of the optical field propagates without gain, and with only a small amount of distributed loss resulting from impurities and crystal imperfections. The true gain of the laser structure is the modal gain, which is defined as the peak material gain of the quantum well γ multiplied by a confinement factor Γ, which is the normalized overlap integral of the optical field with the gain region, given by

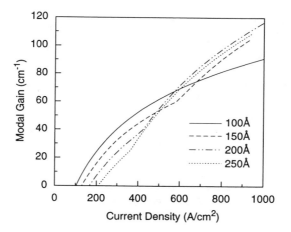

FIGURE 1.7. Peak modal gain vs current density in a GaAs quantum-well laser for quantum well widths of 100, 150, 200, and 150 Å.

$$\Gamma = \frac{\int_{-L_z/2}^{L_z/2}|E|^2 dz}{\int_{-\infty}^{\infty}|E|^2 dz}. \tag{1.28}$$

The confinement factor Γ is approximately proportional to the quantum well width L_z.

A good example of how the peak modal gain versus current density in a GaAs quantum-well laser varies with quantum well width is shown in Fig. 1.7. These data are from calculations made by Chinn[28] based on the models described in Ref. 25. The decrease in the transparency current density with quantum well width is apparent from the intercepts along the ordinate. The contribution from higher order quantum states, for well widths greater than 100 Å, are evident as kinks in the gain curves. It is important to note that, even though the confinement factor decreases with decreasing quantum well width, the slopes of the modal gain curves remain comparable. Thus, increased gain for smaller quantum well widths more than offsets the decreased optical confinement. Differential gain plays a role in the spectral linewidth, discussed by Ohtsu in Chap. 3, and high-frequency characteristics of quantum-well lasers, discussed by Bowers in Chap. 4.

1.6. THRESHOLD CURRENT DENSITY IN QUANTUM-WELL LASERS

The laser threshold is defined as the current density at which the gain equals the losses. For a semiconductor laser this is written as

$$\Gamma \gamma_t = \alpha + \frac{1}{2L} \ln \frac{1}{R_1 R_2}, \tag{1.29}$$

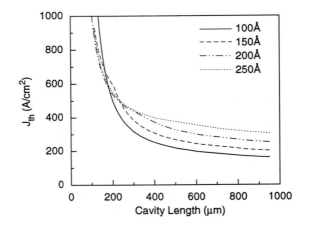

FIGURE 1.8. Threshold current density vs cavity length for a GaAs quantum-well laser.

where the gain is the modal gain $\Gamma\gamma_t$ at threshold, and the losses consist of absorption losses α distributed throughout the structure (typically 2–10 cm^{-1}) and a term for the light lost to transmission at the facets where R_1 and R_2 are the facet reflectivities ($\sim$ 0.32 for uncoated facets). The curves of Fig. 1.7 can be fit nearly perfectly, for lowest order quantum transitions, by an expression having the form[29]

$$\Gamma\gamma = \Gamma\beta J_0 \ln \frac{J}{J_0}, \qquad (1.30)$$

where β is a gain coefficient (cm/A). For the 100 Å quantum well of Fig. 1.7, β has a value of 10.7 cm/A. From Eqs. (1.29) and (1.30), we can derive an expression for the laser threshold current density

$$J_{th} = \frac{J_0}{\eta} \exp \frac{\alpha + \dfrac{1}{L} \ln \dfrac{1}{R}}{\Gamma\beta J_0} \quad (A/cm^2). \qquad (1.31)$$

In Eq. (1.31) a quantum efficiency term η has been added to account for carrier losses that occur independent of the optical processes. The quantum efficiency approaches unity for good semiconductor lasers. The threshold current density versus cavity length for the GaAs quantum well laser material of Fig. 1.7 is shown in Fig. 1.8. These data are calculated assuming $\alpha = 5$ cm^{-1} and $\eta = 1$. The effects of the quantum well are most apparent in the reduction in laser threshold current density with quantum well width at long cavity lengths. Two important conclusions can be drawn from Fig. 1.8. In contrast to a conventional double heterostructure laser, a cavity length of 1000 μm is not the long cavity length limit and the threshold current density is not yet a minimum at this length. Also, at short cavity

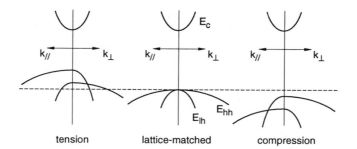

FIGURE 1.9. Schematic diagram of the conduction and valence bands in strained (tension and compression) and unstrained quantum-well heterostructures.

lengths ($< 400~\mu$m) the end losses are large enough to dominate the sublinear gain in quantum wells (Fig. 1.7).

1.7. STRAINED LAYER QUANTUM-WELL LASERS

The conduction and valence energy bands of strained layer lattice mismatched systems are altered[30,31] by the presence of biaxial strain. This includes changes in the effective masses[32,33] of the various energy bands and shifts in energy[34] of the band edges relative to each other. In an unstrained compound semiconductor quantum well heterostructure, the band edges of interest are the conduction-band edge and the degenerate light-hole (spin$\pm\frac{1}{2}$) and heavy-hole (spin$\pm\frac{3}{2}$) valence-band edges. For the purposes of this discussion, the conduction and valence bands can be treated as parabolic near zone center, as in Eq. (1.02), and are shown schematically in Fig. 1.9. In a strained layer heterostructure system, the presence of biaxial tension (or compression) defeats the cubic symmetry of the semiconductor. This results in removal of the degeneracy in the valence-band edge and both heavy-hole and light-hole band edge energies decrease (or increase) in energy with respect to the conduction-band edge with increasing strain. The decrease (or increase) in the light-hole band edge energy is greater, as shown schematically in Fig. 1.9. An orbital strain Hamiltonian for a given band at $k = 0$ can be written[35] in terms of the hydrostatic deformation potential a and the tetragonal shear deformation potential b. Assuming biaxial stress in the growth plane, the Hamiltonian can[36] be simplified and eigenvalues calculated. This results in energy differences between the conduction band and valence bands at $k = 0$ given to first order by

$$\Delta E_{\text{hh}} = -2a\epsilon\left(\frac{C_{11}-C_{12}}{C_{11}}\right) + b\epsilon\left(\frac{C_{11}+2C_{12}}{C_{11}}\right), \tag{1.32}$$

$$\Delta E_{\text{lh}} = -2a\epsilon\left(\frac{C_{11}-C_{12}}{C_{11}}\right) - b\epsilon\left(\frac{C_{11}+2C_{12}}{C_{11}}\right), \tag{1.33}$$

TABLE 1.2. Material parameters for InAs and GaAs[37-41]

		InAs	GaAs
lattice constant (Å)	a_0	6.0585	5.6535
elastic coefficients (10^{12} dyne/cm^2)	C_{11}	0.865	1.188
	C_{12}	0.485	0.538
hydrostatic deformation potential (eV)	a	-6.0	-8.2
shear deformation potential (eV)	b	-1.8	-1.7
electron effective mass	m_e	0.023	0.069
heavy-hole effective mass	m_h	0.41	0.47
light-hole effective mass	m_l	0.027	0.074

where ΔE_{hh} is the shift in the heavy-hole valence-band edge with respect to the conduction-band edge, ΔE_{lh} is the shift in the light-hole valence-band edge with respect to the conduction-band edge, C_{ij} are elastic stiffness coefficients, and the strain ϵ is given by

$$\epsilon = \pm \left| \frac{\Delta a_0}{a_0} \right|, \tag{1.34}$$

where a_0 is the lattice constant. The sign in Eq. (1.34) is chosen to be positive for biaxial compression and negative for biaxial tension.

The parameters necessary to calculate Eqs. (1.32) and (1.33) for In$_x$Ga$_{1-x}$As strained layers can be interpolated from values for the endpoint binary semiconductors, InAs and GaAs and are given[37-41] in Table 1.2. There are two important quantum-well heterostructure laser systems involving strained In$_x$Ga$_{1-x}$As layers. They are the In$_x$Ga$_{1-x}$As-InP heterostructure system which is important for laser wavelengths near 1.5 μm, and the In$_x$Ga$_{1-x}$As-GaAs system which allows for laser emission in the range of 0.9–1.1 μm. The lattice constant for In$_x$Ga$_{1-x}$As determined from Vegard's Law is shown in Fig. 1.10. The dashed portion of the line in Fig. 1.10 corresponds to the normal range for In$_x$Ga$_{1-x}$As-GaAs strained layer heterostructures. Only compressive strain is possible. The solid portion of the line in Fig. 1.10 corresponds to the normal range for In$_x$Ga$_{1-x}$As-InP strained layer heterostructures where both compressive and tensile strain are possible. The strain-induced energy shifts in the heavy-hole and light-hole valence-band edges for the In$_x$Ga$_{1-x}$As-GaAs[31] (dashed lines) and the In$_x$Ga$_{1-x}$As-InP (solid lines) heterostructure materials systems are shown in Fig. 1.11. Under compression, the heavy-hole valence-band edge shifts only slightly away from the conduction-band edge while the light-hole band edge can shift by hundreds of meV. Thus, the energy gap is defined by the conduction band and the heavy-hole valence band. Under tension, the heavy-hole band edge shifts slightly toward the conduction-band edge while the light-hole band edge shows a much greater shift toward the conduction-band edge. Discounting the quantum size effect for a moment, it is apparent that, under tension, the energy gap is defined by the conduction band and the light-hole valence band.

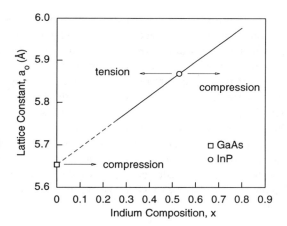

FIGURE 1.10. The lattice constant for $In_xGa_{1-x}As$. The dashed portion of the line corresponds to $In_xGa_{1-x}As$-GaAs strained layer heterostructures and the solid portion of the line corresponds to $In_xGa_{1-x}As$-InP strained layer heterostructures.

1.8. CRITICAL THICKNESS IN STRAINED LAYERS

In lattice-matched heterostructure systems, there is essentially no limit to the number or thickness of quantum wells in the structure. In a strained layer lattice mismatched system, however, the elastic accommodation[42,43] of the strain energy associated with the mismatch, without the formation of misfit dislocations, must be considered. The unit cell of $In_xGa_{1-x}As$ can be more than 3% larger or smaller than InP ($a_0 = 5.868$ Å) as indicated in Fig. 1.10. In the case of a quantum-well layer

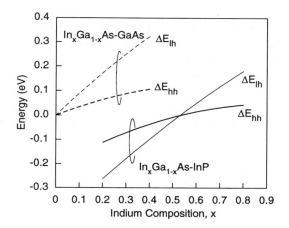

FIGURE 1.11. Strain-induced energy shifts in the heavy-hole and light-hole valence-band edges vs composition for $In_xGa_{1-x}As$-GaAs (dashed lines) and $In_xGa_{1-x}As$-InP (solid lines) heterostructures.

under biaxial tension, the $In_xGa_{1-x}As$ cell is stretched in both directions parallel to the interface, and compressed in the direction normal to the interface. In the case of a quantum-well layer under biaxial compression, the $In_xGa_{1-x}As$ cell is shortened in both directions parallel to the interface, and elongated in the direction normal to the interface (uniaxial tension). The strain energy that results is approximately equal to the misfit and produces forces at the interfaces. If the force at either interface exceeds the tension in a dislocation line, migration of a threading dislocation occurs and results in the formation of a single misfit dislocation.

For a strained quantum-well layer, a critical layer thickness h_c below which the misfit strain is accommodated elastically without formation of misfit dislocations, can be defined in terms of the elastic constants of the materials. Matthews and Blakeslee[42] calculate the critical thickness for layers in a superlattice having equal elastic constants as

$$h_c = \frac{a_0}{2\sqrt{2}\pi f} \frac{(1-0.25\nu)}{(1+\nu)} \left(\ln \frac{h_c\sqrt{2}}{a_0} + 1 \right), \tag{1.35}$$

where h_c is the critical thickness and a_0 is the lattice constant of the strained layer. The misfit f is defined simply as

$$f = \frac{\Delta a_0}{a_0} \tag{1.36}$$

and ν is Poisson's ratio, which is defined as

$$\nu = \frac{C_{12}}{C_{11}+C_{12}}. \tag{1.37}$$

The critical thickness of Eq. (1.35) for $In_xGa_{1-x}As$-InP and $In_xGa_{1-x}As$-GaAs strained quantum wells as a function of indium composition can be solved simply by numerical methods and is shown in Fig. 1.12. For thicknesses below the critical thickness the strain is accommodated elastically and the layer is coherent with the substrate. For thicknesses greater than the critical thickness the layer is relaxed, contains a large number of misfit dislocations, and is generally unsuitable[44,45] for use in a semiconductor laser. The design constraint is that the thickness of the quantum well is bounded by the critical thickness and the practical limitations associated with growing thin epitaxial layers. A further constraint may be imposed by the selection of an emission wavelength.

1.9. THRESHOLD CURRENT DENSITY IN STRAINED LAYER LASERS

Perhaps more important for quantum-well heterostructure lasers are that the changes in the valence-band structure of $In_xGa_{1-x}As$ strained layers result in the expectation of reduced laser threshold current density. In a strained layer lattice-mismatched system, biaxial strain strongly affects more than just the positions of the valence-

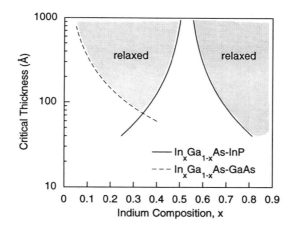

FIGURE 1.12. Critical thickness for $In_xGa_{1-x}As$-InP and $In_xGa_{1-x}As$-GaAs strained quantum wells as a function of indium composition.

band edges at $k = 0$. A salient feature of Fig. 1.9 is that in compression the upper valence band (hh) has the heavy-hole effective mass in the direction perpendicular ($\perp$) to the plane of the interfaces, but is no longer symmetric and has a much lighter effective mass in the direction parallel ($\parallel$) to the growth interfaces. It is the effective masses in the direction parallel ($\parallel$) to the growth interfaces which affect the density of states function and, hence, optical gain and threshold current density in strained layer quantum-well heterostructure lasers.

Adams[46] and, independently, Yablonovitch and Kane[47] have argued that the large asymmetry between the light conduction-band effective mass and the much heavier valence-band effective mass is a fundamental limitation to low threshold laser performance in conventional double heterostructure or quantum well heterostructure lasers. The necessary conditions for transparency have been described above in Sec. 1.4. In unstrained quantum-well heterostructure lasers at transparency, the hole effective mass is so much greater than the electron effective mass that the electron quasi-Fermi level is degenerate while the hole quasi-Fermi level is far from degenerate, as shown in Fig. 1.4. The introduction of compressive strain, however, results in a reduction in the hole effective mass in the direction parallel to the growth interfaces and yields a corresponding reduction in the injected carrier density required to reach threshold. This is illustrated in Fig. 1.13, which is the density of states versus energy for a strained layer quantum-well heterostructure laser. As in Fig. 1.4, the solid lines are the electron and hole density of states versus energy, the dashed lines are the bulk density of states functions, and the dotted lines are the product of the quantum well density of states function and the Fermi–Dirac distribution function for the quasi-Fermi levels shown. At the point of transparency, the areas under the dotted curves in Fig. 1.13 are smaller than in Fig. 1.4. Thus, the lighter valence-band effective mass results in a reduced density of states, and a

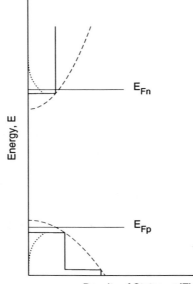

FIGURE 1.13. The density of states vs energy for a strained layer quantum-well heterostructure laser. The solid lines are the electron and hole density of states vs energy, the dashed lines are the bulk density of states functions, and the dotted lines are the product of the quantum-well density of states function and the Fermi–Dirac distribution function for the quasi-Fermi levels shown.

correspondingly smaller value for the injected carrier density at transparency. The lowest transparency carrier densities are obtained when the electron and hole effective masses are equal and as small as possible.

Figure 1.14 shows[48] the injected carrier density nL_z at transparency (dashed line), and the corresponding transparency current density J_0 (solid line), as a function of indium composition for $In_xGa_{1-x}As$-GaAs strained layer quantum-well heterostructure lasers ($L_z = 100$ Å). Even for unstrained $In_xGa_{1-x}As$, the injected carrier density and the transparency carrier density would decrease with increasing composition simply because of the decreasing $In_xGa_{1-x}As$ electron effective mass. With the additional strain-induced reduction of the hole effective mass, the transparency carrier density in the strained layer wells is up to 2.4 times smaller than for comparable GaAs quantum wells. The relationship between injected carrier density and current density is given by Eq. (1.23). The spontaneous lifetime in the well is a function of composition and also increases rapidly with decreasing carrier density. From Fig. 1.14, it can be seen that the reduction in transparency carrier density associated with strain is amplified[48] by the effect of the associated increase in the spontaneous lifetime, and results in a decrease in the transparency current density of up to 5.6 times compared to similar GaAs quantum well structures.

The transparency current density is only part of the threshold current density in

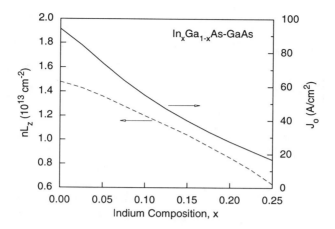

FIGURE 1.14. Injected carrier density nL_z at transparency (dashed line), and transparency current density J_0 (solid line), as a function of indium composition for $In_xGa_{1-x}As$-GaAs strained layer quantum-well heterostructure lasers ($L_z = 100$ Å).

a quantum-well heterostructure laser. In order to completely assess the impact of strain on quantum-well lasers, we must also consider the gain, described above in Sec. 1.5, which must necessarily be strongly influenced by the strain-induced changes in the valence-band structure of these materials. Consider Eq. (1.25). To first order, the bulk momentum matrix element $|M|^2$, the effective index $\bar{n}$, and the reduced effective mass m_r all change with composition and will undergo additional changes associated with strain. The anisotropy factor A_{ij} does not change but plays a much more important role in the summation as the valence-band degeneracy is lifted and recombination to either light-hole or heavy-hole bands is enhanced. Numerical solutions to Eq. (1.25) can be made to include the effects of strain on the valence-band structure. Shown in Fig. 1.15 is the modal gain versus current density[49] for $In_xGa_{1-x}As$-GaAs strained layer quantum well heterostructure lasers ($L_z = 100$ Å, $x = 0.15$, 0.20, and 0.25). The modal gain for a comparable GaAs quantum-well heterostructure ($x = 0.00$) is shown for reference. The gain coefficient β of Eq. (1.30) increases with indium composition and the transparency current density, which is the intercept along the x-axis, decreases. Shown in Fig. 1.16 is the threshold current density versus cavity length[48] of Eq. (1.31) for the $In_xGa_{1-x}As$-GaAs strained layer laser material of Fig. 1.15. These data are calculated assuming $\alpha = 5$ cm^{-1} and $\eta = 1$. The effects of strain are manifested as a large reduction in threshold current density at long cavity lengths, to less than 36% of the unstrained value.

1.10. POLARIZATION AND GAIN IN STRAINED LAYER LASERS

The laser emission in conventional double heterostructure lasers and unstrained quantum-well heterostructure lasers is normally polarized in the TE direction. This

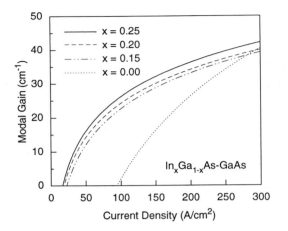

FIGURE 1.15. Modal gain vs current density for $In_xGa_{1-x}As$-GaAs strained layer quantum-well heterostructure lasers ($L_z = 100$ Å, $x = 0.15$, 0.20, and 0.25). The modal gain for a comparable GaAs quantum well heterostructure ($x = 0.00$) is shown for reference.

is because the reflectivity[50] of the cleaved facet cavity slightly favors TE polarization, even though the gain for both TE and transverse magnetic (TM) polarizations is the same. In a strained layer quantum-well heterostructure laser, however, TE and TM gain[51] are different. The physical basis for this has been described nicely in a paper[52] by O'Reilly and co-workers, and we adopt their formalism here. In an unstrained material, the valence-band states are fourfold degenerate and described by the states

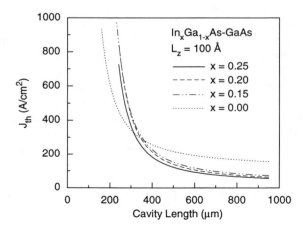

FIGURE 1.16. Threshold current density versus cavity length for $In_xGa_{1-x}As$-GaAs strained layer lasers.

$$|3/2, +3/2\rangle = |(X+iY)\alpha\rangle \frac{1}{\sqrt{2}}, \qquad (1.38a)$$

$$|3/2, +1/2\rangle = \left|-(X+iY)\beta\rangle \frac{1}{\sqrt{6}} + |Z\alpha\rangle \sqrt{\frac{2}{3}}, \qquad (1.38b)\right.$$

$$|3/2, -1/2\rangle = |(X-iY)\alpha\rangle \frac{1}{\sqrt{6}} + |Z\beta\rangle \sqrt{\frac{2}{3}}, \qquad (1.38c)$$

$$|3/2, -3/2\rangle = |(X-iY)\beta\rangle \frac{1}{\sqrt{2}}, \qquad (1.38d)$$

where X is a function defined along the length (x) of the laser, Y is a function in the transverse or lateral plane (y) of the laser, Z is a function normal (z) to the plane of the junction, and α and β are spin states. Electrons recombining with y-like and z-like states contribute to TE and TM gain, respectively. Electrons recombining with x-like states contribute only to spontaneous emission. Averaging over all of the states in Eq. (1.38) results in equal contributions to light polarized along the three directions. In a strained layer structure, however, the fourfold degeneracy is broken and the $\pm 3/2$ states dominate recombination under biaxial compression, while the $\pm 1/2$ states dominate recombination under biaxial tension. The $\pm 3/2$ states in Eq. (1.38) have no z-like component so that TE gain will be enhanced relative to TM gain under biaxial compression. Conversely, the $\pm 1/2$ states in Eq. (1.38) are mainly z-like so that TM gain will be enhanced under biaxial tension.

The $In_xGa_{1-x}As$-GaAs strained layer system is limited to biaxial compression. The reduction in the laser threshold current density shown in Fig. 1.16 is obtained because of both the increase in the TE gain with compressive strain (Fig. 1.15) and the decrease in the transparency current density (Fig. 1.14) that results from the reduction in the valence-band effective mass. The $In_xGa_{1-x}As$-InP strained layer materials system allows for both biaxial compression and biaxial tension (Fig. 1.10). From the discussion above, it would appear that biaxial compression is to be preferred because, even though TM gain will be enhanced under biaxial tension, little or no reduction in the valence-band effective mass is expected. Hence, no large reduction in the transparency current density is expected. Experimental results on comparable strained layer $In_xGa_{1-x}As$-InP quantum-well heterostructure lasers indicate, however, that laser structures with active regions under biaxial tension[53–55] have threshold current densities that are at least as low as those with active regions under biaxial compression. The polarization anisotropy in gain has been calculated for the $In_xGa_{1-x}As$-InP system using realistic models for the valence-band structure including mixing effects and intraband relaxation. The modal gain for $In_xGa_{1-x}As$-InP quantum-well heterostructure lasers under both tension and compression ($\Delta a_0/a_0 = \pm 1.5\%$) is shown[52] in Fig. 1.17. Shown for reference as a dashed line in Fig. 1.17 is the modal gain for an unstrained $In_{0.53}Ga_{0.47}As$-InP quantum well. From Fig. 1.17 we see that the reduction in transparency current

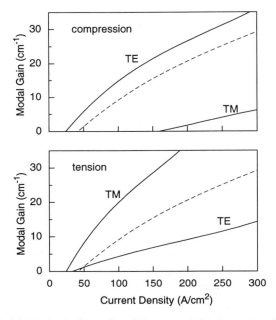

FIGURE 1.17. Modal gain for In$_x$Ga$_{1-x}$As-InP quantum-well heterostructure lasers under both tension and compression ($\Delta a_0/a_0 = \pm 1.5\%$). Shown for reference as a dashed line is the modal gain for an unstrained In$_{0.53}$Ga$_{0.47}$As-InP quantum well.

density is comparable for the lasers under tension and compression and the reduction is much smaller for this material system than for the In$_x$Ga$_{1-x}$As-GaAs strained layer system (Fig. 1.15).

Interestingly, the modal TM gain under tension ($\beta = 17.7$ cm/A) in Fig. 1.17 is 30% larger than the modal TE gain under compression ($\beta = 13.6$ cm/A). This has a distinct effect on the laser threshold current density. Shown in Fig. 1.18 is the threshold current density versus cavity length of Eq. (1.31) for the In$_x$Ga$_{1-x}$As-InP strained layer laser material of Fig. 1.17. The heavy solid line is for 1.5% biaxial tension, the light solid line corresponds to 1.5% biaxial compression, and the unstrained case is shown as a dashed line for reference. These data are calculated assuming $\alpha = 5$ cm^{-1} and $\eta = 1$. At long cavity lengths, the threshold current density is reduced for both compressive and tensile strain with the reduction being larger for tension rather than compression.

1.11. SUMMARY

The semiconductor quantum-well heterostructure laser is already playing a well established and predominant role in present technology. Some of the many applications that have benefited from the use of quantum wells are described by Gray in Chap. 9 and Duan in Chap. 10. In addition, the quantum well and its particular

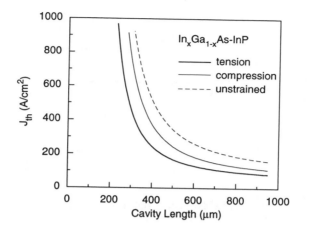

FIGURE 1.18. Threshold current density vs cavity length for $In_xGa_{1-x}As$-InP strained layer lasers. The heavy solid line is for 1.5% biaxial tension, the light solid line corresponds to 1.5% biaxial compression, and the unstrained case is shown as a dashed line for reference.

advantages, such as low threshold current density, high efficiency, high differential gain, and the relationship between size and wavelength, make it the structure of choice for most areas of semiconductor laser development. For example, the quantum-well heterostructure is common to distributed feedback lasers, described by Chinone and Okai in Chap. 2, and semiconductor laser amplifiers, discussed by Agrawal in Chap. 8. In the future, it seems likely that the quantum-well heterostructure will continue to play a key role in the development of vertical cavity semiconductor lasers (VCSEL). These devices, which are the subject of Chap. 5 written by Chang-Hasnain, benefit in two ways since both the periodic gain region and the multilayer reflectors are formed from thin layer quantum-well structures. Laser emission at previously unavailable wavelengths is becoming possible by combining the physics of quantum wells, and particularly strained layer quantum wells, with new materials. These include visible III–V materials, discussed by Hatakoshi in Chap. 6, and II–VI materials, described by Nurmikko and Gunshor in Chap. 7. Finally, there seems to be no fundamental reason why additional reduced dimensionality[56] should not be possible. For example, if we follow the analysis of Sec. 1.3 and introduce a second degree of quantization in the x-direction we can derive the density of states function for a quantum wire, given by

$$\rho(E)dE = \frac{1}{2\pi}\left(\frac{2m_e}{\hbar^2}\right)^{1/2}\frac{1}{L_xL_z}\sum_{nm}(E-E_{n,m})^{-1/2}dE. \qquad (1.39)$$

The density of states function of Eq. (1.39) for a quantum wire is shown in Fig. 1.19. These data are for GaAs and $L_x = L_z = 150$ Å. Shown for reference as a dashed line in each of the figures is the bulk GaAs density of states from Eq. (1.13). This figure should be compared with Fig. 1.3. Note that in theory, the density of

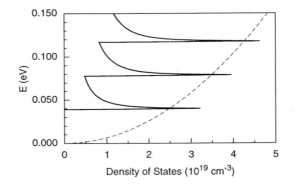

FIGURE 1.19. The conduction-band density of states vs energy for GaAs quantum wire having widths of $L_x = L_z$ = 150 Å. Shown for reference as a dashed line in the figure is the bulk GaAs density of states.

states function has singularities at the confined state energies. These sharp features would, however, be smoothed by the normal random variation in thickness and as a result of the finite interband relaxation times. Even so, higher peak gain can be obtained from the quantum wire structure than from comparable quantum well heterostructure lasers. This is likely to be accompanied by an increase in the differential gain. Both effects contribute to lower laser thresholds and higher modulation frequencies.

The author is grateful to S. R. Chinn, P. D. Dapkus, G. M. Smith, and P. S. Zory, Jr. for helpful discussions and technical assistance. This work was supported by the National Science Foundation (ECD 89-43166 and DMR 89-20538), the Joint Services Electronics Program (N00014-90-J-1270) and SDIO/IST (DAAL03-92-G-0272).

REFERENCES

1. H. C. Casey, Jr. and M. B. Panish, *Heterostructure Lasers-Part A: Fundamental Principles* (Academic, New York, 1978), p. 1.
2. P. D. Dapkus, IEEE J. Quantum Electron. **QE-23**, 650 (1987).
3. *Semiconductor Injection Lasers*, edited by J. K. Butler (IEEE, New York, 1980).
4. *Semiconductor Diode Lasers*, edited by W. Streiffer and M. Ettenberg (IEEE, New York, 1991).
5. *Selected Papers on Semiconductor Diode Lasers*, edited by J. J. Coleman (SPIE-The International Society for Optical Engineering, Bellingham, WA, 1992).
6. C. H. Henry, in *Quantum Well Lasers*, edited by P. S. Zory, Jr. (Academic, San Diego, 1993), p. 1.
7. L. Esaki and R. Tsu, IBM J. Res. Dev. **14**, 61 (1970).
8. R. Dingle, W. Wiegmann, and C. H. Henry, Phys. Rev. Lett. **33**, 827 (1974).
9. A. Y. Cho, R. W. Dixon, H. C. Casey, Jr., and R. L. Hartman, Appl. Phys. Lett. **28**, 501 (1976).
10. W. T. Tsang, Appl. Phys. Lett. **34**, 473 (1979).
11. R. D. Dupuis and P. D. Dapkus, Appl. Phys. Lett. **32**, 473 (1978).
12. R. D. Dupuis, P. D. Dapkus, N. Holonyak, Jr., E. A. Rezek, and R. Chin, Appl. Phys. Lett. **32**, 295 (1978).

13. R. Eisberg and R. Resnick, *Quantum Physics of Atoms, Molecules, Solids, Nuclei, and Particles* (Wiley, New York, 1974) p. G-1.
14. J. M. Luttinger and W. Kohn, Phys. Rev. **97**, 869 (1955).
15. M. Missous, in *Properties of Aluminum Gallium Arsenide*, edited by S. Adachi (INSPEC, The Institute of Electrical Engineers, London, 1993), p. 73.
16. J. S. Blakemore, J. Appl. Phys. **53**, R123 (1982).
17. S. Adachi, J. Appl. Phys. **58**, R1 (1985).
18. S. Adachi, in *Properties of Aluminum Gallium Arsenide*, edited by S. Adachi (INSPEC, The Institute of Electrical Engineers, London, 1993), p. 66.
19. J. T. Verdeyen, *Laser Electronics*, 2nd ed. (Prentice-Hall, Englewood Cliffs, NJ, 1989), p. 366.
20. M. G. A. Bernard and G. Duraffourg, Phys. Status Solidi **1**, 699 (1961).
21. C. B. Su and R. Olshansky, Appl. Phys. Lett. **41**, 833 (1982).
22. R. Olshansky, C. B. Su, J. Manning, and W. Powazinik, IEEE J. Quantum Electron. **QE-20**, 838 (1984).
23. Y. Arakawa, H. Sakaki, M. Nishioka, and J. Yoshino, Appl. Phys. Lett. **46**, 519 (1985).
24. G. P. Agrawal and N. K. Dutta, *Long Wavelength Semiconductor Lasers* (Van Nostrand Reinhold, New York, 1986), p. 74.
25. S. R. Chinn, P. S. Zory, and A. R. Reisinger, IEEE J. Quantum Electron. **24**, 2191 (1988).
26. M. B. Panish, H. C. Casey, Jr., S. Sumski, and P. W. Foy, Appl. Phys. Lett. **22**, 590 (1973).
27. H. C. Casey, Jr. and M. B. Panish, *Heterostructure Lasers-Part A: Fundamental Principles* (Academic, New York, 1978), p. 20.
28. S. R. Chinn (private communication).
29. P. W. A. McIlroy, A. Kurobe, and Y. Uematsu, IEEE J. Quantum Electron. **QE-21**, 1958 (1985).
30. E. P. O'Reilly, in *Quantum Well Lasers*, edited by P. S. Zory, Jr. (Academic, San Diego, 1993), p. 329.
31. J. J. Coleman, in *Quantum Well Lasers*, edited by P. S. Zory, Jr. (Academic, San Diego, 1993), p. 367.
32. G. C. Osbourn, Superlattices and Microstructures **1**, 223 (1985).
33. E. P. O'Reilly and G. P. Witchlow, Phys. Rev. B **34**, 6030 (1986).
34. G. H. Olsen, C. J. Nuese, and R. T. Smith, J. Appl. Phys. **49**, 5523 (1978).
35. G. E. Pikus and G. L. Bir, Sov. Phys. Solid State **1**, 136 (1959).
36. A. Gavini and M. Cardona, Phys. Rev. B **1**, 672 (1970).
37. S. Adachi, J. Appl. Phys. **53**, 8775 (1982).
38. K. F. Huang, K. Tai, S. N. G. Chu, and A. Y. Cho, Appl. Phys. Lett. **54**, 2026 (1989).
39. *Numerical Data and Functional Relationships in Science and Technology*, edited by O. Madelung (Springer, Berlin, 1982).
40. S. Niki, C. L. Lin, W. S. C. Chang, and H. H. Wieder, Appl. Phys. Lett. **55**, 1339 (1989).
41. E. Yablonovitch and E. O. Kane, J. Lightwave Technol. **LT-6**, 1292 (1988).
42. J. W. Matthews and A. E. Blakeslee, J. Cryst. Growth **27**, 118 (1974).
43. F. C. Frank and J. H. van der Merwe, Proc. R. Soc. (London) A **198**, 216 (1949).
44. K. J. Beernink, P. K. York, J. J. Coleman, R. G. Waters, J. Kim, and C. M. Wayman, Appl. Phys. Lett. **55**, 2167 (1989).
45. K. J. Beernink, P. K. York, and J. J. Coleman, Appl. Phys. Lett. **55**, 2585 (1989).
46. A. R. Adams, Electron. Lett. **22**, 249 (1986).
47. E. Yablonovitch and E. O. Kane, J. Lightwave Technol. **LT-4**, 504 (1986).
48. J. J. Coleman, K. J. Beernink, and M. E. Givens, IEEE J. Quantum Electron. **28**, 1983 (1992).
49. S. W. Corzine, R. H. Yan, and L. A. Coldren, Appl. Phys. Lett. **57**, 2835 (1990).
50. T. Ikegami, IEEE J. Quantum Electron. **QE-8**, 470 (1972).
51. T. C. Chong and C. G. Fonstad, IEEE J. Quantum Electron. **25**, 171 (1989).
52. E. P. O'Reilly, G. Jones, A. Ghiti, and A. R. Adams, Electron. Lett. **27**, 1417 (1991).
53. C. E. Zah, R. Bhat, B. Pathak, C. Caneau, F. J. Favire, N. C. Andreakis, D. M. Hwang, M. A. Koza, C. Y. Chen, and T. P. Lee, Electron. Lett. **27**, 1414 (1991).
54. P. J. A. Thijs, L. F. Tiemeijer, P. I. Kuindersma, J. J. M. Binsma, and T. van Dongen, IEEE J. Quantum Electron. **27**, 1426 (1991).
55. T. Tanbun-Ek, N. A. Olsson, R. A. Logan, K. W. Wecht, and A. M. Sergent, IEEE Photon. Technol. Lett. **3**, 103 (1991).
56. E. Kapon, in *Quantum Well Lasers*, edited by P. S. Zory, Jr. (Academic, San Diego, 1993), p. 461, and references cited.

CHAPTER 2

Distributed Feedback Semiconductor Lasers

Naoki Chinone and Makoto Okai

Optoelectronics Research Center, Hitachi Central Research Laboratory,
Kokobunji, Tokyo 185, Japan

2.1. INTRODUCTION

Semiconductor lasers whose output is a single spectral line—that is, lasers that operate in a single longitudinal mode—are required for various applications, such as long-distance optical-fiber communications. Longitudinal-mode control therefore attracts much attention and various single-mode device structures have been proposed. One provides stable single-longitudinal-mode oscillation by using so-called distributed feedback from an internal grating. This distributed feedback (DFB) laser is the most promising single-longitudinal-mode device and is now widely commercialized for use in long-distance communication systems.

In this chapter, review of the DFB laser is described on laser oscillation theory (Sec. 2), device fabrication (Sec. 3), laser performances (Sec. 4), and recent progresses (Sec. 5).

Here, the history of studies on the DFB lasers are briefly reviewed. The basic concepts of the DFB laser were established by Kogelnik and Shank[1] in 1971, who made the first DFB laser by using dyed gelatin on a glass substrate. The first semiconductor DFB laser oscillation was achieved by optical pumping to a corrugated GaAs surface and was reported by Nakamura et al.[2] in 1973. An injection-type DFB laser was then reported by Scifres et al.,[3] who used a GaAlAs/GaAs single-heterojunction structure. The first continuous-wave (cw) oscillation at room temperature was achieved almost simultaneously in 1975 by Casey et al.[4] and Nakamura et al.,[5] who used a GaAlAs/GaAs separate-confinement heterostructure. The structure of the DFB laser reported by Aiki et al.[6] is shown schematically in Fig. 2.1. The grating in this structure was fabricated by the holographic interference method using an ultraviolet (He–Cd) gas laser as a light source. This method is still widely used because the grating period needed is usually beyond the resolution of

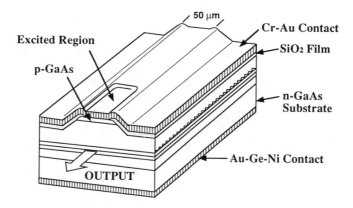

FIGURE 2.1. Schematic structure of a GaAlAs/GaAs distributed feedback laser achieving cw operation at room temperature (after Ref. 6 © 1976 IEEE).

commercially available photolithography equipment. The grating period of the device shown in Fig. 2.1, for the example, is only 0.3813 μm, which is the third order for the 0.8 μm wavelength range. To avoid damaging the active layer during the grating fabrication process, the grating was formed on an optical guide layer grown on the active layer. As a result, this laser was capable of cw operation in a single-longitudinal mode at a wavelength of 0.89 μm.

DFB laser oscillation was also intensively studied theoretically. Kogelnik and Shank[7] proposed a coupled-wave formalism for analyzing plane wave propagation in the periodic structure and obtained simple expressions for the resonant wavelengths and threshold. Yariv[8] extended this coupled-wave formalism to analyze guided-wave propagation in a periodically corrugated waveguide, and this analysis is more suitable for designing heterostructure lasers. Wang,[9] on the other hand, used Bloch wave formalism to analyze wave propagation in periodic structures. Although the equations used in these two formalisms seem to be different, Yariv and Gover[10] showed that the two approaches are equivalent.

After cw operation was achieved in GaAlAs/GaAs DFB lasers, attention was focused on making DFB lasers oscillate at wavelengths of 1.3 and 1.55 μm. This wavelength range is important because the fiber loss is minimum at these wavelengths in optical-fiber communication systems, and the mode partition noise inherent in Fabry–Perot lasers, was becoming a serious problem as the transmission bit rate increased. The first DFB laser oscillating in the 1 μm wavelength range was made from InGaAsP/InP and was reported in 1979 by Doi et al.[11] This laser oscillated at 1.15 μm, but only at low temperatures. Room temperature cw operation at 1.5 μm in InGaAsP/InP DFB lasers was achieved in 1981 by Utaka et al.,[12] and these long-wavelength DFB lasers (Fig. 2.2) showed stable single-mode oscillation. Since an InP substrate is transparent to the laser wavelength, the grating was formed on a surface of the InP substrate. Because the grating pitch for this wavelength

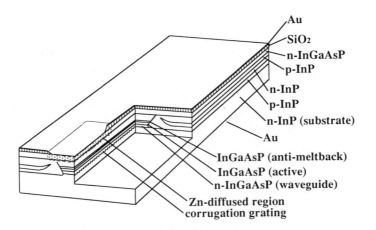

Au
SiO₂
n-InGaAsP
p-InP
n-InP
p-InP
n-InP (substrate)
Au
InGaAsP (anti-meltback)
InGaAsP (active)
n-InGaAsP (waveguide)
Zn-diffused region
corrugation grating

FIGURE 2.2. Schematic structure of an InGaAsP/InP distributed feedback laser achieving cw operation at room temperature (after Ref. 12).

range is larger than that of GaAlAs/GaAs lasers, the first-order grating can be fabricated by ultraviolet laser holography.

The efforts to achieve stable operation by long-wavelength DFB lasers stimulated precise analyses of longitudinal mode characteristics. The DFB laser theory predicted two equivalent oscillation modes in a DFB laser with a uniformly distributed grating. Fortunately, reflection from the laser facets at both ends of the grating resolves the degeneracy of longitudinal modes and selects one of the two modes. The present cleavage process, however, does not allow the position of the facet—and thus the phase of the reflected light—to be controlled precisely. There could, therefore, be devices in which gain difference between the main mode and the biggest submode is not large enough for stable single-mode oscillation. It was becoming clear that in long-distance communication systems, the ratio of the intensity of the main mode to that of the biggest submode should be at least 30 dB, and the mode degeneracy was becoming a serious obstacle to satisfy this requirement. Extensive theoretical and experimental studies of the probability of single-mode oscillation in DFB lasers with various reflectivities at each end revealed that a probability of more than 80% can be achieved by choosing the proper reflectivities.[13] As a result of those studies, long-wavelength DFB lasers usually having a low-reflection-coated front facet and a high-reflection-coated or as-cleaved rear facet became commercially available in the middle of the 1980s.

The degeneracy problem can be solved by introducing a phase shift to the uniform grating. Haus and Shank[14] theoretically proposed in 1976 that single-mode oscillation could be achieved by introducing a quarter-wavelength ($\lambda/4$) shift at the center of the device and Suzuki and Tada proposed a chirped grating structure[15] in 1980. However, because the grating pitch needed is as small as 0.2 μm, the development of suitable fabrication methods have taken several years. The first

$\lambda/4$-shifted DFB laser was reported in 1984 by Sekatedjo et al.[16] who used electron beam lithography, and by Utaka et al.[17] who used simultaneous exposure of positive and negative photoresists. Both groups obtained single-longitudinal-mode oscillation and observed near threshold spectrum structures inherent in $\lambda/4$-shifted DFB lasers. Other methods for producing the $\lambda/4$ shift were reported later,[18-20] but studies of the spectral characteristics of the $\lambda/4$-shifted DFB laser revealed that the oscillation mode frequently became unstable at high output powers. Soda et al.[21] studied this phenomenon by self-consistently analyzing coupled-mode equations and carrier-distribution equations. They found that the concentration of light intensity at the $\lambda/4$-shift region reduces the number of injected carriers and that this reduces the gain difference between the main mode and the biggest submode. They also showed that the coupling coefficient of the grating has an optimum value for stabilizing a longitudinal mode at high output powers. In addition to an appropriately fabricated grating, single-mode operation also requires facet reflectivities below 1% at both ends. Knowledge of the proper design parameters and the development of suitable fabrication method led to $\lambda/4$-shifted DFB lasers becoming commercially available in the late 1980s.

The DFB lasers have been widely used in the field of optical communication systems. The necessity of single-mode operation in digital transmission systems depends on the transmission wavelength as well as the transmission bit rate and distance. The fiber loss is minimal at a wavelength of 1.55 μm, but the wavelength dispersion in conventional single-mode fibers is large. The DFB laser, therefore, plays a more important role at 1.55 μm than it does at 1.3 μm. Single-mode operation is usually preferable when the bit rate is greater than 600 Mb/s and the transmission distance is more than several tens of kilometers. In particular, the systems with bit rates above 1 Gb/s and distances greater than 50 km use the DFB laser exclusively both at 1.3 and 1.55 μm. Analog transmission systems also require single-mode operation whenever mode-partition noise seriously degrades the signal-to-noise ratio. The DFB laser is used, for example, in subcarrier multiplexed analog transmission systems.

The distributed Bragg reflector (DBR) laser is another important single-mode laser. The gratings of DBR lasers, in contrast to those of DFB lasers, are outside of the active region. Laser oscillation takes place in a resonator formed by two grating mirrors or by one grating mirror and a cleaved facet. Single-mode oscillation can be obtained because the grating mirror has wavelength selectivity, but because there are many modes in the resonator, temperature variation can elicit mode jumping between adjacent modes. The first DBR laser oscillation was demonstrated by Reinhart et al.,[22] using GaAlAs/GaAs. Long-wavelength DBR lasers using InGaAsP/InP were intensively studied later.[23,24] An advantage of the DBR laser is that gratings can be fabricated in an area apart from the active layer. The DBR laser is therefore suitable for integration with other optical devices and for making multifunctional light sources such as multielectrode wavelength-tunable lasers.[25] Although this chapter will mainly concentrate on DFB lasers, readers should note that the DBR laser is also a promising single-mode light source.

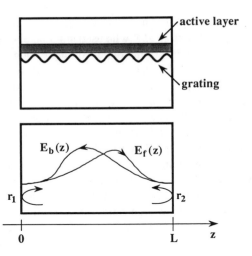

FIGURE 2.3. Schematic structure of a distributed feedback laser and the notations used in the calculations.

2.2. THEORY

2.2.1. Coupled-Wave Equations

DFB lasers oscillate in a stable single-mode because a grating fabricated inside the laser selects a single-longitudinal mode. The longitudinal mode behavior of DFB lasers is analyzed by using coupled-wave equations.[7] First let us derive the coupled-wave equations from the wave equation for the laser field along the laser cavity of z-direction (Fig. 2.3). The laser field is confined in x- and y-directions by waveguide structure. We assume the fundamental waveguide mode for x- and y-directions and focus on the laser field along the laser cavity of z-direction. The wave equation is expressed as

$$\frac{d^2E}{dz^2} + \epsilon k_0^2 E = 0, \qquad (2.1)$$

where E is the electric field and ε is the dielectric constant. The term z is the distance along the laser cavity and k_0 equals ω/c (ω is the mode radian frequency and c is the speed of light in a vacuum). If the dielectric constant ϵ varies periodically along the z-axis with the grating pitch Λ, it can be expanded in the following Fourier series:

$$\epsilon = \bar{\epsilon} + \sum_l \Delta\epsilon_l \exp\left(j\,\frac{2\pi}{\Lambda}\,lz + j\varphi\right), \qquad (2.2)$$

where $\bar{\epsilon}$ is the averaged dielectric constant and $\Delta\epsilon_l$ is the perturbation of the dielectric constant. The term j represents $\sqrt{-1}$ and φ is the grating phase at $z = 0$. The general solution of Eq. (2.1) is expressed as

$$E = A \exp(-j\beta z) + B \exp(j\beta z). \tag{2.3}$$

Here β is the propagation constant and is expressed by

$$\beta = n_{eff}k_0 + j \frac{g_{th}}{2}, \tag{2.4}$$

where n_{eff} is the effective reflective index and g_{th} is the threshold gain expressed for power. By substituting Eqs. (2.2) and (2.3) into Eq. (2.1), we obtain the following relation:

$$\left(\frac{d^2A}{dz^2} \exp(-j\beta z) - 2j\beta \frac{dA}{dz} \exp(-j\beta z) - \beta^2 A \exp(-j\beta z) \right)$$

$$+ \left\{ \frac{d^2B}{dz^2} \exp(j\beta z) + 2j\beta \frac{dB}{dz} \exp(j\beta z) - \beta^2 B \exp(j\beta z) \right\}$$

$$+ \left\{ \bar{\epsilon} + \sum_l \Delta\epsilon_l \exp\left(j \frac{2\pi}{\Lambda} lz + j\varphi \right) \right\} k_0^2 \{A \exp(-j\beta z) + B \exp(j\beta z)\} = 0.$$
$$\tag{2.5}$$

The terms d^2A/dz^2 and d^2B/dz^2 in this equation can be neglected because A and B change slowly. By using the relation $(\bar{\epsilon}k_0^2 - \beta^2)E = 0$, we obtain

$$\frac{dA}{dz} \exp(-j\beta z) - \frac{dB}{dz} \exp(j\beta z) = -j[\kappa A \exp(-2j\Delta\beta z + j\varphi + j\beta z)$$

$$+ \kappa^* B \exp(2j\Delta\beta z - j\varphi - j\beta z)], \tag{2.6}$$

where the coupling coefficient κ is given by

$$\kappa = \frac{\Delta\epsilon_m k_0^2}{2\beta}, \tag{2.7}$$

and κ^* is the conjugate to κ and κ increases with the increase of corrugation depth, and

$$\Delta\beta = \beta - \beta_0. \tag{2.8}$$

The term β_0 (which is equal to $\pi m/\Lambda$, where m is an integer) in Eq. (2.8) is the Bragg propagation constant. When the transform Eq. (2.5) into Eq. (2.6), the Fourier component in Eq. (2.2) we used is the one that nearly satisfies the phase-match condition ($l = m$).

By equating the coefficients of $\exp(-j\beta z)$ and $\exp(j\beta z)$ on both sides of Eq. (2.6), we obtain the coupled-mode equations

$$\frac{dA}{dz} = -j\kappa^*B \exp(2j\Delta\beta z - j\varphi) \tag{2.9}$$

and

$$\frac{dB}{dz} = j\kappa A \exp(-2j\Delta\beta z + j\varphi). \tag{2.10}$$

The coupled-wave equations can be rewritten by using β_0, instead of Eq. (2.3), to express E as

$$E = E_f \exp(-j\beta_0 z) + E_b \exp(j\beta_0 z). \tag{2.11}$$

By comparing Eqs. (2.3) and (2.11), we can see that

$$E_f = A \exp(-j\Delta\beta z) \tag{2.12}$$

and

$$E_b = B \exp(j\Delta\beta z). \tag{2.13}$$

Substituting Eqs. (2.12) and (2.13) into Eqs. (2.9) and (2.10) allows us to transform the coupled-wave equations as follows:

$$\frac{dE_f}{dz} = -j\Delta\beta E_f - j\kappa^*E_b \exp(-j\varphi) \tag{2.14}$$

and

$$\frac{dE_b}{dz} = j\Delta\beta E_b + j\kappa E_f \exp(j\varphi). \tag{2.15}$$

From Eqs. (2.4) and (2.8) we can express $\Delta\beta$

$$\Delta\beta = (n_{\text{eff}}k_0 - \beta_0) + j\frac{g_{\text{th}}}{2} = \delta + j\frac{g_{\text{th}}}{2}, \tag{2.16}$$

where δ express the wave number's deviation from the Bragg condition.

The general solution of the coupled-wave Eqs. (2.14) and (2.15) are expressed as follows:

$$E_f = E_{f1} \exp(-j\gamma z) + E_{f2} \exp(j\gamma z) \tag{2.17}$$

and

$$E_b = E_{b1} \exp(-j\gamma z) + E_{b2} \exp(j\gamma z). \tag{2.18}$$

By substituting Eqs. (2.17) and (2.18) into Eqs. (2.14) and (2.15) and equating the coefficients of $\exp(-j\gamma z)$ and $\exp(j\gamma z)$ on both sides, we obtain the following four equations:

$$(-\gamma + \Delta\beta)E_{f1} = -\kappa^* E_{b1} \exp(-j\varphi), \tag{2.19}$$

$$(\gamma + \Delta\beta)E_{f2} = -\kappa^* E_{b2} \exp(-j\varphi), \tag{2.20}$$

$$(-\gamma - \Delta\beta)E_{b1} = \kappa E_{f1} \exp(j\varphi), \tag{2.21}$$

and

$$(\gamma - \Delta\beta)E_{b2} = \kappa E_{f2} \exp(j\varphi). \tag{2.22}$$

Combining Eqs. (2.19) and (2.21) or Eqs. (2.20) and (2.22) give us the following dispersion relation:

$$\gamma = \sqrt{(\Delta\beta)^2 - \kappa\kappa^*}. \tag{2.23}$$

From Eqs. (2.19) to (2.22), we obtain

$$E_{b1} = r_{G1}E_{f1} \tag{2.24}$$

and

$$E_{f2} = r_{G2}E_{b2}, \tag{2.25}$$

where r_{G1} and r_{G2} are given by

$$r_{G1} = \frac{-\gamma + \Delta\beta}{-\kappa \exp(-j\varphi)} = \frac{\kappa \exp(j\varphi)}{-\gamma - \Delta\beta} \quad \text{and}$$

$$r_{G2} = \frac{-\kappa^* \exp(-j\varphi)}{\gamma + \Delta\beta} = \frac{\gamma - \Delta\beta}{\kappa \exp(j\varphi)}. \tag{2.26}$$

Now let us use the coupled-wave equations to find the oscillation condition of DFB lasers as shown in Fig. 2.3. The boundary conditions are given as

$$E_f(0) = r_1 E_b(0) \tag{2.27}$$

and

$$E_b(L) = r_2 E_f(L), \tag{2.28}$$

where r_1 and r_2 are, respectively, the reflectivities for the electric field at $z = 0$ and at $z = L$ (L is the cavity length). The reflectivity can be written

$$r_i = |r_i|\exp(j\phi_i) \quad (i = 1,2), \tag{2.29}$$

where ϕ_i is the phase shift caused by the reflection at the cavity facets. The value of ϕ_i depends on the corrugation phase at the cavity facets. By substituting Eqs. (2.17) and (2.18) into Eqs. (2.27) and (2.28) and by eliminating E_{f2} and E_{b1} by using Eqs. (2.24) and (2.25), we obtain

$$(1 - r_1 r_{G1})E_{f1} + (r_{G2} - r_1)E_{b2} = 0 \tag{2.30}$$

and

$$(r_{G1} - r_2)\exp(-j\gamma L)E_{f1} + (1 - r_2 r_{G2})\exp(j\gamma L)E_{b2} = 0. \tag{2.31}$$

The values E_{f1} and E_{b2} have nontrivial solutions if

$$\frac{(r_{G2} - r_1)(r_{G1} - r_2)}{(1 - r_1 r_{G1})(1 - r_2 r_{G2})} \exp(-2j\gamma L) = 1. \tag{2.32}$$

This equation expresses the oscillation condition for DFB lasers.

2.2.2. DFB Lasers with a Uniform Grating

Evaluating the transmission characteristics as a function of the normalized optical gain gL less than the normalized threshold gain $g_{th}L$ helps us understand the lasing modes. When the normalized coupling coefficient κL is 2.0 and $r_1 = r_2 = 0$, we can obtain the transmittance T for a uniform grating by using Eqs. (2.24) and (2.25) and the relation $E_b(L) = 0$:

$$T = \left|\frac{E_f(L)}{E_f(0)}\right|^2 = \left|\frac{1 - r_{G1} r_{G2}}{\exp(j\gamma L) - r_{G1} r_{G2} \exp(-j\gamma L)}\right|^2. \tag{2.33}$$

This transmittance is shown in Fig. 2.4 as a function of the normalized optical gain gL. In these lasers two modes oscillate simultaneously, the intensity of the longitudinal modes grows with the increasing values of gL, and the lasing modes acquire infinite amplitude at the normalized threshold gain $g_{th}L$ (1.972).

The lasing longitudinal mode of DFB lasers with a uniform grating is analyzed by using Eq. (2.32). Many pairs of δ and g_{th} satisfy this equation and the pair with the smallest g_{th} represents the main mode. The threshold gain difference Δg_{th} between the main mode and the biggest submode is an index of the stability of single-mode operation, and according to Eqs. (2.32) and (2.29) the value of Δg_{th} is a function of the phase of the corrugation at the cavity facets. Figure 2.5 shows calculated $g_{th}L$ as a function of δL by using Eq. (2.32). In these calculations, power reflectivity at the left facet r_1^2 is assumed to be 0%, and power reflectivity at the right facet r_2^2 is assumed to be 32%, and the normalized coupling coefficient κL is assumed to be 2.0. φ_2 is the phase of the corrugation at the right facet. When φ_2 is 1.0π or 1.5π, single-mode operation is stable ($\Delta g_{th}L$ is larger than 0.60). When φ_2 is 0 or 0.5π, on the other hand, two modes oscillate simultaneously. The phase of the corrugation at the cavity facets cannot be controlled because the grating pitch for 1.55 μm DFB lasers is as small as 240 nm. Thus for DFB lasers with a uniform grating, the

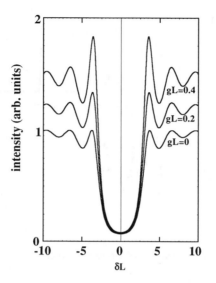

FIGURE 2.4. Transmittance T for a uniform grating as a function of normalized detuning δL.

probability of single-mode operation depends on the phase of the corrugation at the cavity facets. If r_1^2 and r_2^2 in DFB lasers with a uniform grating are controlled to be 0% to eliminate the effect of reflection at the cavity facet, two modes oscillate simultaneously.

2.2.3. Phase-Shifted DFB Lasers

To obtain stable single-mode operation regardless of the phase of the corrugation at the cavity facets, Haus and Shank[14] proposed a $\lambda/4$-shifted DFB lasers. $\lambda/4$-shifted

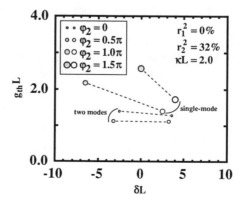

FIGURE 2.5. Normalized threshold gain $g_{th}L$ for DFB lasers with a uniform grating as a function of normalized detuning δL of the main mode and the neighboring submode. φ_2 is the phase of the corrugation at the right facet.

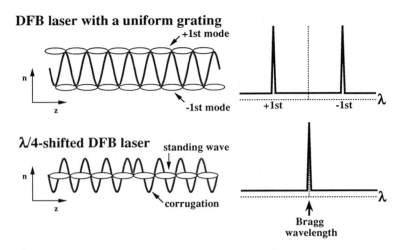

FIGURE 2.6. Schematic representation of longitudinal modes in DFB lasers with a uniform grating and λ/4-shifted DFB lasers. Corrugations and standing waves around the center of the laser axis are shown at the left-hand side of the figure.

DFB lasers operate in a stable single mode when both facet reflectivities are 0%. By considering the standing waves along the laser cavity (Fig. 2.6), we can easily understand why the mode selectivity of λ/4-shifted DFB lasers differs from that of DFB lasers with a uniform grating. In the DFB laser with a uniform grating there are two standing waves with the same threshold gain: The +1st mode and the −1st mode. Because the effective refractive index of the laser cavity differs for each of these modes, they oscillate at different wavelengths. In the λ/4-shifted DFB lasers, however, where the corrugation phase reverses at the center of the cavity, there is only one standing wave. This mode oscillates at the Bragg wavelength, which is midway between the wavelengths of the two mode of DFB lasers with a uniform grating.

Lasers with a nonuniform structure along the laser cavity, such as λ/4-shifted DFB lasers and chirped DFB lasers,[15] can be analyzed by using the F-matrix method.[26] By using Eqs. (2.17), (2.18), (2.24), and (2.25), we obtain the following matrix equations:

$$\begin{pmatrix} E_f(0) \\ E_b(0) \end{pmatrix} = \begin{pmatrix} 1 & r_{G2} \\ r_{G1} & 1 \end{pmatrix} \begin{pmatrix} E_{f1} \\ E_{b2} \end{pmatrix} \tag{2.34}$$

and

$$\begin{pmatrix} E_f(L) \\ E_b(L) \end{pmatrix} = \begin{pmatrix} \exp(-j\gamma L) & r_{G2}\exp(j\gamma L) \\ r_{G1}\exp(-j\gamma L) & \exp(j\gamma L) \end{pmatrix} \begin{pmatrix} E_{f1} \\ E_{b2} \end{pmatrix}. \tag{2.35}$$

Combining Eqs. (2.34) and (2.35), we obtain the following F-matrix for $E_f(0)$, $E_b(0)$, $E_f(L)$, and $E_b(L)$:

$$\begin{pmatrix} E_f(L) \\ E_b(L) \end{pmatrix} = \begin{pmatrix} F_{11} & F_{12} \\ F_{21} & F_{22} \end{pmatrix} \begin{pmatrix} E_f(0) \\ E_b(0) \end{pmatrix}.$$ (2.36)

The elements of this F-matrix can be written as

$$F_{11} = \frac{1}{1-r_{G1}r_{G2}} \left[\exp(-j\gamma L) - r_{G1}r_{G2} \exp(j\gamma L) \right],$$ (2.37)

$$F_{12} = \frac{-r_{G2}}{1-r_{G1}r_{G2}} \left[\exp(-j\gamma L) - \exp(j\gamma L) \right],$$ (2.38)

$$F_{21} = \frac{r_{G1}}{1-r_{G1}r_{G2}} \left[\exp(-j\gamma L) - \exp(j\gamma L) \right],$$ (2.39)

and

$$F_{22} = \frac{1}{1-r_{G1}r_{G2}} \left[r_{G1}r_{G2} \exp(-j\gamma L) - \exp(j\gamma L) \right].$$ (2.40)

The cavity of a DFB laser can be divided into N small segments along the z-axis, and the kth segment is then represented by the kth F-matrix. The total F-matrix can be given by multiplying the kF-matrices:

$$\begin{pmatrix} F_{11} & F_{12} \\ F_{21} & F_{22} \end{pmatrix} = \prod_{k=1}^{N} \begin{pmatrix} F_{11}^{(k)} & F_{12}^{(k)} \\ F_{21}^{(k)} & F_{22}^{(k)} \end{pmatrix}.$$ (2.41)

The effect of the facet reflectivities r_1 and r_2 can be included as follows:

$$\begin{pmatrix} F_{11} & F_{12} \\ F_{21} & F_{22} \end{pmatrix} = \frac{1}{\sqrt{(1-r_1^2)(1-r_2^2)}} \begin{pmatrix} 1 & -r_1 \\ -r_1 & 1 \end{pmatrix} \prod_{k=1}^{N} \begin{pmatrix} F_{11}^{(k)} & F_{12}^{(k)} \\ F_{21}^{(k)} & F_{22}^{(k)} \end{pmatrix} \begin{pmatrix} 1 & r_2 \\ r_2 & 1 \end{pmatrix}.$$ (2.42)

The transmittance T is $|1/F_{22}|^2$ and the oscillation condition is given by

$$F_{22} = 0.$$ (2.43)

The transmittance T of a $\lambda/4$-shifted grating and the oscillation condition of $\lambda/4$-shifted DFB lasers can be given by using the F-matrix method. The cavity is divided into two sections as shown by the inset in Fig. 2.7. When both facet reflectivities are 0, the matrix element F_{22} is expressed as

$$F_{22} = F_{21}^{(1)}F_{12}^{(2)} + F_{22}^{(1)}F_{22}^{(2)}.$$ (2.44)

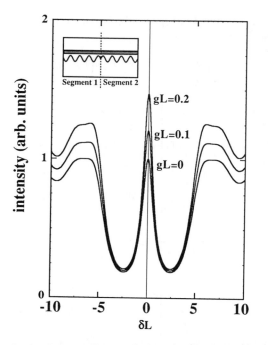

FIGURE 2.7. Transmittance T for a $\lambda/4$-shifted grating as a function of normalized detuning δL.

By using Eqs. (2.38), (2.39), and (2.40), we can rewrite Eq. (2.44) as

$$
\begin{aligned}
F_{22} = &-\frac{r_{G1}(\varphi = 0)r_{G2}(\varphi = \pi)}{[1-r_{G1}(\varphi = 0)r_{G2}(\varphi = 0)][1-r_{G1}(\varphi = \pi)r_{G2}(\varphi = \pi)]} \\
&\times\left[\exp\left(-\frac{j\gamma L}{2}\right)-\exp\left(\frac{j\gamma L}{2}\right)\right]^2 \\
&+\frac{1}{[1-r_{G1}(\varphi = 0)r_{G2}(\varphi = 0)][1-r_{G1}(\varphi = \pi)r_{G2}(\varphi = \pi)]} \\
&\times\left[r_{G1}(\varphi = 0)r_{G2}(\varphi = 0)\exp\left(-\frac{j\gamma L}{2}\right)-\exp\left(\frac{j\gamma L}{2}\right)\right] \\
&\times\left[r_{G1}(\varphi = \pi)r_{G2}(\varphi = \pi)\exp\left(-\frac{j\gamma L}{2}\right)-\exp\left(\frac{j\gamma L}{2}\right)\right]. \qquad (2.45)
\end{aligned}
$$

The transmittance T of a $\lambda/4$-shifted grating is shown in Fig. 2.7 as a function of detuning γL. Stable single-mode lasing occurs at the Bragg wavelength ($\delta L = 0$) where the transmittance is maximum in $\lambda/4$-shifted DFB lasers, whereas the lasing occurs on the two sides of the stopband in DFB lasers with a uniform grating as shown in Fig. 2.4. Solutions of $F_{22} = 0$ for the normalized coupling coefficient κL

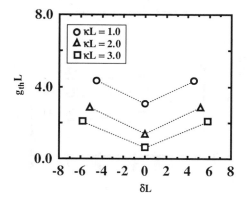

FIGURE 2.8. Normalized threshold gain $g_{th}L$ for $\lambda/4$-shifted DFB lasers as a function of normalized detuning δL of the main mode and the neighboring submodes.

of 1.0, 2.0, and 3.0 are shown in Fig. 2.8. Stable single-mode operation at the Bragg wavelength can be obtained in $\lambda/4$-shifted DFB lasers whose facets have reflectivities of 0. The $g_{th}L$ of their main mode decreases with increasing of κL and single-mode operation is more stable for larger values of κL.

Despite the theoretical stabilities of this single-mode operation, mode jumping occurs at high output powers. This mode instability results from the spatial hole-burning effect,[21] which is shown schematically in Fig. 2.9. The light intensity increases near the center of the laser cavity, where the $\lambda/4$-phase-shift is. This increased intensity decreases the carrier density and, by the plasma effect, increases the refractive index near the center of the laser cavity. The increased refractive index increases the original $\lambda/4$-phase shift and degrades the stability of the single-

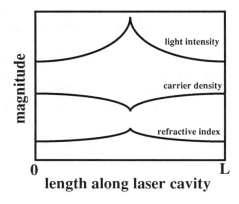

FIGURE 2.9. The spatial hole-burning effect.

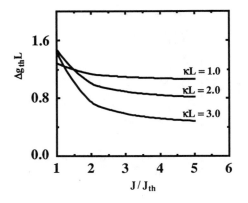

FIGURE 2.10. Normalized threshold gain difference $\Delta g_{th}L$ between the main mode and the biggest submode as a function of the normalized current J/J_h for $\lambda/4$-shifted DFB laser.

mode operation. The spatial hole-burning can be analyzed by combining the rate equation and the coupled-wave equations[27,28] and self-consistently solving for the distributions of carrier density, refractive index, and mode gain along the laser cavity. The threshold gain difference of the main mode and the biggest submode is shown in Fig. 2.10 as a function of the normalized threshold current J/J_{th} for κL values of 1.0, 2.0, and 3.0. Stable single-mode operation is obtained near the threshold, but because of the spatial hole-burning effect the stability is degraded drastically with increasing driving current. This degradation of the single-mode stability is more severe when κL is large.

2.2.4. Gain-Coupled DFB Lasers

Index-coupled DFB lasers have been investigated for a long time and are already commercially available. And gain-coupled DFB lasers have recently received a great deal of attention because of their stable single-mode operation regardless of facet reflectivity and their high-speed and low-chirp performance. Although ideally only the gain is periodically distributed along the laser cavity in gain-coupled lasers,[1] actual devices also have a periodic distribution of refractive index. To make this analysis more general, we will also consider this latter distribution. The periodic distributions of gain and refractive index are expressed as

$$g_{th} = \bar{g}_{th} + \Delta g \, \cos\left(\frac{2\pi}{\Lambda} + \varphi + \theta\right) \tag{2.46}$$

and

$$n = n_{eff} + \Delta n \, \cos\left(\frac{2\pi}{\Lambda} + \varphi\right). \tag{2.47}$$

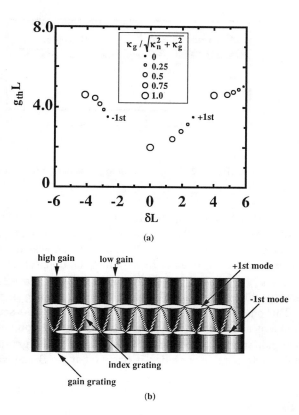

FIGURE 2.11. (a) Normalized threshold gain $g_{th}L$ as a function of normalized detuning δL of the main mode and the neighboring submodes for a gain-coupled DFB laser with an antiphase index grating. (b) Standing waves in a gain-coupled DFB laser with an antiphase index grating.

The lasing mode of gain coupled lasers can be analyzed by using the coupled-wave equations with coupling coefficient modified as follows:

$$\kappa = \kappa_n + j\kappa_g \exp(j\theta), \qquad (2.48)$$

where κ_n is the coupling coefficient of refractive index and κ_g is the coupling coefficient of gain. These are expressed as

$$\kappa_n = k_0 \frac{\Delta n}{2} \qquad (2.49)$$

and

$$\kappa_g = \frac{\Delta g}{2}. \qquad (2.50)$$

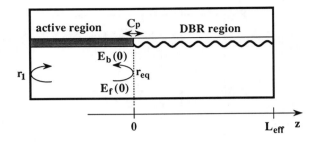

FIGURE 2.12. Distributed Bragg reflector (DBR) laser.

The normalized threshold gain $g_{th}L$ calculated as a function of normalized detuning δL is shown in Fig. 2.11(a). When these values were calculated, the reflectivity of both cavity facets was assumed to be 0%, $\sqrt{\kappa_g^2 + \kappa_n^2}L$ was fixed to be 1.0 and the phase difference θ was assumed to be π radians (antiphase). In a laser with a pure index grating ($\kappa_g / \sqrt{\kappa_g^2 + \kappa_n^2} = 0$), two modes (+1st and −1st modes) oscillate simultaneously. When the ratio of gain grating increases, $g_{th}L$ for the +1st mode decreases. This phenomenon can be understood as follows: As shown in Fig. 2.11(b) the +1st mode is selected by the gain grating because the loop of the +1st mode is at the high-gain region. In a laser with a pure gain grating ($\kappa_g / \sqrt{\kappa_g^2 + \kappa_n^2} = 1$), stable single-mode operation is obtained at the Bragg wavelength.

2.2.5. DBR Lasers

Distributed Bragg reflector (DBR) lasers have a passive corrugated waveguide region (Fig. 2.12). The oscillation condition is given by an equation similar to that for Fabry–Perot lasers:

$$r_1 r_{eq} \exp(2j\beta L) = 1, \tag{2.51}$$

where r_1 is the reflectivity from the left cavity facet and r_{eq} is the equivalent reflectivity from the DBR region. The wave number β given by Eq. (2.4) for DFB lasers should be changed for DBR lasers by substituting the power absorption coefficient α_G for g_{th}:

$$\beta = n_{eff}k_0 - j\,\frac{\alpha_G}{2}. \tag{2.52}$$

The equivalent reflectivity r_{eq} in Eq. (2.51) is given by

$$r_{eq} = C_p r_{DBR}, \tag{2.53}$$

where C_p is the efficiency of coupling between the active region and DBR region. The term r_{DBR} is the reflectivity from the DBR region and is given by

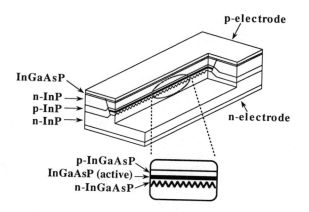

FIGURE 2.13. The structure of a buried-heterostructure (BH) DFB laser.

$$r_{\text{DBR}} = \frac{E_b(0)}{E_f(0)}. \tag{2.54}$$

By substituting Eqs. (2.17) and (2.18) into Eq. (2.54) and eliminating E_{b1} and E_{f2} by using Eqs. (2.24) and (2.25), we can rewrite r_{DBR} as

$$r_{\text{DBR}} = \frac{r_{G1}E_{f1} + E_{b2}}{E_{f1} + r_{G2}E_{b2}}. \tag{2.55}$$

When $E_b(L_{\text{eff}}) = 0$, the equivalent reflectivity is given as

$$r_{\text{eq}} = C_p \frac{r_{G1}[1 - \exp(-2j\gamma L_{\text{eff}})]}{1 - r_{G1}r_{G2} \exp(-2j\gamma L_{\text{eff}})}. \tag{2.56}$$

Thus, the oscillation condition for DBR lasers is given by

$$r_1 C_p \frac{r_{G1}[1 - \exp(-2j\gamma L_{\text{eff}})]}{1 - r_{G1}r_{G2} \exp(-2j\gamma L_{\text{eff}})} \exp(2jn_{\text{eff}}k_0L - \alpha_G L) = 1. \tag{2.57}$$

2.3. FABRICATION

2.3.1. Overview

The process used to fabricate DFB lasers is almost the same as that used to fabricate Fabry–Perot lasers, and the difference between these fabrication processes is due to the corrugated structure of DFB and DBR lasers. Figure 2.13 shows the structure of an index-coupled buried-heterostructure (BH) DFB laser. The corrugated structure in the index-coupled DFB lasers is below or above the active layer, whereas the corrugated structure in a DBR laser is in the passive waveguide regions.

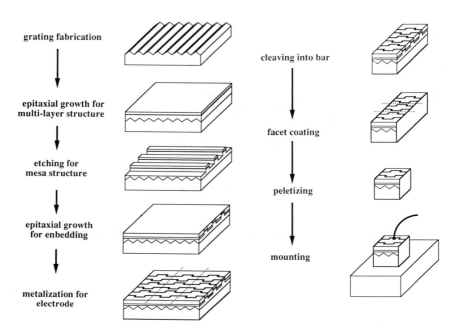

FIGURE 2.14. Process flow.

The relation between the lasing wavelength λ_L and the grating pitch Λ is expressed as follows:

$$\lambda_L = \frac{2n_{\text{eff}}\Lambda}{m}, \tag{2.58}$$

where n_{eff} is the effective refractive index and m is an integer. The first-order ($m = 1$) grating is effective for the edge-emitting lasers, which emits light parallel to the surface of the substrate. When n_{eff} is 3.25, the grating pitch Λ is 238 nm for a λ_L of 1.55 μm and 200 nm for a λ_L of 1.3 μm. The second-order ($m = 2$) grating is effective for surface-emitting lasers, which emits light perpendicular to the surface of the substrate.

The fabrication of BH DFB lasers from InP/InGaAsP materials shown in Fig. 2.13 is illustrated in Fig. 2.14. The corrugated structure is below the active layer and a buried heterostructure is used to obtain lateral confinement. First the grating on the surface of the n-InP substrate is fabricated. Because the grating structure and shape are closely related to the stability of the longitudinal single-mode operation, the methods for fabricating and preserving the grating during the epitaxial growth are described in detail in the next subsection. Multiple layers are grown on the substrate by liquid-phase epitaxy (LPE), metal-organic vapor-phase epitaxy (MOVPE), or molecular beam epitaxy (MBE). The multilayers are the n-InGaAsP lower-guide layer, the InGaAsP active layer, the p-InGaAsP upper-guide layer, the *p*-InP cladding layer, and the InGaAsP cap layer. The mesa structure for optical and electrical confinements is fabricated by using a wet or dry etching technique. Wet etching results in little damage on the semiconductor crystal, and many kinds of etchants have been developed for selective etching. A certain crystal facet appears after selective wet etching because the etching speed is different for different crystal facets. Scales can be better controlled by using a dry etching technique like reactive ion etching (RIE), and dry etching is also more uniform than wet etching. The region damaged by dry etching can be removed by wet etching or it can be repaired by thermal treatments. To obtain the fundamental lateral mode the width of the active layer in the mesa structure should be controlled to be about 1 μm. Lateral mode control is important for introducing the laser light into optical fibers efficiently. So that carriers can be injected selectively into the active region in the mesa structure, the mesa structure is embedded epitaxially with *p*- and *n*-InP blocking layers, or with semi-insulating blocking layer. The *p*-side and *n*-side electrodes are fabricated by an evaporation or plating technique, and the Schottky barriers are generated in the metal-semiconductor interface. To obtain a low-resistance contact with the tunneling effect, the Schottky barrier should be narrow, and heavy doping in the semiconductor contacting reduces the barrier width effectively. The electrodes are alloyed by thermal treatments, and the processed wafer is cleaved into bars. Using a diamond cutter to scribe small indentations near the edge of the wafer and then slightly flexing the wafer results in a cleavage that gives mirror facets perpendicular to the surface. The facets are coated with dielectric thin film by sputtering or by an electron beam evaporation method, and the facet reflectivity is controlled by adjusting the thickness of this coating film. The stability of longitudinal single-mode operation strongly depends on the facet reflectivity: To obtain single-mode operation in $\lambda/4$-shifted DFB lasers, the reflectivity of both facets should be less than 1.0%. Facet coating is also effective for passivation. The bars are then sawed, perpendicular to the mirror facets and with a diamond cutter, into chips. Typical cavity lengths are 200–400 μm and typical widths are 200–400 μm. With spring contacts the chips are preliminary tested under pulsed conditions, and the selected chips are mounted on the heat sink of metalized SiC or diamond. Injection current causes a temperature increase that exponentially increases the threshold current density J_{th} according to the following equation:

$$J_{th} = J_0 \exp\left(\frac{T_j}{T_0}\right), \tag{2.59}$$

where J_0 and T_0 are empirical parameters and T_j is the temperature at the active region. T_0 is called the characteristic temperature. One contact is obtained by soldering to the metalized heat sink and the other contact is obtained by bonding an Au wire to the opposite surface of the laser chip.

2.3.2. Grating Fabrication

The spectral characteristics of DFB lasers are closely related to their grating structure, and although single-mode oscillation can be obtained with a uniform grating, sophisticated structures such as $\lambda/4$-shifted and chirped gratings result in more stable single-mode operation. A $\lambda/4$-shifted grating is one in which the corrugation phase is inverted, and a chirped grating is one in which the corrugation pitch changes gradually. The original grating shape is deformed thermally during the crystal growth, and the final grating shape (after crystal growth) is important because it determines the coupling coefficient κ.

A two-beam holographic exposure method, shown in Fig. 2.15(a), is generally used to fabricate the uniform gratings. The laser beam is expanded and collimated by a pinhole and a pair of lenses, and the collimated beam is divided by a beam splitter into two beams that meet again in the sample stage. Grating patterns generated by the interference between the two collimated beams are transferred photolithographically onto the surface of a semiconductor substrate spin-coated with a photoresist to a thickness of about 100 nm. The grating patterns of the developed photoresist are transferred to the substrate by selective wet etching or dry etching. The fabricated grating pitch Λ is expressed as

$$\Lambda = \frac{\lambda_s}{2 \sin \theta \cos \delta}, \tag{2.60}$$

where λ_s is the wavelength of the light source and the angles θ and δ are those shown in Fig. 2.15(a). By adjusting the angles θ and δ, the average accuracy of the grating pitch can be controlled to within 0.01 nm.

$\lambda/4$-shifted gratings and chirped gratings as well as uniform gratings can be obtained with the two-beam holographic exposure method. $\lambda/4$-shifted gratings are fabricated by multilayer photoresist processing or by placing a phase-mask contact with the substrate during the holographic exposure. In the positive and negative photoresist method[29] shown in Fig. 2.15(b), some parts of the substrate are covered with a positive photoresist, other parts are covered with a negative photoresist, and both parts are exposed simultaneously. Because of the inverse optical characteristics of the photoresists, the grating phase reverses between the positive and negative photoresist regions. In the phase-shift layer method[19] shown in Fig. 2.15(c), a patterned phase-shift layer is fabricated on the photoresist thin film. A $\lambda/4$-shift between the covered and uncovered regions can be obtained by this method because of the optical length difference produced by the phase-shift layer. In the phase-modulating mask method[18] shown in Fig. 2.15(d), a pattered photomask is placed in contact with the substrate during the exposure: The thickness difference in the mask gives the $\lambda/4$ shift. As shown in Fig. 2.15(e) chirped gratings are fabricated by using the spherical beam method,[30] which uses divergent beams instead of collimated beams. The grating period along the z axis is given by

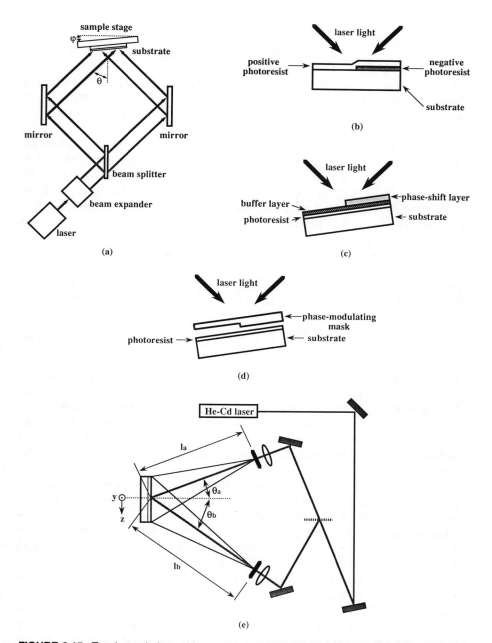

FIGURE 2.15. Two-beam holographic exposure method. (a) Conventional setup. (b) Positive and negative photoresist method. (c) Phase-shifted layer method. (d) Phase-modulating mask method. (e) Spherical laser beam method (after Ref. 30).

$$\Lambda(z) = \lambda_s \left\{ \frac{l_a \sin \theta_a + z}{\sqrt{l_a^2 + 2zl_a \sin \theta_a + z^2}} + \frac{l_b \sin \theta_b - z}{\sqrt{l_b^2 - 2zl_b \sin \theta_b + z^2}} \right\}^{-1}, \quad (2.61)$$

where λ_s is the wavelength of the laser light source. Lengths l_a and l_b and angles θ_a and θ_b are shown in Fig. 2.15(e). When $l_a = 800$ mm, $l_b = 200$ mm, and $\theta_a = \theta_b = 28.1°$, the grating pitch variation over a 400 μm distance in the z-direction is within 0.1%.

The photomask self-interference method[20] can be used to obtain high-performance gratings with any kind of structure. Grating photomasks are fabricated by using a mechanical ruling machine and the successive replica process. The mechanical ruling machine has an average pitch accuracy of 0.01 mm and can change the grating pitch for each groove. High-performance structure such as $\lambda/4$-shifted grating can be obtained when fabricating the master gratings. The exposure setup for the photomask self-interference method is shown in Fig. 2.16(a). Laser light is used as the light source for the contact exposure: The collimated laser light irradiates the grating photomask from the Bragg angle to generate the transmission waves and the first-order diffraction waves under the photomask. No higher-order diffraction waves are generated under the following condition:

$$\frac{\lambda_s}{2} < \Lambda < \lambda_s, \quad (2.62)$$

where Λ is the grating pitch on the photomask and λ_s is the wavelength of the laser light source. Interference between the transmission waves and the first-order diffraction waves generates fringe patterns under the grating photomask. Figure 2.16(b) shows scanning electron microscope (SEM) views of the gratings fabricated on the surface of InP substrates. This is a combination of two pictures: One of a uniform grating and the other of a $\lambda/4$-shifted grating (generated by shortening the pitch of six grooves by 20 nm).

High-performance gratings can also be obtained by electron beam lithography.[16] The pitch accuracy of this method depends on the stability of electron beam scanning systems, and subnanometer controllability has been demonstrated in fabricating a laser array.[31] Although the throughput of the electron beam lithography method is low, it can be increased by using x-ray lithography to transfer the grating pattern on the photomasks fabricated by electron beam lithography.[31]

During crystal growth the grating shape is deformed by mass transport due to heating that occurs during epitaxy growth. And because the grating shape determines the coupling coefficient κ, longitudinal single-mode operation cannot be obtained unless we somehow preserve the grating during crystal growth. In the LPE method, mass transport can be prevented by covering a GaAs substrate.[32] Arsenic from this substrate is incorporated into the vacancy of the P site of the InP substrate, and this incorporation suppresses the evaporation of P and the migration of In. The growth rate in the concave region of the grating is also suppressed by the incorporation of Ga and As.[33] Increasing the PH$_3$ gas pressure is also an effective way to

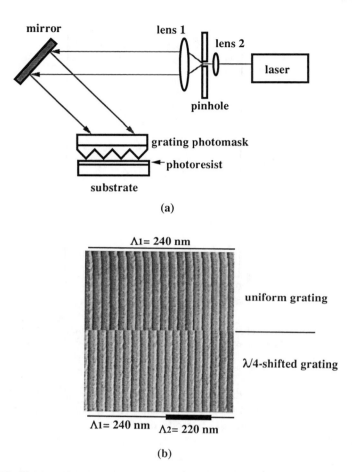

FIGURE 2.16. Photomask self-interference method. (a) Exposure system (see Ref. 20). (b) SEM view of grating fabricated on the surface of InP substrate (after Ref. 20).

suppress mass transport.[34] In the MOCVD method, AsH_3 and PH_3 gas supplied while increasing the temperature before crystal growth suppresses the evaporation of P from the InP substrate.[35]

2.4. PERFORMANCE

2.4.1. Spectral Characteristics

2.4.1.1. DFB Laser with a Uniform Grating

The light-current characteristics of an InGaAsP/InP buried-heterostructure (BH) DFB laser with a uniform grating are shown in Fig. 2.17.[36] The threshold current,

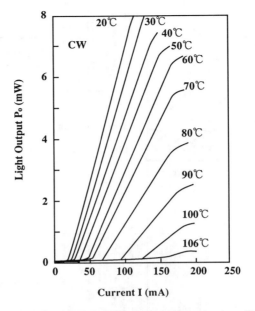

FIGURE 2.17. Light-current characteristics of a buried-heterostructure (BH) distributed feedback (DFB) laser. Parameter is ambient temperature (after Ref. 36 © 1987 IEEE).

differential efficiency, and their temperature dependence are almost the same as those of conventional BH Fabry–Perot lasers. The spectra of the DFB laser are shown in Fig. 2.18.[37] Stable single-mode oscillation was observed. The spectral characteristics of the DFB laser cavity are clearly seen in amplified spontaneous emission spectra produced at a subthreshold bias. At the center of the spectrum, spontaneous emission is suppressed because of the stopband of the grating, as theoretically predicted in Fig. 2.4. Intensity of one of the two modes adjacent to both sides of the stopband increases with increasing current. At high output powers the side-mode suppression ratio (SMSR, that is, the ratio of the intensity of the main mode to that of the biggest submode) of more than 30 dB is obtained. The output power of the sub modes is thus substantially less than that of the main mode.

Polarization of the laser output is usually parallel to the junction plane [transverse electric (TE) mode] like that in Fabry–Perot lasers, because the coupling coefficient to the grating is slightly larger for the TE mode than that for the transverse magnetic (TM) mode and also confinement factor of the laser field to the active layer is usually larger for the TE mode than that for the TM mode. Reflection from the facet is also larger for the TE mode, which enhances the polarization selectivity.

The temperature dependencies of oscillation wavelength and of threshold current in DFB lasers are shown in Fig. 2.19.[36] The absence of mode hopping through a temperature range of −20–80 °C shows that temperature dependence of the wavelength is much smaller than that of conventional Fabry–Perot lasers. The tempera-

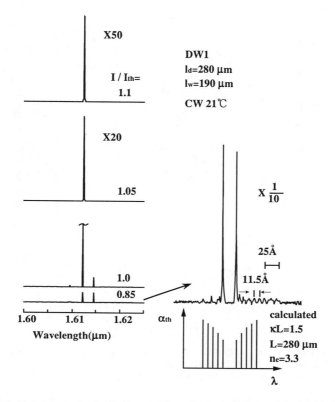

FIGURE 2.18. Spectra of a buried-heterostructure distributed feedback laser. *I* is the operating current, and I_{th} is the threshold current (after Ref. 37 © 1983 IEEE).

ture dependence of 0.08–0.1 nm/°C in DFB lasers is due to that of the refractive index of the semiconductor material. At very low or very high temperatures, the gain at the DFB mode can be lower than that of Fabry–Perot modes formed between two facets, since the temperature dependence of the gain-peak wavelength is much larger than that of the DFB mode wavelength. At extreme temperatures, the DFB mode can therefore hop toward one of the Fabry–Perot modes. The temperature range in which DFB mode oscillation is stable depends on the allocation of the DFB mode wavelength in the wavelength bandwidth of the laser gain and on the gain of Fabry–Perot modes determined by the facet reflectivities. Proper design of DFB lasers can usually result in stable single-mode oscillation over a range of more than 100 °C.

DFB lasers are studied not only in the 1.5 μm wavelength range described above, but also in the 1.3 and 0.8 μm wavelength ranges. In both wavelength ranges, stable single-mode oscillation was achieved.

In DFB lasers with a uniform grating, reflection from the facets selects one of the two DFB modes nearest the stopband as the main oscillation mode. It has been

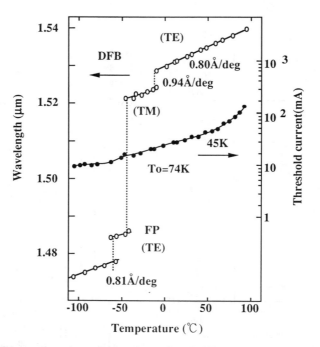

FIGURE 2.19. Temperature dependence of wavelength (○) and threshold current (●) of a distributed feedback laser. DFB is the distributed feedback mode, FP is the Fabry–Perot mode, TE is the transverse electric mode, and TM is the transverse magnetic mode (after Ref. 36 © 1987 IEEE).

experimentally confirmed, by precisely controlled shallow etching of a laser facet, that the oscillation mode is determined by the phase of the light reflected from the facets.[38] Hopping between two DFB modes was clearly indicated by changing the phase of reflected light from one of the facets. Because such a precise control of the facet position is not possible in the usual cleavage process of facets, 100% reproducibility of the single-mode oscillation cannot be achieved in the DFB lasers with a uniform grating. Probability of the single-mode oscillation was studied in the DFB lasers with various facet reflectivities.[39] It was found that more than 80% reproducibility can be obtained by proper choice of facet reflectivities. To improve differential efficiency, the front facet is usually low-reflection coated and the rear facet is as-cleaved or high-reflection coated.

2.4.1.2. Phase-Shifted DFB Lasers

To get reproducible single-mode oscillation, the λ/4-shifted DFB laser is one of the most promising light sources. The emission spectra of BH λ/4-shifted DFB lasers with a multiple-quantum-well (MQW) active layer (a) and with a bulk active layer (b) are shown in Fig. 2.20.[40] The MQW active layer has ten InGaAs well layers and

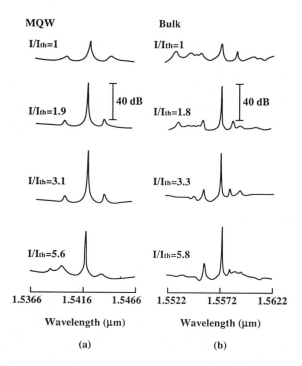

FIGURE 2.20. Spectra of (a) a multiquantum well (MQW) and (b) a bulk λ/4 DFB laser. *I* is the operating current and I_{th} is the threshold current (after Ref. 40 © 1990 IEEE).

InGaAsP barrier layers grown by MOCVD. The spectral characteristics inherent in λ/4-shifted DFB lasers are clearly seen in the amplified spontaneous emission spectra at a threshold bias. The main oscillation mode is at the center of the stopband and it increases as the bias current increases. Stable single-mode oscillation is obtained, and the SMSR is above 40 dB. At higher output powers, however, the oscillation-mode wavelength tends to move to longer-wavelength side of the stopband and the biggest submode with a shorter wavelength grows in the DFB laser with a bulk active layer. As described in Sec. 2.2, this is due to so-called spatial hole burning in the phase-shifted region. This effect eventually causes the single-mode stability to deteriorate even in λ/4-shifted DFB lasers. To suppress this spatial hole-burning effect, several methods were proposed: (1) Coupling coefficient of the grating is optimized,[21] (2) a multiple-quantum-well structure is introduced to the active layer, and (3) a corrugation-pitch-modulated grating structure is utilized,[41] in which the phase shift corresponding to λ/4 is produced along a certain length of a grating region having a slightly different corrugation pitch. This structure reduces the concentration of the light field into the phase-shift region and thereby reduces the spatial hole burning effect. Combination of those methods greatly reduces the spatial hole burning effect, which will be described in detail in Sec. 2.5.

Polarization selectivity is smaller in $\lambda/4$-shifted DFB lasers than that in DFB lasers with a uniform grating, because the reflectivity of both facets should be minimized to get reproducible single-mode oscillation in $\lambda/4$-shifted DFB lasers, resulting in smaller enhancement of the TE mode by the facet reflection. The SMSR is frequently limited by the intensity of the TM mode. Such a problem can be solved by utilizing the MQW structure in the active layer, because the MQW structure has polarization dependence of the gain and the gain of the TE mode is much larger than that of the TM mode. This advantage of the MQW structure has been experimentally confirmed.

2.4.2. Dynamic Characteristics

2.4.2.1. Intensity Modulation

Conventional Fabry–Perot lasers tend to oscillate in several modes, especially when the laser is directly modulated. This is because when the laser current is modulated from below threshold, the gain transiently exceeds the threshold gain for other modes over a certain wavelength range. This transient multimode oscillation disappears in several nanoseconds corresponding to the carrier lifetime and the laser oscillates in a steady state. The average number of oscillation modes therefore depends on the modulation speed. In Fabry–Perot lasers, this kind of spectral broadening degrades transmission characteristics because optical pulse deformation and also mode-competition noise are caused by wavelength dispersion of refractive index in optical fibers. In DFB lasers, however, the single mode is maintained even when the laser is directly modulated. Such single-mode lasers are therefore called dynamic single-mode (DSM) lasers, since in a steady state, conventional Fabry–Perot lasers also tend to oscillate in a single mode at high output powers. The spectra of directly modulated DFB lasers are shown in Fig. 2.21,[36] where it can be seen that a single-mode oscillation is maintained even when the laser is directly modulated. Although side modes are intensified by the modulation due to the transient overgrowth of the gain as those in Fabry–Perot lasers, a SMSR above 30 dB can be realized.

2.4.2.2. Spectral Broadening

As shown in Fig. 2.21, the spectral width of the oscillating mode tends to broaden as the modulation speed is increased, especially when the bias current is below threshold. This broadening is due to variation of refractive index, resulting from variation of the injected carrier density around threshold level at the transient. The spectral width of modulated DFB lasers depends on the modulation speed and on a parameter α defined as

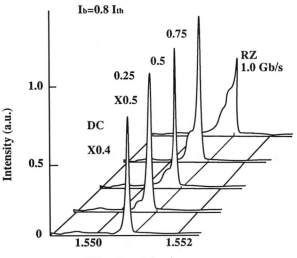

FIGURE 2.21. Spectra of a directly modulated bulk DFB laser. Parameter is the modulation speed in return-to-zero (RZ) format. I_b is the dc bias current, and I_{th} is the threshold current (after Ref. 36 © 1987 IEEE).

$$\alpha = -\frac{4\pi}{\lambda}\frac{(dn/dN)}{(dg/dN)}, \tag{2.63}$$

where λ is the wavelength, n is the refractive index, g is the gain, and N is the carrier density. The α value for bulk InGaAsP/InP lasers is estimated to be 5–8. This parameter is frequently called the "spectral linewidth enhancement factor" because, as described in the next subsection, it plays an important role in defining the spectral linewidth of semiconductor lasers. It was originally defined as indicated in Eq. (2.63), but recently the α parameter is also frequently expressed as the ratio of phase variation to intensity variation in a transmitted laser light. This kind of spectral broadening is becoming serious as the transmission speed and distance in fiber communication systems are increased. Because the wavelength dispersion of refractive index in conventional single-mode fibers is large in the 1.5 μm wavelength range, it is especially important that the α parameter should be reduced for long-distance transmissions in this wavelength range.

The α parameter can be reduced by increasing the so-called differential gain dg/dN. Increasing the value of dg/dN simultaneously improves the relaxation oscillation frequency inherent in semiconductor lasers, which limits their high-speed performance. The differential gain dg/dN can be increased by detuning of the oscillation wavelength from the gain peak wavelength, since dg/dN is larger on the shorter-wavelength side of the gain-wavelength profile. This detuning is accomplished by designing the grating pitch for the oscillation wavelength of a DFB mode

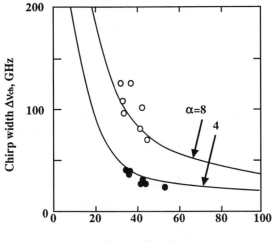

FIGURE 2.22. Chirping width as a function of pulse width in gain-switched DFB lasers with an MQW (●) and a bulk (○) active layer. α is the linewidth enhancement factor (after Ref. 45).

to be within this wavelength range. The concept of detuning has been theoretically proposed[42] and experimentally confirmed.[43] The differential gain can also be improved by incorporating an MQW structure into the active layer, where two-dimensional confinement of electrons into thin quantum wells enhances dg/dN.[44] The values of α parameters were experimentally evaluated by measuring wavelength chirping in gain switched bulk and MQW DFB lasers, and these values are shown in Fig. 2.22.[45] An α parameter of 3–4 has been reached in the MQW lasers, whereas that of the bulk DFB lasers is 8. As shown in Fig. 2.22, the spectral broadening of a directly modulated MQW DFB laser was much reduced compared with that of a bulk DFB laser.

The α parameter can be further reduced by making strained MQW structures and by making multiple quantum wire structures. For very long distance transmissions above several 100 km at bit rates above several Gb/s, however, the α parameter must be controlled below 1.0. For such applications, external optical modulators combined with a DFB laser as a light source are being studied intensively, because their α parameter is usually below 1.0.[46]

2.4.2.3. Frequency Modulation

A frequency-shift keying (FSK) modulation scheme is being intensively studied for use in coherent communication systems. This kind of modulation is accomplished by directly modulating DFB lasers, since their wavelength (that is, their optical frequency) can be modulated directly by refractive index variation. Small signal pulse modulation is used to minimize amplitude modulation, but the wavelength in the low-frequency range is thermally modulated and the direction of the wavelength shift is opposite to that resulting from refractive index variation. The efficiency of frequency modulation is therefore reduced in the low-frequency range, around several MHz. To overcome this problem, improved devices, such as multielectrode DFB lasers,[47] and improved modulation schemes, such as alternate-mark-inversion modulations,[48] are being studied.

2.4.3. Spectral Linewidth

The spectral linewidth of single-mode semiconductor lasers is important in coherent communication systems and in measurement systems. The spectral linewidth $\Delta \nu$ of semiconductor lasers is theoretically expressed as

$$\Delta \nu = \frac{v_g^2 h \nu g \, \alpha_m n_{\rm sp}}{8 \pi P} \, (1 + \alpha^2), \tag{2.64}$$

where v_g is the velocity of light in the active layer, $h \nu$ is the photon energy, g is the threshold gain, α_m is the mirror loss, $n_{\rm sp}$ is the spontaneous emission factor, α is the linewidth enhancement factor, and P is the output power. This formula is the same as the Shallow–Towns formula except for the addition of an α parameter.[49] Since the α parameter for semiconductor lasers is substantially larger than 1, this factor dominates the estimation of the spectral linewidth. Experimental results of the linewidth are shown in Fig. 2.23 as a function of inverse output power for bulk and MQW DFB lasers. The spectral linewidth difference between bulk and MQW DFB lasers is due to the different α parameter of those two kinds of lasers. As stated in the preceding subsection, the α parameter of MQW lasers is reduced by increasing the differential gain. In both kinds of lasers the spectral linewidth narrows as the output power increases, as predicted by Eq. (2.64). It saturates at a certain output power, however, and rebroadens above this power. The reasons for the linewidth rebroadening are not yet fully understood. One is degradation of SMSR caused by spatial hole burning at high output powers. Mode partition noise resulting from the reduction of SMSR can increase the spectral linewidth. Another discrepancy between experimental results and theoretical predictions is that the linewidth does not extrapolate to 0 at an inverse output power of 0. The reason for the residual linewidth is also not clear. One explanation suggested is that the residual linewidth is caused by $1/f$ noise from measurement circuits and from the series resistance of the semiconductor laser itself.[50,51]

2.4.4. Other Characteristics

2.4.4.1. Laser Noise

With conventional Fabry–Perot mode lasers, there is a large amount of mode-partition noise, when wavelength-selective components are used in optical systems. Because the optical fibers used in optical transmission systems have wavelength

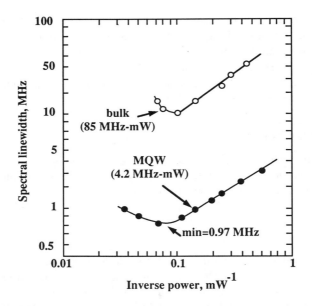

FIGURE 2.23. Spectral linewidth as a function of inverse output power of DFB lasers with a MQW (●) and a bulk (○) active layer (after Ref. 45).

selectivity due to their wavelength dispersion, mode-partition noise is a serious problem in high-speed digital and analog transmission systems. The DFB laser with a high SMSR eliminates this mode-partition noise, and its noise level is determined mostly by the quantum noise associated with spontaneous emission. But because laser light is very coherent, optical feedback from external optical components—such as the reflection from optical connectors—causes feedback noise. The DFB laser is therefore usually used with an optical isolator between the laser and optical fiber. Intensity modulation systems typically need more than 30 dB isolation. In coherent communication systems, on the other hand, fluctuation of optical frequency directly affects transmission characteristics and complete elimination of the feedback noise in those systems requires an isolation of at least 60 dB.

2.4.4.2. Reliability

Reliability is one of the most important issues for DFB lasers as light sources in optical communication systems. Because the active layer of a DFB laser is usually grown on a corrugated substrate, the quality of the epitaxial layers has been carefully studied from the viewpoint of reliability. Transmission electron micrographs have revealed that there is no serious dislocation in the epitaxial layers grown on the corrugated substrate, since the corrugation height necessary for DFB laser oscillation is only several tens of nanometers and the active layer is separated from the corrugated surface by an optical-guide layer.[33] And for DFB lasers, in contrast to the

TABLE 2.1. Properties of various kinds of high-speed lasers.

Laser structure	Active layer	Wavelength	Bandwidth	Reference
Semi-insulator buried DFB	Bulk	1.3 μm	18 GHz	53
Polyimide buried DFB	Bulk	1.55 μm	17 GHz	54
Constricted mesa DFB	Strained 4 QW	1.53 μm	17 GHz	55
Ridge waveguide FP	4 QW	1.10 μm	30 GHz	56
FP	Strained 7 QW	1.55 μm	25 GHz	57

conventional Fabry–Perot lasers, we must be concerned with whether or not the stability of single-mode oscillation deteriorates with aging. Studies of aging characteristics have confirmed, however, that single-mode stability does not seriously deteriorate as long as other device characteristics do not. Recent studies of the reliability of DFB lasers for coherent communication systems have found that increases of the threshold current result in increased spectral linewidth,[52] due mainly to increased cavity loss. The DFB laser is nonetheless reliable enough for use in coherent communication systems.

2.5. RECENT PROGRESS

2.5.1. High-Speed Lasers

High-speed lasers are necessary for high-bit-rate data transmission systems, and the properties of different kinds of high-speed lasers are summarized in Table 2.1. The modulation bandwidth of semiconductor lasers is limited by relaxation oscillation and frequency rolloff. The relaxation oscillation frequency is proportional to the square root of differential gain and it can be increased by introducing quantum-well or strained-quantum-well structures into the active region. Frequency rolloff is caused by parasitic capacitance and resistance. The parasitic capacitance mainly comes from the reverse p-n junction of embedding layers and it can be reduced to few pF by embedding the active stripe with Fe-doped semi-insulating semiconductor or polyimide. The resistance is typically 2–6 Ω and it consists of contact resistance and series bulk resistance. The series bulk resistance is dominant and it can be reduced by heavily doping the cladding layer.

2.5.2. Narrow Spectral Linewidth Lasers

Narrow spectral linewidth semiconductor lasers are necessary for coherent transmission systems, and the exact spectral linewidth required depends on the

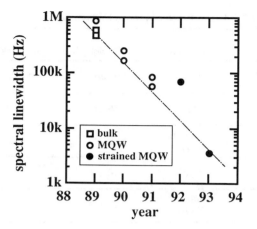

FIGURE 2.24. Trend for narrowing the spectral linewidth.

modulation/detection schemes and the transmission speed. Schemes which have higher receiver sensitivity require narrower spectral linewidth and the linewidth required becomes broader with increasing transmission speed. Frequency modulation/heterodyne detection schemes with transmission speed of 10 Gbit/s require a spectral linewidth of 10 MHz, and phase modulation/homodyne detection schemes, which have the highest receiver sensitivity, require a spectral linewidth less than 100 kHz for 10 Gbit/s data transmission.

Although the spectral linewidth of semiconductor lasers is theoretically proportional to the inverse power according to the modified Shawlow–Townes equations,[49] the linewidth rebroadening and the linewidth floor are observed in actual devices as described in Sec. 2.4.3. The recent trend in the narrowing of spectral linewidth is shown in Fig. 2.24. A spectral linewidth of less than 1 MHz was first obtained by introducing multiquantum-well (MQW) structure into the active layer.[58] This structure can reduce the spectral linewidth enhancement factor to half the value of the bulk structure, and this is effective for reducing the linewidth-power product and the spatial hole-burning. Spectral linewidths less than 100 kHz were obtained by making a long-cavity DFB laser with a corrugation-pitch-modulated (CPM) grating structure[59] that suppresses the spatial hole-burning effect and also by making a long-cavity DBR laser with an optimized grating structure.[60] And in 1993 a spectral linewidth of only 3.6 kHz[61] was obtained by introducing strained multiquantum-well structure into a long-cavity CPM-DFB laser (Fig. 2.25). This combination of CPM structure and the reduction of spectral linewidth enhancement factor by a strained MQW structure very efficiently reduces the spatial hole-burning.

2.5.3. Wavelength Tunable Lasers

Wavelength tunable lasers are required as a local oscillator for selecting a channel in coherent FDM systems and as a wavelength converter in wavelength division

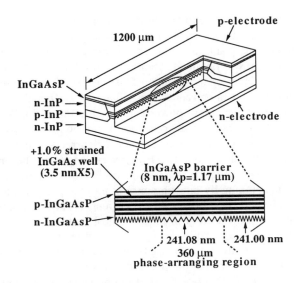

FIGURE 2.25. Structure of strained MQW corrugation-pitch-modulated DFB laser (after Ref. 61).

multiplex (WDM) and frequency division multiplex (FDM) systems. Many kinds of wavelength tunable lasers based on DFB and DBR lasers have been proposed for obtaining a wide wavelength tuning range, and the structures and characteristics of representative wavelength tunable lasers are summarized in Table 2.2.

Multisection DFB lasers[62] can tune the wavelength by changing the ratio of the

TABLE 2.2. Properties of various kinds of wavelength tunable lasers.

Laser structure	Tuning range	Spectral linewidth	Tuning speed	Output power
Multi-section DFB	Continuous several nm	Less than few MHz	0.1 ns	more than 10 mW
Multi-section DBR	Continuous several nm	Less than several 10 MHz	1 ns	more than 10 mW
Heater-integrated DFB	Continuous several nm	Less than few MHz	several ms	more than 10 mW
TTG*	Continuous several nm	Less than several 10 MHz	1 ns	several mW
SSG** DBR SG*** DBR	Non-continuous 100 nm	Less than several 10 MHz	0.1 ns	more than 10 mW

TTG*; Tunable twin guide
SSG**; Super structure grating
SG***; Sampled grating

currents injected into each electrode, but the continuous tuning range is limited to a few nanometers. The Bragg wavelength and the phase of the optical wave are controlled simultaneously and the tuning mechanism is complicated. A continuous tuning range of 7.2 nm was obtained recently by introducing strained multiquantum wells into the active layer of multisection DFB lasers.[63] The spectral linewidth of multisection DFB lasers can be reduced less than a few MHz, their tuning speed is as fast as 0.1 ns, and their output power can be more than 10 mW.

Multisection DBR lasers,[64] on the other hand, can be given a continuous tuning of several nanometers by controlling the ratio of currents injected into the DBR section and phase control section. The Bragg wavelength can be tuned shorter by injecting current into the DBR section and the optical wave phase can be controlled to obtain continuous tuning by injecting current into the phase control section. The spectral linewidth becomes broad when current is injected into the passive regions—that is, the DBR section or phase control section.

Although a laser with a thin-film heater integrated near the active layer[65] can be tuned to longer wavelengths, its response time is several milliseconds—which is six orders of magnitude slower than multisection DFB lasers.

A tunable twin guide (TTG) laser[66] has a tuning layer parallel to the active layer and its wavelength can be shortened by injecting the current into the tuning layer. This current injection broadens the spectral linewidth and increases the optical loss, thereby reducing the output power.

Wavelength tuning ranges of about 100 nm have been reported for superstructure grating DBR lasers[67] and sampled grating DFB lasers.[68] The continuous wavelength tuning range of these lasers is only few nanometers, but their discrete modes cover about 100 nm. These lasers are useful for wavelength switching in optical cross connections.

2.5.4. Gain-Coupled DFB Lasers

As stated in Sec. 2.2.4, gain-coupled DFB lasers with a uniform grating can oscillate in a stable single mode at the Bragg wavelength. The single-mode operation of gain-coupled DFB lasers is more resistant to instabilities caused by facet reflection and reflected light returning into the laser cavity than is the single-mode operation of index-coupled DFB lasers. This is because gain-coupled lasers have a larger threshold gain difference between the main mode and the biggest submode. The first cw operation of a gain-coupled DFB laser at room temperature[69] was reported for the structure shown in Fig. 2.26(a). This gain-coupled DFB laser, made by compensating the index grating with the buffer layer, was a GaAs/AlGaAs laser emitting at 880 nm. Its averaged threshold current was 20 mA, its cavity length was 200 μm, and it operated in a single mode without facet coating. Simpler gain-coupled DFB lasers structures have been since reported. In the structure shown in Fig. 2.26(b), for example, a loss grating[70] is introduced to produce the gain-coupled grating. A gain-coupled grating and an index-coupled grating are combined and they are antiphase in this structure. The duty factor of the loss grating should be controlled

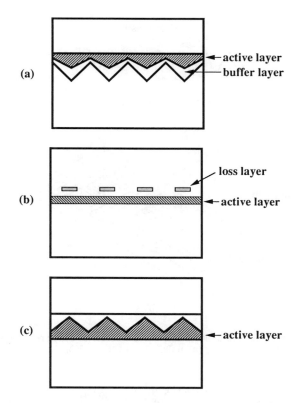

FIGURE 2.26. Structure of gain-coupled DFB lasers. (a) Pure gain coupling. (b) Gain-coupling with antiphase index grating. (c) Gain-coupling with inphase index coupling.

to be less than 0.2 to suppress the increase of the threshold current caused by the loss grating. From viewpoint of reliability, this structure is very promising because the active layer is not processed directly. And it has been pointed out recently that the spectral linewidth enhancement factor could be theoretically reduced to half of the material value with this antiphase structure.[71] In the structure shown in Fig. 2.26(c), the active layer is corrugated to produce the gain-coupled grating. In this structure, the combined gain-coupled grating and index-coupled grating are in-phase.

2.5.5. Grating Surface-Emitting Lasers

Surface-emitting lasers are suitable for integration and are promising for optical interconnections and optical parallel processing. The lasing light of DFB or DBR lasers can be taken out perpendicular to the surface by using the second-order grating. The condition for Bragg refraction is given as follows:

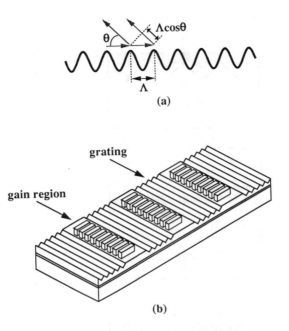

(a)

(b)

FIGURE 2.27. Grating surface-emitting laser. (a) Condition for Bragg reflection. (b) Structure of grating surface-emitting laser array (after Ref. 77 © 1989 IEEE).

$$\Lambda(1 + \cos \ \theta) \ = \ m n_{\text{eff}} \lambda_0, \tag{2.65}$$

where Λ is the period of the grating and θ is the refraction angle as shown in Fig. 2.27(a), m is the order number, n_{eff} is the effective refractive index, and λ_0 is the wavelength in vacuum. The second-order grating ($m = 2$) provides the light for optical feedback ($\theta = 0$) generating laser oscillation, and it provides the light that is emitted perpendicular to the surface ($\theta = 90°$).

High output power (30–150 mW) and improved differential quantum efficiency (10%–30%) have been obtained by introducing multiquantum wells into the active layer.[72–74] Further increases in power and reductions of beam divergence have been achieved with array structure.[75,76] A grating surface-emitting laser array is illustrated in Fig. 2.27(b).[77] The gratings between the gain sections provide feedback for laser oscillation and they provide the light radiated perpendicular to the surface. Part of the light is passed through each grating section and phase locking is achieved in the longitudinal direction. This structure achieved a maximum output power of more than 3 W and a differential quantum efficiency of 20%–46%.

2.5.6. Integrated Lasers

Lasers integrated with other functional devices, such as modulators and optical amplifiers, are reported to result in performance higher than that obtained when

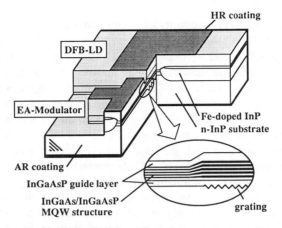

FIGURE 2.28. Structure of integrated light source composed of a DFB laser and an electroabsorption modulator (after Ref. 80).

using isolated lasers, and it is important for this integration that the active layer of lasers and the waveguides of other functional devices be connected smoothly. Conventional processing techniques require etching and regrowth to fabricate regions that have different band gaps in the lateral direction. Selective epitaxial growth, however, allows these regions to be fabricated in one-step epitaxial growth. The thickness of a selectively grown layer depends on the mask pattern because most of the source species incident on the masked region do not deposit there; instead they diffuse laterally to the growth window. Layer thickness increases when the mask is narrower. The band-gap energy of selectively grown multiquantum-well structures therefore decreases with increases in the stripe width of the mask—because of the increased well thickness.[78,79]

Figure 2.28 shows the structure of a laser integrated with electroabsorption modulator and fabricated by using a selective epitaxial growth technique.[80] The laser active layer and the modulator absorption layer consist of one continuous InGaAs/InGaAsP MQW structure with slightly different thicknesses and compositions. The band-gap energy of the laser active layer and that of modulator waveguide layer were controlled by selective epitaxial growth: The photoluminescence peak wavelength of laser active layer is 1.56 μm and that of modulator waveguide layer is 1.48 μm. The transition regions between the two sections is less than 50 μm long. The 3 dB bandwidth of this integrated device is 14 GHz and under 10 Gbit/s modulation the chirping is less than 0.1 nm.

2.6. CONCLUDING REMARKS

This chapter reviewed the theory and performance of the DFB laser, which operates in a stable single-longitudinal mode and has been widely used as a light source in high-speed optical communication systems. One crucial problem yet to be solved, however, is that of the reproducibility of oscillation wavelength. Wavelength-division-multiplexed (WDM) transmission systems, for example, have recently been intensively studied because of their potential value in increasing transmission capacity and in constructing multiwavelength network architectures. Each channel of a WDM system is allocated a different wavelength and requires a DFB laser oscillating at this wavelength. Therefore, multiple DFB lasers with different wavelengths should be prepared for this system. DFB lasers are also used as the local oscillators of the receivers in coherent communication systems, where the wavelength of the local oscillator laser must be precisely tuned to that of the transmitter laser. Frequency-division-multiplexed (FDM) transmission systems are also being studied using the coherent communication technology in the attempt to further increase transmission capacity. An advantage of the FDM system is that the wavelength allocated to each channel can be very close to that of the adjacent channels because each channel can be selected by precisely tuning the wavelength of the local oscillator laser in the receiver. Both WDM and FDM systems thus require DFB lasers with precisely predetermined wavelengths. The wavelength of a DFB laser is determined not only by the grating pitch but also by the effective refractive index of the waveguide, which depends on the width of the active region as well as on the thicknesses of the layers. Production of DFB lasers for these systems will therefore require the precision of fabrication processes to be much improved.

Another open problem in DFB laser technology is that of extending the range of wavelength available. In addition to having applications in communication systems, DFB lasers are useful in optical measurement systems requiring laser light that is coherent or wavelength stable or both. The DFB laser can, for instance, be used to detect a gaseous ingredient in the air, since any gas has a series of sharp absorption lines at specific wavelengths. The DFB lasers that are presently available commercially are made for communication applications and operate at wavelength ranges near 1.3 and 1.55 μm. Application fields in measurement systems would be widely extended if DFB lasers were available for other wavelength ranges.

Another problem still to be solved is the fabrication of gratings for DFB lasers. Because DFB lasers need grating pitches as small as 200 nm even in the 1.3 wavelength range, their gratings can so far be made by using the ultraviolet laser interference method, which is less stable and therefore has a lower yield than usual lithographic methods. Lithography technologies, including electron beam lithography, are being gradually improved for other semiconductor devices, such as Si-LSIs, and these more stable, industrial technologies will also be applied to fabricate DFB laser gratings.

Finally, the DFB (and DBR) lasers have an advantage over the Fabry–Perot laser in that they can be monolithically integrated with other optical devices, since mirror facets are not necessary for laser oscillation. Such integrated devices are being intensively studied, and optical integrated circuits using the DFB or DBR laser as a light source will be soon commercially available for communication systems. One of the most promising of these circuits is, as described in Sec. 2.5, a DFB laser integrated with an optical modulator.

REFERENCES

1. H. Kogelnik and C. V. Shank, Appl. Phys. Lett. **18**, 152 (1971).
2. M. Nakamura, A. Yariv, H. W. Yen, and S. Somekh, Appl. Phys. Lett. **22**, 515 (1973).
3. D. R. Scifres, R. D. Burnham, and W. Streifer, Appl. Phys. Lett. **25**, 203 (1974).
4. H. C. Casey, Jr., S. Somekh, and M. Ilegems, Appl. Phys. Lett. **27**, 142 (1975).
5. M. Nakamura, A. Aiki, J. Umeda, and A. Yariv, Appl. Phys. Lett. **27**, 403 (1975).
6. K. Aiki, M. Nakamura, and J. Umeda, IEEE J. Quantum Electron. **QE-12**, 597 (1976).
7. H. Kogelnik and C. V. Shank, Appl. Phys. Lett. **43**, 2327 (1972).
8. A. Yariv, IEEE J. Quantum Electron. **QE-9**, 919 (1973).
9. S. Wang, IEEE J. Quantum Electron. **QE-10**, 413 (1974).
10. A. Yariv and A. Gover, Appl. Phys. Lett. **26**, 537 (1975).
11. A. Doi, T. Fukuzawa, M. Nakamura, R. Ito, and K. Aiki, Appl. Phys. Lett. **35**, 441 (1979).
12. K. Utaka, S. Akiba, K. Sakai, and Y. Matsushima, Electron. Lett. **17**, 961 (1981).
13. T. Matsuoka, Y. Yoshikuni, and G. Motosugi, Electron. Lett. **21**, 1151 (1985).
14. H. A. Haus and C. V. Shank, IEEE J. Quantum Electron. **QE-12**, 532 (1976).
15. A. Suzuki and K. Tada, Proc. SPIE **239**, 10 (1980).
16. K. Sekartedjo, N. Eda, K. Furuya, Y. Suematsu, F. Koyana, and T. Tanbun-Ek, Electron. Lett. **20**, 80 (1984).
17. K. Utaka, S. Akiba, and Y. Matsushima, Electron. Lett. **20**, 1008 (1984).
18. M. Shirasaki, H. Soda, S. Yamaguchi, and H. Nakajima, 11th ECOC/5th IOOC Dig. (Venezia), pp. 25–28, 1985.
19. T. Numai, M. Yamaguchi, I. Mito, and K. Kobayashi, Jpn. J. Appl. Phys. **26**, L1910 (1987).
20. M. Okai, S. Tsuji, N. Chinone, and T. Harada, Appl. Phys. Lett. **55**, 415 (1989).
21. H. Soda, Y. Kotaki, H. Sudo, H. Ishikawa, S. Yamakoshi, and H. Imai, IEEE J. Quantum Electron. **QE-23**, 804 (1987).
22. F. K. Reinhart, R. A. Logan, and C. V. Shank, Appl. Phys. Lett. **27**, 45 (1975).
23. K. Abe, K. Kishino, T. Tanbun-Ek, S. Arai, F. Koyana, K. Matsumoto, T. Watanabe, and Y. Suematsu, Electron. Lett. **18**, 410 (1982).
24. O. Mikami, T. Saitoh, and H. Nakagome, Electron. Lett. **18**, 458 (1982).
25. Y. Tohmori, Y. Suematsu, H. Tsushima, and S. Arai, Electron. Lett. **19**, 656 (1983).
26. M. Yamada and K. Sakuda, Appl. Opt. **26**, 3474 (1987).
27. Y. Nakano, O. Kamatani, and K. Tada, 11th IEEE Int. Semiconductor Laser Conf. (Boston), paper H-2, 1988.
28. M. Aoki, K. Uomi, T. Tsuchiya, S. Sasaki, M. Okai, and N. Chinone, IEEE J. Quantum Electron. **27**, 1782 (1991).
29. K. Utaka, S. Akiba and Y. Matsushima, Electron. Lett. **20**, 1008 (1984).
30. A. Suzuki and K. Tada, Thin Solid Films **72**, 419 (1980).
31. M. Nakao, K. Sato, T. Nishida, T. Tamamura, A. Ozawa, Y. Sato, I. Okada, and H. Yoshihara, Electron. Lett. **25**, 148 (1989).
32. J. Kinoshita, J. Okuda, and Y. Uematsu, Electron. Lett. **19**, 215 (1983).
33. H. Nakamura, S. Tsuji, A. Ohishi, M. Hirao, H. Kakibayashi, and H. Matshumura, Int. Symp. GaAs and Related Compounds (Kanazawa), pp. 169–174, 1985.
34. H. Nagai, Y. Noguchi, T. Matsuoka, and Y. Suzuki, Jpn. J. Appl. Phys. **22**, L291 (1983).
35. M. Ohishi, Y. Itaya, and Y. Imamura, 10th IEEE Int. Semiconductor Laser Conf. (Kanazawa), paper G-5, 1986.
36. S. Tsuji, A. Ohishi, H. Nakamura, M. Hirao, N. Chinone, and H. Matsumura, J. Lightwave Technol., **LT-5**, 822 (1987).
37. S. Akiba, K. Utaka, K. Sasaki, and Y. Matsushima, IEEE J. Quantum Electron. **QE-19**, 1052 (1983).
38. T. Matsuoka, Y. Yoshikuni, and H. Nagai, IEEE J. Quantum Electron. **QE-21**, 1880 (1985).
39. T. Matsuoka, Y. Yoshikuni, and G. Motosugi, Electron. Lett. **21**, 1151 (1985).
40. M. Aoki, K. Uomi, S. Sasaki, and N. Chinone, IEEE Photon. Technol. Lett. **2**, 617 (1990).
41. M. Okai, N. Chinone, H. Taira, and T. Harada, IEEE Photon. Technol. Lett. **1**, 200 (1989).
42. K. Vahala and A. Yariv, Appl. Phys. Lett. **45**, 501 (1984).
43. K. Kamite, H. Sudo, M. Yano, H. Ishikawa, and H. Imai, IEEE J. Quantum Electron. **QE-23**, 1054 (1987).
44. Y. Arakawa and A. Yariv, IEEE J. Quantum Electron. **QE-21**, 1666 (1985).
45. K. Uomi, S. Sasaki, T. Tsuchiya, and M. Okai, Electron. Lett. **26**, 52 (1990).

46. M. Suzuki, H. Tanaka, H. Taga, S. Yamamoto, and Y. Matushima, IEEE J. Lightwave Technol. **LT-10**, 90 (1992).
47. Y. Yoshikuki and G. Motosugi, J. Lightwave Technol. **LT-5**, 516 (1987).
48. H. Tsushima, S. Sasaki, R. Takeyari, and K. Uomi, J. Lightwave Technol. **9**, 666 (1991).
49. C. H. Henry, "Theory of the linewidth of semiconductor lasers," IEEE J. Quantum Electron., **18**, 259 (1982).
50. K. Kikuchi, IEEE J. Quantum Electron. **25**, 684 (1989).
51. M. Okai, T. Tsuchiya, K. Uomi, N. Chinone, and T. Harada, IEEE Photon. Technol. Lett. **3**, 427 (1991).
52. M. Fukuda, F. Kano, T. Kurosaki, and J. Yoshida, IEEE Photon. Technol. Lett. **4**, 305 (1992).
53. S. J. Wang, T. M. Shen, and N. K. Dutta, IEEE Phton. Technol. Lett. **1**, 258 (1989).
54. K. Uomi, H. Nakano, and N. Chinone, Electron. Lett. **25**, 668 (1989).
55. Y. Hirayama, M. Morinaga, M. Onomura, M. Tanimura, M. Tohyama, M. Funemizu, M. Kushibe, N. Suzuki, and M. Nakamura, IEEE J. Light. Technol. **10**, 1272 (1992).
56. S. Weisser, J. D. Ralton, E. C. Larkins, I. Esquivias, P. J. Tasker, J. Fleissner, and J. Rosenzweig, Electron. Lett. **28**, 2141 (1992).
57. P. A. Morton, R. A. Logan, T. Tanbun-Ek, P. F. Sciortino, Jr., A. M. Sergent, R. K. Montgomery, and B. T. Lee, Electron. Lett. **28**, 2156 (1992).
58. S. Takano, T. Sasaki, H. Yamada, M. Kitamura, and I. Mito, Electron. Lett. **25**, 356 (1989).
59. M. Okai, T. Tsuchiya, A. Takai, and N. Chinone, IEEE Photon. Technol. Lett. **4**, 526 (1992).
60. T. Kunii, Y. Matsui, H. Horikawa, T. Kamijyo, and T. Nonaka, Electron. Lett. **27**, 691 (1991).
61. M. Okai, M. Suzuki, and T. Taniwatari, Electron. Lett. **29**, 1696 (1993).
62. Y. Yoshikuni, K. Oe, G. Motosugi, and T. Matsuoka, Electron. Lett. **22**, 1153 (1986).
63. P. I. Kaindersma, W. Scheepers, J. H. M. Cnoops, P. J. A. Thijs, G. L. A. v. d. Hofstad, T. V. Dongen, and J. J. M. Binsma, 12th IEEE Int. Semiconductor Laser Conf. (Davos), paper M-4, 1991.
64. S. Murata, I. Mito, and K. Kobayashi, Electron. Lett. **23**, 403 (1987).
65. S. Sakano, T. Tsuchiya, M. Suzuki, S. Kitajima, and N. Chinone, IEEE Photon. Technol. Lett. **4**, 321 (1992).
66. S. Illek, W. Thulke, C. Chanen, H. Lang, and M. C. Amann, Electron. Lett. **26**, 46 (1990).
67. Y. Tomori, Y. Yoshikuni, H. Ishii, F. Kano, T. Tamamura, and Y. Kondo, Electron. Lett. **29**, 352 (1993).
68. V. Jayaraman, A. Mathur, L. A. Coldren, and P. D. Dupkus, 13th Int. Semiconductor Laser Conf. (Takamatsu), post-deadline paper PD-11, 1992.
69. Y. Luo, Y. Nakano, K. Tada, T. Inoue, H. Hosomatsu, and H. Iwaoka, IEEE J. Quantum Electron. **27**, 1724 (1991).
70. Y. Nakano, Y. Luo, and K. Tada, Appl. Phys. Lett. **55**, 1606 (1989).
71. K. Kudo, J. I. Shim, K. Komori, and S. Arai, IEEE Photon. Technol. Lett. **4**, 531 (1992).
72. K. Kojima, S. Noda, K. Mitsunaga, K. Kyuma, and K. Hamanaka, Appl. Phys. Lett. **50**, 1705 (1987).
73. G. A. Evans, N. W. Carlson, J. M. Hammer, M. Lurie, J. K. Butler, S. L. Palfrey, L. A. Carr, F. Z. Hawrylo, E. A. James, C. J. Kaiser, J. B. Kirk, and W. F. Reichert, Appl. Phys. Lett. **51**, 1478 (1987).
74. G. A. Evans, N. W. Carlson, J. M. Hammer, M. Lurie, L. K. Bulter, L. A. Carr, F. Z. Hawrylo, C. J. Kaiser, J. B. Kirk, W. F. Reichert, S. R. Chinn, J. R. Sheady, and P. S. Zory, Appl. Phys. Lett. **52**, 1037 (1988).
75. N. W. Carlson, G. A. Evans, J. M. Hammer, M. Lurie, L. A. Carr, F. Z. Hawrylo, E. A. James, C. J. Kaiser, J. B. Kirk, W. F. Reichert, and D. A. Truxal, Appl. Phys. Lett. **52**, 939 (1988).
76. G. A. Evans, N. W. Carlson, J. M. Hammer, M. Lurie, L. K. Bulter, S. L. Palfery, R. Amantea, L. A. Carr, F. Z. Hawrylo, E. A. James, C. J. Kaiser, J. B. Kirk, W. F. Reichert, S. R. Chinn, J. R. Shealy, and P. S. Zory, Appl. Phys. Lett. **53**, 2123 (1988).
77. G. A. Evans, N. W. Carlson, J. M. Hammer, M. Lurie, L. K. Bulter, S. L. Palfery, R. Amantea, L. A. Carr, F. Z. Hawrylo, E. A. James, C. J. Kaiser, J. B. Kirk, and W. F. Reichert, IEEE J. Quantum Electron. **25**, 1525 (1989).
78. T. Kato, T. Sasaki, N. Nida, K. Komatsu, and I. Mito, 17th ECOC (Paris), paper WeB7-2, 1991.
79. M. Aoki, H. Sano, M. Suzuki, M. Takahashi, K. Uomi, and A. Takai, Electron. Lett. **27**, 2138 (1991).
80. M. Aoki, M. Suzuki, M. Takahashi, H. Sano, T. Ido, T. Kawano, and A. Takai, Electron. Lett. **28**, 1157 (1992).

CHAPTER 3

Phase Noise and its Control in Semiconductor Lasers

M. Kourogi and M. Ohtsu

*Tokyo Institute of Technology, 4259 Nagatuta-tyo, Midori-ku,
Yokohama-shi, Kanagawa 226, Japan and the Kanagawa Academy of
Science & Technology, 3-2-1 Sakado, Takatsu-ku, Kawasaki-shi,
Kanagawa 213, Japan*

3.1. INTRODUCTION

In recent years, the techniques of molecular beam epitaxy, chemical beam epitaxy, and metal organic-chemical vapor deposition have been used to deposit homogeneous and reproducible thin films on a substrate for semiconductor laser devices, and the technique of large-scale laser fabrication and integration has been developed.[1] Using these techniques, a structure of strained quantum wells has been employed to realize high-quality laser devices which cover a wide lasing wavelength range.[2] Although the semiconductor laser is inherently a multilongitudinal mode laser due to its wide gain bandwidth and its strong coupling between longitudinal modes originating from fast intraband relaxation of carriers, an advanced laser device, regarded for practical purposes as a single-mode laser,[3] has been successfully fabricated by utilizing the quarter-wavelength shifted distributed feedback (DFB) structure to suppress the unwanted vestigial modes.[4] In addition to these advanced devices, development of high-quality optical isolators enables laser devices to lase stably without suffering from the injection effects of external lightwaves. Furthermore, the frequency of a semiconductor laser can be swept by sweeping the injection current. The range of frequency sweep can be further extended if the technique of nonlinear optical frequency conversion is utilized by injecting the light of semiconductor lasers into nonlinear optical crystals. With the improving performance of these devices both active and passive semiconductor lasers and semiconductor laser-based coherent light sources have recently been used widely not only in the field of photonics but also in the study of basic physics such as atomic physics, the verification of quantum mechanics, and quantum optics.

However, the spectral linewidth and of the semiconductor laser device is still large, due to the large magnitude of frequency fluctuation [i.e., frequency modulation (FM) noise] and its large noise bandwidth. This FM noise must be reduced if the laser is to be used for the applications described above. For this purpose, it is advantageous to utilize a high efficiency and wide bandwidth of direct FM characteristics, which is due to the modulation of the refractive index of the active layer induced by plasma oscillation of carriers. That is, FM noise can be efficiently reduced by controlling the injection current of the laser. Moreover, control of carrier density by injecting external lightwaves into the laser active waveguide has also been used as a convenient optical feedback technique to reduce the FM noise so as to control the laser frequency. Studies on frequency control of semiconductor lasers have been actively carried out recently, and the resulting low FM noise semiconductor lasers have been used as the light sources for some advanced application systems.

To achieve a narrow linewidth, wide-band tunable semiconductor laser system, the minimum requirements listed below should be met.

(1) FM noise reduction of the field spectrum.

(2) Stable and wide-band frequency sweep.

If a high-quality light source meeting these requirements is made available, the following requirements should also be met to fully utilize the low FM noise properties.

(3) Accurate measurement of the optical frequency and center frequency stabilization of the field spectrum of the laser.

In this chapter, we review the recent progress made in satisfying these requirements. Our previous work on this topic has been summarized in Refs. 5 and 6.

3.2. PRINCIPLE OF FREQUENCY/PHASE CONTROL

In regard to requirements (1) and (2) above, this section describes the principle of FM noise reduction by negative electrical feedback and optical feedback, and the principle of the optical phase-locking loop.

3.2.1. Negative Electrical Feedback

Negative electrical feedback techniques using an external Fabry–Perot (FP) cavity have been widely used for reducing FM noise, thereby meeting requirement (1) above. Although the reproducibility of the resonance frequency of an FP cavity is lower than that of the atomic or molecular spectral lines,[5,7] a simpler, accurate, low FM noise, and low drift frequency reference can be achieved by using materials with extremely low expansion characteristics, such as Zerodure (a product of Schott Glasswork Co.) or ULE (a product of Corning Glass Co.) as a spacer for a super cavity with a finesse value as high as 1×10^6.[8] In negative electrical feedback techniques using the super cavity, the minimum value of FM noise attained based on the principle of FM noise reduction is less than that attainable by any other tech-

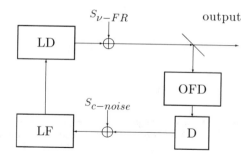

FIGURE 3.1. Block diagram of negative electrical feedback. LD: semiconductor laser. OFD: optical frequency discriminator. D: photodetector. LF: loop filter. $S_{\nu\text{-FR}}$: frequency noise of free-running laser. $S_{c\text{-noise}}$: noise in the photodetector.

nique. Described here is the principle behind the methods for stabilizing the laser frequency to the FP cavity using negative electrical feedback, and shown next are the theoretical limits of the method. Then, the theoretical limit of FM noise is compared in some different negative electrical feedback methods using an FP cavity as the optical frequency discriminator.

3.2.1.1. Principle of Negative Electrical Feedback

In negative electrical feedback control of the semiconductor laser frequency, the injection current of the laser is controlled by the demodulated signal after the magnitude of FM noise is determined by an optical frequency discriminator, as is shown by the block diagram in Fig. 3.1. The FM noise of a free-running laser is generated in the laser cavity and corresponds to both the quantum FM noise of the coherent state which is due to vacuum fluctuation and has been called the Schawlow–Townes' limit,[9-11] and the FM noise generated by fluctuations in injection current and in ambient temperature. The Schawlow–Townes' limit $S_{\nu\text{-FR-lim}}$ of the frequency noise power spectrum density in the free-running semiconductor laser is $(1+\alpha^2)h\nu_0/(2\pi\tau_p)^2 P_{\text{laser}}$, where the quantity α represents the linewidth enhancement factor, h is Planck's constant, ν_0 is the average optical frequency of the laser, τ_p is the photon lifetime of the laser cavity, and P_{laser} is the output power from the laser cavity. From this block diagram, the frequency noise power spectrum density under feedback condition $S_{\nu\text{-electrical}}$ is given by

$$S_{\nu\text{-electrical}} = \left|\frac{1}{1+H_{\text{open}}}\right|^2 S_{\nu\text{-FR}} + \left|\frac{H_{\nu/c}H_{\text{loop}}}{1+H_{\text{open}}}\right|^2 S_{c\text{-noise}}, \qquad (3.1)$$

where $S_{c\text{-noise}}$ is the power spectrum density of the current noise which is generated in the photodetector, $S_{\nu\text{-FR}}$ is the power spectrum density of the FM noise, $H_{\text{open}}(=H_{\nu/c}H_{P/\nu}H_{c/P}H_{\text{loop}})$ is the open loop transfer function, $H_{\nu/c}$ Hz/A is the current-frequency transfer function, $H_{P/\nu}$ W/Hz is the laser frequency-power transfer function of the optical frequency discriminator, $H_{c/P}$ A/W is the power-

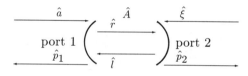

Fabry-Perot cavity

FIGURE 3.2. Internal field and the external field of FP cavity. $\hat{a}$: laser field incident on port 1. $\hat{\xi}$: vacuum fluctuation incident on port 2. $\hat{A}$: internal field of the cavity. $\hat{r},\hat{l}$: running wave along the right and the left directions in the cavity. $\hat{p}_1,\hat{p}_2$: output running wave from ports 1 and 2.

current transfer function of the detector, and H_{loop} is the transfer function of the loop filter. This equation shows that the value of $S_{\nu\text{-electrical}}$ approaches the value of $(S_{c\text{-noise}})/|H_{P/\nu}H_{\nu/c}|^2$ if feedback gain is sufficiently large. Furthermore, the magnitude of the residual FM noise under feedback can be smaller than the Schawlow–Townes' limit of the solitary laser if $(S_{c\text{-noise}})/|H_{P/\nu}H_{c/P}|^2 < [(1 + \alpha^2)h\nu]/(2\pi\tau_p)^2 P_{\text{laser}}$. This state of light, with very low FM noise falling below the Schawlow–Townes' limit, is referred to as the hyper-coherent state.

Then, to what extent can FM noise can be decreased? The intrinsic noise generated when the laser frequency is detected and fed back to the laser is a quantum noise which is called a "shot noise" in the photodetector. Therefore, if control is performed ideally, FM noise can be reduced to the FM noise minimum determined by the shot noise. The FM noise level when ideal control is performed can be obtained by replacing $S_{c\text{-noise}}$ in Eq. (3.1) by the power spectrum density of the shot noise $S_{c\text{-noise-lim}}$ ($= 2eI$ A^2/Hz, where e is the electron charge value, and I is the average photocurrent flowing in the photodetector). However, to understand the intrinsic nature of the limit to FM noise reduction, a more detailed discussion is required, and as will be given next.

3.2.1.2. The Limit of FM Noise Reduction by Negative Electrical Feedback

Detailed theoretical analysis of electrical feedback technique was presented in Ref. 12 in which a Mach–Zehnder interferometer was used as the optical frequency discriminator. In this subsection, the more important case where the FP cavity is used for the optical frequency discriminator is considered and the theoretical control limit is studied.

First, in order to study the FP cavity which is used for the optical frequency discriminator, the relationship between the internal and external fields of the FP cavity must be obtained. An analysis of the internal and external fields of the cavity was given previously in Ref. 13. Figure 3.2 shows how the vacuum fluctuations are injected into the FP cavity used for the feedback laser system. In this figure, the two mirror portions where the internal field of the FP cavity is coupled to the external field are called port 1 and 2. The complex amplitude of the electrolytic field of the laser beams, which are running waves incident to ports 1 and 2, are expressed by the photon annihilation operators $\hat{a}$ and $\hat{\xi}$, respectively. The internal field of the cavity is represented by the two operators: One propagating to the right and the other to the left, $\hat{r}$ and $\hat{l}$, respectively. Laser beams, which are running wave! s outgoing from ports 1 and 2, are expressed by the operators $\hat{p}_1$ and $\hat{p}_2$, respectively. Ignoring the optical loss inside the FP cavity, the relation between the internal and external field of the FP cavity is derived as follows:

$$\sqrt{T}\hat{a}(t) + \sqrt{R}\hat{l}(t) = \hat{r}(t), \qquad (3.2)$$

$$\sqrt{T}\hat{\xi}(t) + \sqrt{R}\hat{r}(t - \tau_l/2)\exp(-\pi\nu_0\tau_l) = \hat{l}(t + \tau_l/2)\exp(\pi\nu_0\tau_l), \qquad (3.3)$$

$$\hat{p}_1 = -\sqrt{R}\hat{a} + \sqrt{T}\hat{l}(t), \qquad (3.4)$$

$$\hat{p}_2 = -\sqrt{R}\hat{\xi} + \sqrt{T}\hat{r}(t - \tau_l/2)\exp(-\pi\nu_0\tau_l), \qquad (3.5)$$

where the value of $\hat{a}^\dagger\hat{a}$, $\hat{\xi}^\dagger\hat{\xi}$, $\hat{r}^\dagger\hat{r}$, $\hat{l}^\dagger\hat{l}$, $\hat{p}_1{}^\dagger\hat{p}_1$ and $\hat{p}_2{}^\dagger\hat{p}_2$ are normalized to the photon flux, R and T ($= 1-R$) are the reflectivity and transmission of the mirrors, respectively, and τ_l is the round-trip time of the laser beam inside the FP cavity. A small signal analysis, using Eqs. (3.2) and (3.3), leads to

$$\frac{d\hat{A}}{dt} = -\left\{\frac{1}{2\tau_{FP}} - j2\pi(\nu_c - \nu_0)\right\}\hat{A} + \sqrt{\frac{1}{2\tau_{FP}}}(\hat{\xi} + \hat{a}), \qquad (3.6)$$

where ν_c is the resonant frequency of the FP cavity to which the laser frequency is stabilized, τ_{FP} [$\cong (\tau_l/2T)$] is the photon lifetime of the FP cavity, and $\hat{A}$ ($= \sqrt{\tau_l}(\hat{r} + \hat{l})/2$) is the operator composed of two running waves of the internal field of the FP cavity. Here $\hat{A}^\dagger\hat{A}$ is normalized to the total photon number in the cavity. This equation represents Langevin's equation of the internal field of the FP cavity. As R can be approximated to unity, the following transform is made with for $\hat{p}_1$ and $\hat{p}_2$ by using $\hat{A}$:

$$\hat{p}_1 = -\hat{a} + \sqrt{\frac{1}{2\tau_{FP}}}\hat{A}, \qquad (3.7)$$

$$\hat{p}_2 = -\hat{\xi} + \sqrt{\frac{1}{2\tau_{FP}}}\hat{A}. \qquad (3.8)$$

Next, taking $\hat{a}$ and $\hat{A}$ to be the field having a frequency of ν_0, they can be approximated as

$$\hat{a} = (a_0 + \Delta\hat{a})\exp(j\hat{\phi}_a) \cong a_0 + \Delta\hat{a} + ja_0\hat{\phi}_a, \qquad (3.9)$$

$$\hat{A} = (A_0 + \Delta\hat{A})\exp(j\hat{\phi}_A) \cong A_0 + \Delta\hat{A} + jA_0\hat{\phi}_A, \qquad (3.10)$$

where a_0 and A_0 are the average amplitude of $\hat{a}$ and $\hat{A}$, and where $\Delta\hat{a}, \Delta\hat{A}$ and $\hat{\phi}_a, \hat{\phi}_A$ are fluctuations in amplitude and phase of $\hat{a}$ and $\hat{A}$, respectively. By

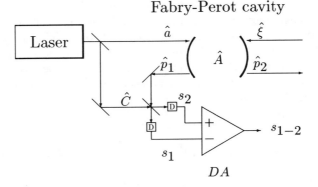

FIGURE 3.3. Construction of a frequency discriminator. $\hat{C}$: reference light for obtaining the imaginary part from $\hat{p}_1$. DA: differential amplifier. D's: photodetectors.

assuming that ν_0 is locked to ν_c for simplicity, Eqs. (3.9) and (3.10) are substituted into Eq. (3.6) to analyze these equations. As a result, the relations among phase, amplitude, and time are obtained, which are

$$A_0 = \sqrt{2\tau_{\mathrm{FP}}}a_0, \tag{3.11}$$

$$\frac{d\Delta\hat{A}}{dt} = -\frac{1}{2\tau_{\mathrm{FP}}}\Delta\hat{A} + \sqrt{\frac{1}{2\tau_{\mathrm{FP}}}}[\mathrm{Re}(\hat{\xi}) + \Delta\hat{a}], \tag{3.12}$$

$$\frac{d\hat{\phi}_A}{dt} = \frac{1}{2\tau_{\mathrm{FP}}}\left(-\hat{\phi}_A + \frac{\mathrm{Im}(\hat{\xi})}{a_0} + \hat{\phi}_a\right), \tag{3.13}$$

where Re() and Im() represent real part and imaginary part, respectively. The real and imaginary parts of the light amplitudes taken from the cavity are obtained by substituting Eqs. (3.9), (3.10), and (3.11) into Eqs. (3.7) and (3.8). They are

$$\mathrm{Re}(\hat{p}_1) = \sqrt{\frac{1}{2\tau_{\mathrm{FP}}}}\Delta\hat{A} - \Delta\hat{a}, \tag{3.14}$$

$$\mathrm{Im}(\hat{p}_1) = (\hat{\phi}_A - \hat{\phi}_a)a_0, \tag{3.15}$$

$$\mathrm{Re}(\hat{p}_2) = a_0 + \sqrt{\frac{1}{2\tau_{\mathrm{FP}}}}\Delta\hat{A} - \mathrm{Re}(\hat{\xi}), \tag{3.16}$$

$$\mathrm{Im}(\hat{p}_2) = \hat{\phi}_A a_0 - \mathrm{Im}(\hat{\xi}). \tag{3.17}$$

Next, the frequency error signal is discussed. Among various methods of using the FP cavity as an optical frequency discriminator, the simplest one is shown in Fig. 3.3 in which ν_0 is locked to ν_c without modulating the laser frequency. That is, in this method, a part of the laser is used as a reference [its operator is expressed as

$\hat{C} = (C_0 + \Delta\hat{C})\exp(j\hat{\phi}_c)$ as was the case for $\hat{a}$] to be interfered with by $\hat{p}_2$. In this interference, if the average phase difference between $\hat{p}_1$ and $\hat{C}$ is set at 90°, the output of the photodetector $\hat{s}_1, \hat{s}_2$ and the output of the differential amplifier $\hat{s}_{1-2}$ in Fig. 3.3 are expressed as follows:

$$\hat{s}_1 = \frac{C_0^2}{2} + C_0[\Delta\hat{C} + \text{Im}(\hat{p}_1)], \tag{3.18}$$

$$\hat{s}_2 = \frac{C_0^2}{2} + C_0[\Delta\hat{C} - \text{Im}(\hat{p}_1)], \tag{3.19}$$

$$\hat{s}_{1-2} = \hat{s}_1 - \hat{s}_2 = 2C_0 \, \text{Im}(\hat{p}_1). \tag{3.20}$$

It is found from these equations that the $\hat{s}_{1-2}$ is proportional to $\text{Im}(\hat{p}_1)$ and that it is free from intensity noise $\Delta\hat{C}$. From Eqs. (3.13), (3.15), and (3.20), the Fourier transform $F_{\hat{s}_{1-2}}$ of $\hat{s}_{1-2}$ can be obtained as follows:

$$F_{\hat{s}_{1-2}} = \frac{2C_0}{j4\pi f\tau_{FP} + 1} F_{\text{Im}(\hat{\xi})} - \frac{j8\pi f\tau_{FP}C_0 a_0}{j4\pi f\tau_{FP} + 1} F_{\hat{\phi}_a}, \tag{3.21}$$

where $F_{\text{Im}(\hat{\xi})}$ and $F_{\hat{\phi}_a}$ are Fourier transforms of $\text{Im}(\hat{\xi})$ and $\hat{\phi}_a$, respectively. The first term of the right side is a vacuum fluctuation injected from port 2 and it is the noise component. The second term is a signal component. As the first term, i.e., the noise term has the characteristics of a low-pass filter and the second term has those of a high-pass filter to the phase signal, high-frequency components are more sensitively detected. From this equation, $H_{P/\nu}$ is obtained by dividing the second term of the right side by Fourier spectrum $jfF_{\hat{\phi}_a}$ of instantaneous frequency and multiplying by $h\nu_0$. It is

$$H_{P/\nu} = \frac{8C_0 a_0 \pi\tau_{FP}h\nu_0}{j4f\pi\tau_{FP} + 1}. \tag{3.22}$$

From this equation, it is found that the optical frequency discriminator has first-order low-pass characteristics with a cutoff frequency of $(1/4\pi\tau_{FP})$.

In the laser without feedback, the power spectrum density of $\hat{s}_{1-2}$ is obtained as below:

$$S_{\hat{s}_{1-2}} = \frac{4C_0^2 S_{\text{Im}(\hat{\xi})}}{4(2\pi f\tau_{FP})^2 + 1} + \frac{16C_0^2(2\pi f\tau_{FP}a_0)^2 S_{\hat{\phi}_a}}{4(2\pi f\tau_{FP})^2 + 1}, \tag{3.23}$$

where $S_{\text{Im}(\hat{\xi})}$ represents the power spectrum density of the imaginary part of $\hat{\xi}$, and $S_{\hat{\phi}_a}$ represents the power spectrum density of $\hat{\phi}_a$ respectively. If light incident to the FP cavity is in the ideal coherent state (i.e., $a_0^2 S_{\hat{\phi}_a} = (a_0\hat{\phi}_a, a_0\hat{\phi}_a^\dagger) = \frac{1}{2}$, and $S_{\text{Im}(\hat{\xi})} = [\text{Im}(\hat{\xi}), \text{Im}(\hat{\xi}^\dagger)] = \frac{1}{2}$), the sum of the first and the second terms becomes white noise $2C_0^2$, generally called shot noise. If in the future phase-squeezed light is realized (i.e., if the relation of $a_0^2 S_{\hat{\phi}_a} < \frac{1}{2}$ is realized), phase detection of the laser below the shot noise limit can be made at high frequencies.

The power spectrum density $S_{\nu\text{-discriminator-lim}}$ representing the FM noise detection limit of the optical frequency discriminator can be obtained by dividing the first term of Eq. (3.23) by $[16C_0^2(2\pi f\tau_{FP}a_0)^2]/4(2\pi f\tau_{FP})^2+1$, which gives

$$S_{\nu\text{-discriminator-lim}} = \frac{S_{\text{Im}(\hat{\xi})}}{8\pi^2\tau_{FP}A_0^2}.$$ (3.24)

As the vacuum fluctuation gives a coherent state, the power spectrum density $S_{\text{Im}(\hat{\xi})}$ of the imaginary part of $\hat{\xi}$ is equal to $\frac{1}{2}$. By substituting it into Eq. (3.24)

$$S_{\nu\text{-discriminator-lim}} = \frac{1}{8(2\pi\tau_{FP})^2a_0^2} = \frac{h\nu_0}{8(2\pi\tau_{FP})^2P_{FP}},$$ (3.25)

or alternatively,

$$S_{\nu\text{-discriminator-lim}} = \frac{1}{16\pi^2\tau_{FP}A_0^2},$$ (3.26)

where P_{FP} is the laser power coupled into the FP cavity, i.e., $a_0^2h\nu$.

These equations represent the limit of FM detection using the optical frequency discriminator, and the control limit when ideal control is applied. However, this value is not the FM noise of the output lightwave but is the FM noise contained in the lightwave of the closed feedback loop. This is because, when controlling the laser frequency, the lightwave from the laser is separated in two parts: One is used for laser frequency control and the other is used for the output, and the two lightwaves are quantum mechanically uncorrelated. One may ask which part of the lightwave may be extracted for output and which part is used for control so as to maximize the output power and obtain minimum FM noise? The answer is that $\hat{p}_2$ must be used as output. With this method, optimum control can be realized without losing power. The laser power lost to control is the $C_0^2h\nu$ received by the photodetector, but this value can be assumed to be small enough. Since the value of $\hat{s}_{1-2}$ is maintained at zero with such a control method, then the relation of $\hat{\phi}_A = \hat{\phi}_a$ is given. With this control method, the phase fluctuation of the internal field of the FP cavity and that of the laser of the closed feedback loop are equalized, and the FM noise level of the internal field of the FP cavity becomes equal to the value in Eq. (3.25). From this result and Eq. (3.13), the FM noise limit of a field of output $\hat{p}_2$ under ideal electrical feedback conditions is given as follows:

$$S_{\nu\text{-electrical-lim}} = f^2S_{\hat{\phi}_A} = \frac{h\nu_0}{8(2\pi\tau_{FP})^2P_{FP}} + \frac{h\nu_0f^2}{2P_{FP}}.$$ (3.27)

The first term on the right side is the FM noise in the control loop. The second term is the FM noise generated by interference with the vacuum fluctuation from port 2, and this value determines the limit for the high frequency components of the noise level of the output lightwave.

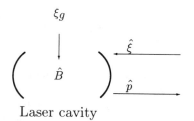

Laser cavity

FIGURE 3.4. External field, internal field of laser cavity. $\hat{\xi}$: vacuum fluctuation from the mirror. ξ_g: vacuum fluctuation from the gain medium. $\hat{B}$: internal field of the laser cavity. $\hat{p}$: output field of the laser cavity.

3.2.1.3. Comparison Between the Limit of FM Noise Reduction by Negative Electrical Feedback and Schawlow–Townes' Limit

The theoretical limit to laser FM noise in a free-running laser was first shown by Schawlow and Townes,[9] and then it was corrected by a factor of $\frac{1}{2}$ (Ref. 10) to give the present theory. Schawlow–Townes' limit $S_{\nu\text{-FR-lim}}$ to the FM noise level of a free-running laser is known as below if the α parameter is ignored, as

$$S_{\nu\text{-FR-lim}} = \frac{h\nu_0}{(2\pi\tau_p)^2 P_{\text{laser}}}. \tag{3.28}$$

When the equivalent optical cavity used for the laser is used as an FP cavity for an optical frequency discriminator for negative electrical feedback, the minimum of FM noise is determined by the photon lifetime of the FP cavity used and the incident laser power.

A comparison between Eqs. (3.28) and (3.27) shows that the value of the white noise level of $S_{\nu\text{-electrical-lim}}$ given by Eq. (3.27) has the same mathematical form as the Schawlow–Townes' limit given by Eq. (3.28) (i.e., the formula for vacuum fluctuations generated in the cavity of the solitary free-running laser), if τ_{FP} and if P_{FP} are replaced by τ_p and P_{laser}, respectively. However, these equations differ by the factor of $\frac{1}{8}$ that appears in Eq. (3.27), which is due to the difference in the way the vacuum fluctuations, i.e., the spontaneous emission fluctuations, are injected into the laser cavity and into the FP cavity. Though all the laser power P_{FP} incident upon the FP cavity can be extracted as output power as shown before, the FM noise level differs depending on whether an optical cavity is used as the free-running laser cavity or as the FP cavity for electrical negative feedback control.

Since no detailed discussions have been made of the constant $\frac{1}{8}$, we discuss it here now.

We first consider the contribution of vacuum fluctuations to the FM noise of the free-running laser. Figure 3.4 shows how the vacuum fluctuations are injected into the free-running laser cavity. If it is assumed that the laser oscillates at a sufficiently high bias level, Langevin's equation for the internal field of the laser oscillator can be shown to be as follows:

$$\frac{d\hat{B}}{dt} = -\left(\frac{1/\tau_p - G}{2} - j2\pi(\nu_p - \nu_0)\right)\hat{B} + \sqrt{1/\tau_p}\hat{\xi} + \sqrt{G}\hat{\xi}_g, \qquad (3.29)$$

where $\hat{B}$ is the operator representing the internal field of the laser cavity, ν_p is the resonant frequency of the laser cavity, G is the gain coefficient of the gain medium, $\hat{\xi}$ and $\hat{\xi}_g$ are the operators representing the vacuum fluctuations injected from the output mirror and the gain medium. As in the case for $\hat{A}$, the operator $\hat{B}$ can be expressed as $\hat{B} = (B_0 + \Delta\hat{B})\exp(j\hat{\phi}_B) \cong B_0 + \Delta\hat{B} + jB_0\hat{\phi}_B$. By substituting it into Eq. (3.29), the power spectrum density of the FM noise of the laser is obtained as follows:

$$S_{\nu\text{-FR-lim}} = \frac{S_{\text{Im}(\hat{\xi})} + S_{\text{Im}(\hat{\xi}_g)}}{4\pi^2\tau_p B_0^2} = \frac{1}{4\pi^2\tau_p B_0^2}, \qquad (3.30)$$

where $S_{\text{Im}(\hat{\xi})} = \frac{1}{2}$ and $S_{\text{Im}(\hat{\xi}_g)} = \frac{1}{2}$ are the power spectrum density of the imaginary parts of $\hat{\xi}$ and $\hat{\xi}_g$, respectively. By substituting the relation of $P_{\text{laser}} = h\nu_0 B_0^2/\tau_p$ for laser power into the above equation, Eq. (3.28) can be obtained.

When the above discussion is compared with the discussion of negative electrical feedback, in which only the vacuum fluctuation incident upon port 2 is the cause of the noise generated in negative electrical feedback control as shown in Eq. (3.24), one notes that in the case of the laser, vacuum fluctuation incident upon all the ports as well as a further term representing vacuum fluctuation $\hat{\xi}_g$ from the gain medium are the cause of noise. This means that if the values of the photon lifetime τ_{FP} and the photon number A_0 of the FP cavity used for negative electrical feedback are equal to those of the laser cavity (τ_p and B_0, respectively) the power spectrum density of the FM noise of the Schawlow–Townes' limit is four times as large as that of the limit of negative electrical feedback as shown by comparing the constant of Eq. (3.26) and that of Eq. (3.30). Furthermore, in the case of negative electrical feedback control, the photon number in the FP cavity is $A_0^2 = (2\tau_{\text{FP}}P_{\text{FP}}/h\nu_0)$, but in the case of the laser, $B_0^2 = (\tau_p P_{\text{laser}}/h\nu_0)$ and thus the coefficient becomes different by a factor of 2. Since the ratio between the internal energy of the cavity and the power spectrum density of the fluctuation of the internal field of the cavity corresponds to the FM noise, the FM noise differs by a factor of 8 between the limit of negative electrical feedback and the Schawlow–Townes' limit of the laser.

3.2.1.4. Comparison of Signal and Noise by the Difference in Controlling Methods

There are many ways to get a feedback error signal from a cavity for use in negative electrical feedback, to wit:

(1) Transmission-type FP cavity:[14–19] The slope of the transmitted resonance fringe is used.

(2) Reflection-type FP cavity:[20] The slope of the reflected resonance fringe, i.e., the directly reflected light at the input mirror and the transmitted light from the FP cavity, are used.

(3) Phase-sensitive detection using reflection-type FP cavity:[21-29] Laser light is frequency modulated prior to its input to the FP cavity, and the reflected resonance is phase-sensitively detected at the modulation frequency. Using this signal the laser frequency can be locked to a resonance frequency of the FP cavity. This method has been used to stabilize microwave signal sources.[30] This is called the method of FM spectroscopy, Pound–Drever locking, and so on.

(4) Resonance peak locking without any frequency or phase modulation:[31,32] By placing a polarizer or a wavelength plate inside the FP cavity, dispersion-shaped resonance can be detected with a polarization analyzer. Then the dispersion-shaped resonance can give a feedback error signal. Various practical examples have been demonstrated.[32] This can also be applied to optical feedback method.[32,33] A similar idea has been introduced for a power build-up cavity for application to atomic beam collimation using a standing wave in the build-up cavity.[34]

Each method has its own merits and demerits depending on its application, and one cannot say which is best, but when comparing them on the basis of simplicity of construction of an optical frequency discriminator, (1) is simplest and structures become more complicated as one goes down the list from (2)–(4). The common merit of (2), (3), and (4) is that their control bandwidth is not limited by the photon lifetime of the FP cavity. They all use a reflected light of FP cavity and when considering the theoretical extreme limit of FM noise reduction, the same result is obtained. The common merit of (3) and (4) is that the laser frequency can be locked to a resonance frequency of the FP cavity. In (3) especially, the circuit becomes complicated for FM modulation and detection but the effects of noise at low frequencies (such noise includes the intensity noise of the laser and noise from the amplifier) injected during signal processing can be reduced. As for (4), the effect of intensity noise can be completely removed, at least in theory, and a wide locking range can be achieved.[32]

Then, what is the minimum value of FM noise in these methods? In the above subsection, the reduction limit of FM noise in the ideal case for (4) is shown. In the following, we discuss the minimum value of the power spectrum density of the FM noise to be obtained by using a negative electrical feedback technique with various methods to get a feedback error signal from the FP cavity.

Figures 3.5(a)–3.5(e) show five methods of laser control using FP cavities for negative electrical feedback. Among them, Figs. 3.5(a)–3.5(d) correspond to (1)–(4), and Fig. 3.5(e) shows a method in which Fig. 3.5(d) is developed. The limits of the FM noise reduction levels $S_{\nu\text{-electrical}}$ are given in Table 3.1. In this table, (a)–(e) correspond to (a)–(e) of Fig. 3.5, η_{PD} is the quantum efficiency of the photodetector, η_T and $1-\eta_R$ are the transmission and reflection efficiencies of the FP cavity, $S'_{\nu\text{-electrical-lim}}$ is $(h\nu_0)/[8(2\pi\tau_{FP})^2 P_{\text{discriminator}}]$, where $P_{\text{discriminator}}$ is the entire laser power fed into the optical frequency discriminator $[(C_0^2+A_0^2)h\nu]$. The reason for not not normalizing by $S_{\nu\text{-electrical-lim}}$ but by $S'_{\nu\text{-electrical-lim}}$ is that in actual

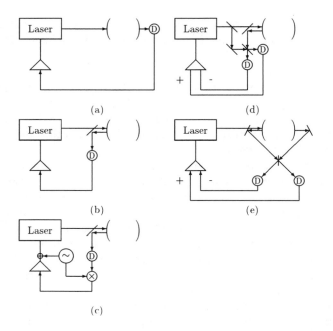

FIGURE 3.5. Optical frequency discriminators based on various methods. D's: photodetectors. (a) Use of the slope of the transmitted resonance fringe of FP cavity. (b) Use of the slope of the reflected resonance fringe of FP cavity. (c) Use of phase-sensitive detection of the reflected resonance of FP cavity. (d) Use of the interference between reflected light of FP cavity and a part of the incident light. (e) Use of the interference between reflected light and reflected light of FP cavity.

applications, light power wholly incident onto the optical frequency discriminator $P_{\text{discriminator}}$ is more important than power coupling to the FP cavity P_{FP}. In (a) and (b) of the table, $\Delta \nu_{\text{lock}}$ is $\nu_c - \nu_0$ and in (c)–(e), $\Delta \nu_{\text{lock}}$ is 0. When $\eta_{\text{PD}} = \eta_R = \eta_T = 1$ is applied to those equations, $S_{\nu\text{-electrical}}$ calculated in the minimum FM noise condition and $S_{\nu\text{-electrical}}$ calculated when the relative amplitude of signal is at its maximum condition are shown in Tables 3.2 and 3.3, respectively. (a)–(e) in the tables corresponds to (a)–(e) of Fig. 3.5. $\Delta \nu_{\text{lock}}$ of (a) and (b), m of (c), and ϵ of (e) in the Table 3.2 show conditions for minimum FM noise. $\Delta \nu_{\text{lock}}$ of (a) and (b), m of (c), and ϵ of (e) in Table 3.3 show the case where the relative signal amplitude reaches its maximum. As found in Table 3.2, $S_{\nu\text{-electrical}}$ does not reach $S'_{\nu\text{-electrical-lim}}$ only in (a), it means that the precise observation of the laser FM noise shown in the previous subsection cannot be done only with the transmission-type FP cavity. The transfer function of frequency discrimination for the case in Fig. 3.5(a) shows second-order low-pass characteristics different from the other methods which show first-order low-pass characteristics, and it is disadvantageous to high-frequency control.[32] As shown in Table 3.1, in the methods (b)–(e), the value of $S_{\nu\text{-electrical}}$ can become $S'_{\nu\text{-electrical-lim}}$ at the limit where laser power $C_0^2 h \nu$ split off as

TABLE 3.1. Formulas of $S_{\nu\text{-electrical}}$.

Fig.	$\dfrac{S_{\nu\text{-electrical}}}{S'_{\nu\text{-electrical-lim}}}$
(a)	$\dfrac{1+4\Delta\nu_{\text{lock}}^2(2\pi\tau_{\text{FP}})^2}{4\Delta\nu_{\text{lock}}^2(2\pi\tau_{\text{FP}})^2\,\eta_T\,\eta_{\text{PD}}}$
(b)	$\dfrac{[1-\eta_R+4\Delta\nu_{\text{lock}}^2(2\pi\tau_{\text{FP}})^2][1+4\Delta\nu_{\text{lock}}^2(2\pi\tau_{\text{FP}})^2]^3}{4\Delta\nu_{\text{lock}}^2(2\pi\tau_{\text{FP}})^2\,\eta_R^2\,\eta_{\text{PD}}}$
(c)	$\dfrac{1-\eta_R J_0(m)^2}{2J_0(m)^2 J_1(m)^2(1-\sqrt{1-\eta_R})^2\,\eta_{\text{PD}}}$
(d)	$\dfrac{1-(1-\epsilon)\eta_R}{\epsilon(1-\epsilon)(1-\sqrt{1-\eta_R})^2\,\eta_{\text{PD}}}$
(e)	$\dfrac{1-\eta_R+\eta_T}{\eta_T\,\eta_{\text{PD}}}$

reference light approaches zero, i.e., $P_{\text{discriminator}} \cong P_{\text{FP}}$. In the methods of (b), (c), and (d), the whole power of the laser can be used as output. This method is advantageous for achieving high stability and high output power.

In the case of the signal maximum condition shown in Table 3.3, in (b), (c), and (d) the FM noise level reaches about twice the value of $S'_{\nu\text{-electrical-lim}}$. This is because P_{FP} is reduced. But in (e), the maximum signal can be obtained at the FM noise limit. That is because P_{FP} becomes equal to the power $C_0^2 h\nu_0$ input into the photodetector. Therefore, in this method, the signal level in (e) is larger than for any other method.

3.2.2. Optical Feedback

The carrier density in a laser active waveguide can be modulated by the injected light. The laser frequency is thus modulated through the refractive index modulation

TABLE 3.2. $S_{\nu\text{-electrical}}$ under minimum $S_{\nu\text{-electrical}}$ in a different method.

Fig.	$\dfrac{S_{\nu\text{-electrical}}}{S'_{\nu\text{-electrical-lim}}}$	Condition	Relative amplitude of signal
(a)	$\dfrac{27}{4}$	$\Delta\nu_{\text{lock}}=\dfrac{1}{4\sqrt{2}\,\pi\tau_{\text{FP}}}$	$\sqrt{\dfrac{8}{81}}$
(b)	1	$\Delta\nu_{\text{lock}}\to 0$	$\cong 4\pi\Delta\nu_{\text{lock}}\tau_{\text{FP}}$
(c)	1	$m\to 0$	$\cong \dfrac{m}{\sqrt{2}}$
(d)	1	$\epsilon\to 0$	$\cong\sqrt{\epsilon}$
(e)	1		1

TABLE 3.3. $S_{\nu\text{-electrical}}$ under maximum signal amplitude in a different method.

Fig.	$\dfrac{S_{\nu\text{-electrical}}}{S'_{\nu\text{-electrical-lim}}}$	Condition	Relative amplitude of signal
(a)	$\dfrac{64}{9}$	$\Delta \nu_{\text{lock}} = \dfrac{1}{4\sqrt{3}\,\pi\tau_{\text{FP}}}$	$\sqrt{\dfrac{27}{256}}$
(b)	$\dfrac{64}{27}$	$\Delta \nu_{\text{lock}} = \dfrac{1}{4\sqrt{3}\,\pi\tau_{\text{FP}}}$	$\sqrt{\dfrac{27}{256}}$
(c)	2.05	$m \cong 1.08$	0.48
(d)	2	$\epsilon = 0.5$	0.5
(e)	1		1

which is induced by the optically induced carrier density modulation. This phenomenon has been used to control the laser frequency by the self-injection of laser light after it is reflected from an external surface. In utilizing this self-injection locking phenomenon, an optical feedback is constructed to reduce the FM noise of semiconductor laser. For optical feedback methods, there are several kinds of external reflectors, e.g., one mirror[35-38] two mirrors (with semiconductor laser located between them),[39] an optical grating,[40-42] a fiber FP cavity,[43-45] a tilted confocal FP cavity,[46,47] a photorefractive phase conjugate mirror,[48] and so on. Semiconductor lasers optically frequency stabilized with a confocal FP cavity, show higher spectral purity with relatively long-term frequency stability when a high-finesse and frequency-stable confocal FP cavity are used, and frequency modulation capability without any AR coating on one facet of the semiconductor laser. Any commercially available semiconductor lasers can thus be directly used in this method without requiring change within the laser device itself.

Figure 3.6 shows a schematic diagram of a confocal FP-cavity-coupled semiconductor laser. The tilted confocal FP cavity acts as an optical band pass filter for a laser light field which is spectrally filtered in the cavity. A part of the light field filtered within the cavity returns to the laser through the cavity input mirror at resonance. The oscillation frequency of the laser is then locked to the resonance of

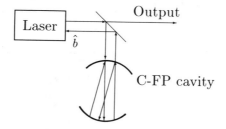

FIGURE 3.6. Schematic diagram of the confocal FP-cavity-coupled semiconductor laser. C-FP cavity: confocal FP cavity. $\hat{b}[= (b_0/B_0)(\hat{B}*K)]$: feedback light.

this confocal FP cavity, i.e., optical self-locking takes place. By this process, the frequency noise of the laser becomes very low, and the center frequency is also stabilized substantially. We discuss here the FM noise reduction characteristics and the frequency response of the direct frequency modulation of an FP-cavity-coupled semiconductor laser.

When the term representing optical feedback is used in Eq. (3.29), the following equation can be obtained:

$$\frac{d\hat{B}}{dt} = -\left(\frac{1/\tau_p - G}{2} - j2\pi(\nu_p - \nu_0)\right)\hat{B} + \sqrt{1/\tau_p}\left[\hat{\xi} + \frac{b_0}{B_0}(\hat{B}*K)\right] + \sqrt{G}\hat{\xi}_g$$

(3.31)

where $*$ shows the convolution integral, b_0 is a loot of the average photon flux of the feedback light, and K is the impulse response of the FP cavity and is obtained as follows:

$$K = \begin{cases} \dfrac{1}{2\tau_{FP}}\exp\left(-\dfrac{t}{2\tau_{FP}} + j2\pi(\nu_c - \nu_0)t + j\theta\right), & \text{for } 0 \le t, \\ 0, & \text{otherwise}, \end{cases}$$

(3.32)

where θ represents the difference between phases of the internal field of the laser and that of the feedback light from the FP cavity which is determined from the path length between the FP cavity and the laser. For simplicity of the calculations, the value of θ is assumed to be a constant. Development of carrier number N over time is shown below:

$$\frac{dN}{dt} = -\frac{N}{\tau_c} - G|\hat{B}|^2 + P,$$

(3.33)

where τ_c is carrier lifetime by natural recoupling, and P is the injection carrier number per unit time. Considered below as linear approximations, the following equations are used:

$$N = N_0 + \Delta N,$$

(3.34)

$$G = G_0 + \frac{dG}{dN}\Delta N,$$

(3.35)

$$\nu_p = \nu_{p0} + \frac{d\nu_p}{dN}\Delta N,$$

(3.36)

where N_0, G_0, and ν_{p0} are the average values of N, G, and ν_p, respectively. ΔN is the fluctuation of N. For simplicity, it is assumed that the laser frequency ν_0 has been locked to the resonant frequency ν_c of the FP cavity by adjusting the distance between the laser and the FP cavity. In small signal analysis, the power spectrum density of the FM noise limit $S_{\nu\text{-optical-lim}}$ in a frequency range less than the relax-

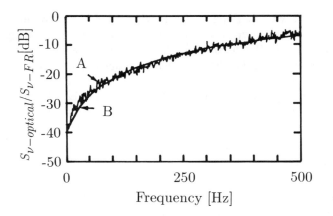

FIGURE 3.7. Measured (curve A) and calculated (curve B) power spectral densities ($S_{\nu\text{-optical}}$) of FM noise under optical feedback condition normalized by free-running condition ($S_{\nu\text{-FR}}$) (after Ref. 49 © 1993 IEEE).

ation oscillation frequency is obtained as follows:

$$S_{\nu\text{-optical-lim}} = \frac{\alpha^2 + 1}{\left| 2\pi B_0 \sqrt{\tau_p} + \dfrac{2b_0 \sqrt{1 + \alpha^2}\,\cos[\theta + \arctan(\alpha)]}{\dfrac{1}{2\pi\tau_{FP}} + j2\,f} \right|^2}, \qquad (3.37)$$

where the α parameter is represented as

$$\alpha = 4\pi \frac{d\nu_p/dN}{dG/dN}. \qquad (3.38)$$

Without optical feedback, i.e., $b_0 = 0$, the following is obtained and the well-known Schawlow–Townes' limit results:

$$S_{\nu\text{-FR-lim}} = \frac{(\alpha^2 + 1)h\nu_0}{(2\pi\tau_p)^2 P_{\text{laser}}}. \qquad (3.39)$$

The power spectral density of the frequency noise of the laser under optical feedback condition is shown in Fig. 3.7. Curve A of Fig. 3.7 shows the power spectral density of the frequency noise normalized to that of free-running conditions. The result calculated by using the equation corresponding to the experiment is also shown by curve B. This shows good agreement.

When the effect of optical power feedback is large, the FM noise limit of the FP-cavity-coupled semiconductor laser is approximated from Eq. (3.37) as

$$S_{\nu\text{-optical-lim}} = \frac{h\nu_0}{4(2\pi\tau_{FP})^2 P_{\text{optical-feedback}}} + \frac{f^2 h\nu_0}{P_{\text{optical-feedback}}}. \qquad (3.40)$$

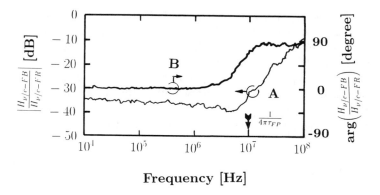

FIGURE 3.8. Measured current-frequency transfer function $H_{\nu/c\text{-FB}}$ in the confocal FP-cavity-coupled semiconductor laser normalized by the current-frequency transfer function $H_{\nu/c\text{-FR}}$ in the free-running semiconductor laser. Curve A is the frequency responses of the FM efficiency, and curve B is the phase (after Ref. 50 © 1990 IEEE).

Notice that the α parameter does not appear in this equation, and the FM noise level is determined only by the lifetime τ_{FP} of the FP cavity and the feedback power $P_{\text{optical-feedback}}$ from the FP cavity. When Eqs. (3.40) are compared with Eq. (3.27) and when the optical feedback power equals to the optical power which is used for the negative electrical feedback, it is found that the limit of the FM noise level of the optical feedback is twice that of the negative electrical feedback because vacuum fluctuations from the optical gain medium have an effect in the case of optical feedback.

The characteristics of frequency response of the direct frequency modulation of an FP-cavity-coupled semiconductor laser must be known to design a servo loop for several applications, e.g., frequency stabilization by negative electrical feedback, optical phase-locking loops, etc. For this purpose, it is sufficient to know the direct frequency modulation characteristics for a small modulation index. Since frequency modulation efficiency is reduced like FM noise is reduced, the ratio of the current-frequency transfer function $H_{\nu/c\text{-FB}}$ in an FP-cavity-coupled semiconductor laser and the current-frequency transfer function $H_{\nu/c\text{-FR}}$ in a free-running laser is given as follows by the ratio of Eqs. (3.37) and (3.39):

$$\frac{H_{\nu/c\text{-FB}}}{H_{\nu/c\text{-FR}}} = \frac{2\pi B_0 \sqrt{\tau_p}}{2\pi B_0 \sqrt{\tau_p} + \dfrac{2b_0 \sqrt{1+\alpha^2}\,\cos[\,\theta + \arctan(\alpha)\,]}{(2\pi\tau_{\text{FP}})^{-1} + j2\,f}}. \tag{3.41}$$

Figure 3.8 shows the measured frequency response of FM efficiency with the transfer function $H_{\nu/c\text{-FB}}$ of an FP-cavity-coupled semiconductor laser. The value of $H_{\nu/c\text{-FR}}$ is normalized to the value under free-running condition $H_{\nu/c\text{-FR}}$. It was measured by an rf network analyzer. The curves in this figure show that the present optical feedback system has high-pass characteristics. The 3 dB cut-off frequency is determined by $1/4\pi\tau_{\text{FP}}$. For $f < 1/4\pi\tau_{\text{FP}}$, FM efficiency and its phase are constant. The phase changes by 90° at $f = 1/4\pi\tau_{\text{FP}}$. On the other hand, for $f > 1/4\pi\tau_{\text{FP}}$, FM efficiency increases at a rate of 20 dB/decade. And then, $\arg[H_{\nu/c\text{-FB}}/H_{\nu/c\text{-FR}}]$ returns to zero at the limit of $f = \infty$.

The FP-cavity-coupled semiconductor laser shows wide-band frequency noise suppression characteristics with a narrow full width of half-maximum linewidth (FWHM) less than 10 kHz as shown in Refs. 46 and 47. The power spectrum density profile $\text{PSD}_{\text{laser}}(\nu)$ of the laser can be calculated by

$$\text{PSD}_{\text{laser}}(\nu) = 2\int_0^\infty \exp\left(-2(\pi\nu_0\tau)^2\frac{1}{\nu_0^2}\int_0^\infty S_\nu\left(\frac{\sin(\pi f\tau)}{\pi f\tau}\right)^2 df\right)$$

$$\times \cos[2\pi(\nu-\nu_0)\tau]d\tau. \tag{3.42}$$

From this calculation, FWHM is estimated πS_ν when S_ν can be approximated as white noise.[51] By using an FP cavity with long photon lifetime as the reflector, the FM noise level and the FWHM are reduced further. However, the suppression of $1/f$ noise and $1/f^2$ noise at low frequencies in the FM noise of an FP-cavity-coupled semiconductor laser is not sufficient because the FM noise suppression ratio of the optical feedback is usually lower than that of the electrical feedback. Therefore, it is better that optical feedback be used to suppress FM noise at high frequencies, while low frequencies are controlled by electrical feedback methods. The spectral linewidth of 7 Hz is obtained by the combined methods of electrical feedback and light feedback.[52]

The problem with the optical feedback method is that oscillation instability, and even chaotic instabilities, can be induced depending on the value of the phase of the injection light and the injection current fed to the laser, which means that additional control to the laser current and the distance between the laser and the cavity is required to maintain a stable optical feedback. In order to avoid the instability of the optical feedback, the value of $\nu_p - \nu_c$ and $\theta + \arctan(\alpha)$ must be controlled to be $\cong 0$. However, a method of control in which the above two parameters are made 0 simultaneously has not yet been realized. The method commonly used is to lock $\nu_0 - \nu_c$ to be 0 by controlling the value of θ. This method is effective and supplies sufficient stability if the lock-in range (the allowable range of $\nu_p - \nu_c$ in which ν_0 can be locked to ν_c) of the FP-cavity-coupled semiconductor laser is wide enough. But if not, mode hop to another FP cavity resonant mode or even chaotic instabilities may occur when the laser frequency leaves from the lock-in range due to frequency drift of semiconductor laser. To solve this problem, if both parameters are controlled by utilizing a new method (e.g., the dependency of the FM noise levels on the distance of the laser and external cavity may yet be successfully employed), stable operation of the FP-cavity-coupled semiconductor laser can be achieved. If so, a compact semiconductor laser light source which is stable over both the long and short term can be obtained by combining the above with negative electrical feedback techniques.

3.2.3. Optical Phase Locking

To meet requirement (2) described in the Introduction, it is necessary to achieve a continuous, accurate, and fine sweep of laser frequencies. This can be achieved by utilizing the heterodyne optical phase-locking loop, i.e., the slave laser frequency is accurately fine tuned to the frequency-stabilized master laser by sweeping the microwave frequency of the local oscillator used for the phase-locking loop. Here, the standard laser of the phase-locking loop is called the main laser, and the laser which is controlled by phase-locking loop is called the slave laser. An AlGaAs laser diode coupled to the external cavity laser frequency has been swept for 64 GHz by this technique,[53] which was limited only by the mode-hopping phenomenon of this conventional laser device. A wider range of continuous, accurate, and fine frequency sweeps is expected by using advanced laser devices such as a DFB laser and/or semiconductor lasers with optical feedback using an optical grating as an external reflector.[40–42] The authors have obtained a heterodyne signal frequency stability of 1×10^{-18} at an integration time of 70 s and a phase error variance of 0.02 rad^2.[54] For homodyne optical phase locking, a bandwidth as wide as 134 MHz (Ref. 55) and a phase error variance of 0.02 rad^2 have been achieved. The basic principles of the optical phase-locking loop and its limit are discussed next.

Schematic diagrams of two kinds of optical phase-locking loop, one a homodyne phase-locking loop and the other a heterodyne phase-locking loop, are shown in Fig. 3.9(a) and an analysis model of the optical phase-locking loop is shown in Fig. 3.9(b). In the homodyne phase-locking loop, a photodetector is used as a phase comparator and its signal is fed back to the frequency of the slave laser, and the frequency/phase of the slave laser is locked to that of the master laser. In the heterodyne phase-locking loop, a beat signal at the frequency difference between the slave and master lasers is detected with the photodetector and then the phase difference between the beat signal and the microwave oscillator is detected by the phase comparator used in the microwave band. By using it as feedback to the slave laser, the frequency/phase of the slave laser is locked to that of the master laser with the offset frequency of the microwave frequency. By utilizing this method, the slave laser frequency can be accurately fine tuned to the master laser by sweeping the microwave frequency of the local oscillator.

First, we will describe the process in which the photodetector becomes a phase comparator for the homodyne phase-locking loop, and describe the noise characteristics. The interference signal I between the main and slave laser which is detected by an optical balanced mixer receiver is obtained as follows:

$$I = \frac{\eta_{PD}e}{h\sqrt{\nu_M \nu_S}} 2\sqrt{P_M P_S} \sin[2\pi(\nu_M - \nu_S)t + (\phi_M - \phi_S)], \qquad (3.43)$$

where P_M and P_S are the laser power of the master and slave lasers, ν_M and ν_S are the frequency average of the master and slave lasers, and ϕ_M and ϕ_S are their phases, respectively. As a homodyne phase-locking loop, the phase error signal can be obtained by substituting $\nu_S = \nu_M$ in Eq. (3.43) shown below:

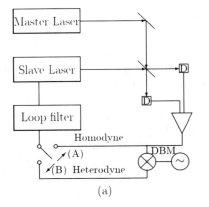

(a)

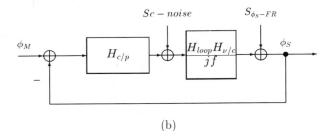

(b)

FIGURE 3.9. (a) Schematic diagram of the optical phase-locking loop (OPLL). D's: photodetectors. DBM: double balanced mixer. (b) An analysis model of OPLL. $H_{\nu/c}$: current-frequency transfer function of the slave laser. H_{loop} : transfer function of the loop filter. $H_{c/p}$: phase-current transfer function of the photodetector. $S_{c\text{-noise}}$: noise in the photodetector. $S_{\phi_S\text{-FR}}$: phase noise of the slave laser under free-running condition.

$$S = \frac{2 \eta_{\text{PD}} e \sqrt{P_M P_S} \, \sin(\phi_M - \phi_S)}{h\nu} \tag{3.44}$$

and, in the range of linear approximation, the result becomes

$$S \cong \frac{2 \eta_{\text{PD}} e \sqrt{P_M P_S}(\phi_M - \phi_S)}{h\nu}. \tag{3.45}$$

Then, the transmission function of the photodetector $H_{c/p}$ is

$$H_{c/p} = \frac{2e \, \eta_{\text{PD}} \sqrt{P_M P_S}}{h\nu}. \tag{3.46}$$

In an optical phase-locking loop, the intrinsic noise that disturbs phase detection is shot noise. The power spectrum density $S_{c\text{-noise-lim}}$ as the current signal of shot noise detected by the light detector is shown below:

$$S_{c\text{-noise-lim}} = 2\,\eta_{\text{PD}}e^2(P_M+P_S)/h\nu_M. \tag{3.47}$$

Therefore, the power spectrum density $S_{\phi_M\text{-}\phi_S\text{-lim}}$ of the measurement limit (suppression limit) of phase difference $\phi_M - \phi_S$ of the phase-locking loop is given as follows by dividing $S_{c\text{-noise-lim}}$ by $H_{c/p}^2$:

$$S_{\phi_M\text{-}\phi_S\text{-lim}} = \frac{(P_M+P_S)h\nu}{2\,\eta_{\text{PD}}P_M P_S}. \tag{3.48}$$

In the result of the case of the homodyne phase-locking loop, the noise component is increased in the process where frequency is transformed in the microwave circuit and the following equation is shown:

$$S_{\phi_M\text{-}\phi_S\text{-lim}} = \frac{(P_M+P_S)h\nu}{\eta_{\text{PD}}P_M P_S}. \tag{3.49}$$

To execute the optimum design of the control loop, the calculation of the power spectrum density $S_{\phi_M\text{-}\phi_S}$ of the phase error shall be made by taking the FM noise of the main and slave lasers into consideration. If the FM noise is assumed to be white noise for simplification, the equation below is obtained:

$$S_{\phi_M\text{-}\phi_S} = \frac{\text{FWHM}_M+\text{FWHM}_S}{\pi f^2}\left|\frac{1}{1+H_{\text{open}}}\right|^2 + \left|\frac{H_{\text{open}}}{1+H_{\text{loop}}}\right|^2 S_{\phi_M\text{-}\phi_S\text{-lim}}, \tag{3.50}$$

where the open loop transfer function H_{open} is given as

$$H_{\text{open}} = \frac{H_{c/p}H_{\nu/c}H_{\text{loop}}}{jf}, \tag{3.51}$$

where H_{loop} is the transfer function of the loop filter and $H_{\nu/c}$ Hz/A is the current-laser frequency transfer function of the laser, respectively, FWHM_S and FWHM_M are spectral linewidths of the slave and master lasers and FM noise is assumed to be white noise. When considering the minimum value of phase error variance $\delta_{\phi_M\text{-}\phi_S}^2$ under conditions of ideal control, assume the following function for H_{open} for simplification of discussion:

$$H_{\text{open}} = \begin{cases} \infty, & \text{for } f \leq f_B, \\ 0, & \text{otherwise.} \end{cases} \tag{3.52}$$

Here, f_B is the control bandwidth of the phase-locking loop. Equations (3.50) and (3.54) are substituted in Eq. (3.50) and phase error variance is calculated. In this case, the optimum value of f_B is obtained as follows:

$$f_B = \sqrt{\frac{2\,\eta_{PD}P_M P_S(\text{FWHM}_M + \text{FWHM}_S)}{\pi(P_M + P_S)h\nu}}. \tag{3.53}$$

The minimum value of phase error variance when f_B is at the optimum value is

$$\delta^2_{\phi_M - \phi_S\text{-lim}} = \sqrt{\frac{2h\nu(\text{FWHM}_M + \text{FWHM}_S)(P_M + P_S)}{\eta_{PD}P_M P_S \pi}}. \tag{3.54}$$

Since in an actual situation, the loop filter in Eq. (3.54) cannot be used, the characteristics of the loop filter which will be used shall be substituted in place of Eq. (3.52). The value obtained from Eq. (3.54) shows the minimum value obtainable in principle.

3.3. EXPERIMENT

This section reviews an example of progress in recent experiments.

3.3.1. Pound–Drever Locking

In order to meet requirement (1) described in the Introduction, laser devices have been investigated and developed. A 150 kHz spectral linewidth semiconductor laser has been reported.[56] Although the FM noise of recently fabricated narrow linewidth semiconductor lasers has been sufficiently low for a high bit rate coherent optical transmission system, further reduction of this noise is still required for advanced and precise coherent optical systems. That is, linewidth reduction (FM noise reduction) of the semiconductor laser should be carried out even for this advanced laser device. For this purpose, the FP cavity, especially the reflection type, has been used as a wideband and high gain frequency demodulator of a negative electrical feedback system because of its high sensitivity to phase demodulation. By measuring the power fluctuations of the light reflected from the FP cavity for FM noise detection, a negative electrical feedback loop can be optimally designed for an AlGaAs laser. A gain-bandwidth product of the feedback loop higher than 10 MHz has been achieved.[20] Similar frequency control for a sub-MHz linewidth segmented electrode multiquantum well distributed feedback (MQW-DFB) laser[57] at 1.5 μm wavelength has also successfully attained a 250 Hz linewidth.[58] Although the fundamental factor for limiting the sensitivity of FM noise detection is quantum noise, laser power fluctuations usually limit the sensitivity of feedback systems in practice. The Pound–Drever locking has been employed to achieve quantum noise-limited sensitivity by rejecting the contribution of the power fluctuations.[25]

Figure 3.10 shows an experimental setup of the Pound–Drever locking using a segmented-electrode MQW-DFB laser whose wavelength is 1.5 μm. Figures 3.11(a) and 3.11(b) show the experimental result of the power spectrum density of FM noise. By comparing this with the experimental result using a conventional reflection-type FP cavity,[58] considerable FM noise reduction can be achieved in the

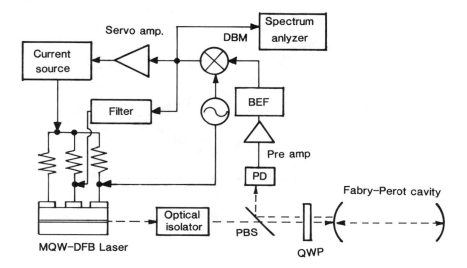

FIGURE 3.10. Experimental setup of the Pound–Drever locking using a segmented-electrode MQW-DFB laser. DBM: double balanced mixer. BEF: band elimination filter at 2 f_m (f_m : modulation frequency). PBS: polarization beam splitter. PD: photodetector. QWP: quarter-wave plate (after Ref. 59).

frequency region below 100 kHz. The power spectral density in this low frequency region was reduced to less than 25 Hz^2/Hz, by which it was estimated that the 3 dB spectral linewidth was reduced to less than 80 Hz.[59]

If a commonly used medium-class FP cavity, for example, with an 1 MHz resonance linewidth, corresponding to a finesse of 150 with a cavity length of 1 m, is used as a frequency demodulator for a 1.5 μm-wavelength semiconductor laser, the field spectral linewidth limited by quantum noise is estimated to be as narrow as 30 μHz assuming that $P_{\text{discriminator}} = 1$ mW. Furthermore, if the recently developed supercavity with a finesse as high as 1×10^6 (Ref. 8) is used, the corresponding quantum-noise-limited linewidth is expected to be 5 nHz, assuming again a cavity length of 1 m. Thus, it is possible in principle to realize an ultralow noise semiconductor laser usable as a coherent light source even with the long interferometers in gravitational wave detection experiment.

3.3.2. Novel Optical Frequency Discriminator

To satisfy requirement (1) described in the Introduction, there is a method to lock the laser frequency to a cavity resonance frequency without any modulation as shown in Figs. 3.5(d) and 3.5(e). In particular, the optical frequency discriminator described by Fig. 3.5(e) is excellent in gain and noise performance. However, performance of the optical frequency discriminator is inadequate due to mechanical instability of the interference between reflection and transmission of an FP cavity.

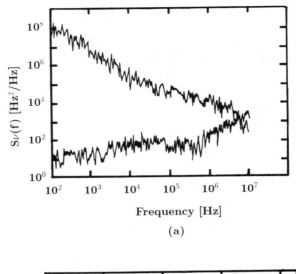

(a)

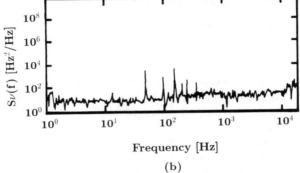

(b)

FIGURE 3.11. Power spectral densities S_ν of FM noise of a segmented electrode MQW-DFB laser of 1.5 μm wavelength. (a) Results for a lower gain and narrower bandwidth (curve A) and a higher gain and wider bandwidth (curve B) controls measured by an rf spectrum analyzer. (b) Extension of curves B in (a) for a lower Fourier frequency range measured by a fast Fourier transform spectrum analyzed (after Ref. 59).

This can be solved by inserting an element with polarization selectivity (such as a Brewster plate) into the FP cavity[31] or by inserting an element such as a wavelength plate or Faraday rotor into the FP cavity.[32] Described here is the latter method, which is equivalent to the method described in Fig. 3.5(e).

The present optical frequency discriminator is schematically explained in Figs. 3.12(a) and 3.12(b). To explain the operation of these optical frequency discriminators, let us consider the setup illustrated in Fig. 3.12(a). Here, linearly polarized light from a laser is fed into a ring cavity in which a 90° polarization rotator is installed. When light from the laser is incident on such a cavity, along with reflec-

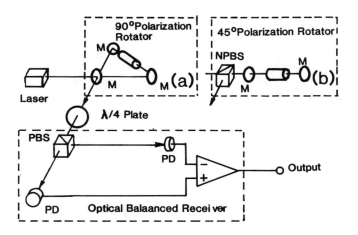

FIGURE 3.12. Present optical frequency discriminator. (a) Traveling-wave type. (b) Standing-wave type. M's: mirrors. PBS: polarization beam splitter. PD's: photodetector. NPBS: nonpolarization beam splitter (after Ref. 32).

tion light output from the incident port, reflection-type resonant output can be obtained in the polarization component identical with the polarization of the incident light and transmission-type resonant output can be obtained in the polarization component having an angle of 90° with the polarization of the incident light. When observing interference of this output light with an assembly of a 1/4λ wave plate, a polarization beam splitter, two photodetectors and differential amplifier, the same result as Fig. 3.5(e) can theoretically be obtained. In this case, unfavorable effects of mechanical instability of the optical distance in Fig. 3.5(e) do not exist. Figure 3.13 shows the output signal from the optical frequency discriminator having specific dispersive spectral shapes, which provides an ideal error signal to servo-lock the laser frequency to one of the FP cavity resonant frequencies without any frequency modulation. This shape originates from interference between two degenerated orthogonal linearly polarized modes of the FP cavity. The laser frequency is locked to the position represented by point P in this figure. This point is the center of a resonant spectral profile of the FP cavity, and the contribution of the laser

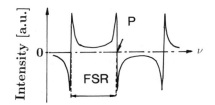

FIGURE 3.13. Resonant profile of the present optical frequency discriminator. FSR: free spectral range. *P*: position to which laser frequency is locked (after Ref. 32).

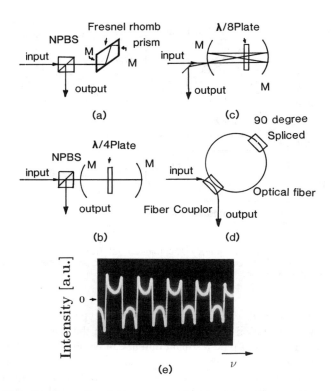

FIGURE 3.14. Various examples of the present optical frequency discriminators created in our laboratory. (a) A Fresnel rhomb λ/4 plate with both ends high reflection coated. (b) An FP cavity on-axis with an intracavity λ/4 plate. (c) A confocal FP cavity off axis with an intracavity λ/8 plate. (d) A ring type fiber cavity with an 90° spliced fiber. (e) The signal obtained by the setup of (a) (free spectral range = 1.7 GHz). Similar signal have been also obtained by (b), (c), (d), and (e). M's: mirrors. NPBS: nonpolarization beam splitter.

intensity modulation noise to the frequency discrimination is suppressed by the differential amplifier. The additional advantage is that even though some electrical surges can throw the laser frequency out of lock, recovery is possible because the dispersive shape has a couple of nonzero wings reaching to ±FSR (free spectral range). The problem with the present method is that the finesse of the FP cavity is not high, since internal losses of the FP cavity are raised by an element installed in the FP cavity. But in the future, when internal losses are minimized by utilizing the anisotropic characteristic by polarization of high reflection mirrors as a relief of the element installed in the external cavity, the finesse of the external cavity should be high. In conclusion, Fig. 3.14 shows some modified stable and compact devices of the optical frequency discriminator created in our laboratory.

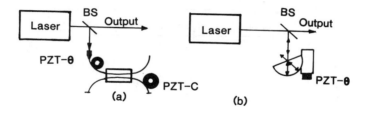

FIGURE 3.15. A fiber cavity (a) and a hemispherical microlens cavity (b), which were used as the external cavity of compact modules. BS's: beam splitters. PZT's: piezoelectric transducers (after Ref. 50 © 1990 IEEE).

3.3.3. Compact Modules of External-Cavity Coupled Semiconductor Laser

FP-cavity-coupled semiconductor lasers can easily be attained so as to high bandwidth FM noise suppression regardless of laser type, and spectral linewidth can be narrowed to 10 kHz or less to satisfy requirement (1) described in the Introduction. This system assembled into a compact module will be widely applied to optical communications, cavity-type fiber gyroscopes, etc. Described are some trials for actualizing compact modules.

3.3.3.1. Fiber Cavity

Figure 3.15(a) shows a semiconductor laser module coupled to a fiber cavity made of a single mode optical fiber. The fiber used for the cavity was aluminum coated at both ends, and its free spectral range and finesse were 100 MHz and 20, respectively. The laser light was coupled to the fiber cavity by a 96:4 fiber coupler, of which the coupling end to the laser and its opposite end were cut to angles of 4° and 8°, respectively, for preventing optical feedback caused by direct reflection from the fiber ends. The operating principle of the fiber cavity coupled semiconductor laser is the same as that of the confocal FP-cavity-coupled semiconductor laser. Here, laser light is locked to the resonance of the fiber optics cavity.

Figure 3.16 shows the measured results of the spectral linewidth at various feedback power levels. The linewidth reduction factor is proportional to 1/(actual feedback power into the laser cavity) as predicted by the theoretical analysis of the FM noise for optical feedback system, i.e., the linewidth reduction factor. At stronger feedback levels, optical feedback locking became unstable, and the linewidth could not be measured exactly. The laser field spectral linewidth was 20 kHz at the maximum feedback level under stable locking conditions.

The feedback phase of the fiber cavity coupled semiconductor laser could be controlled stably by the voltage control applied to the PZT-θ in Fig. 3.14. The oscillation frequency can be swept by PZT-C. The frequency stability of fiber cavity coupled semiconductor lasers can be improved by packaging the fiber cavity on a temperature-controlled copper block, on which the laser is mounted.

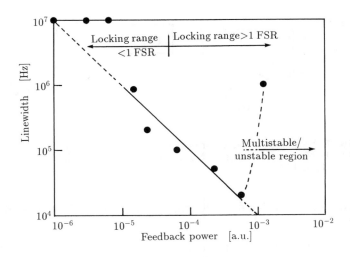

FIGURE 3.16. Measured spectral linewidth of the fiber cavity coupled semiconductor laser as a function of the feedback power level.

3.3.3.2. Hemispherical MicroCavity

Figure 3.15(b) shows a hemispherical microlens cavity, which was made of BK7 glass. Its diameter and reflectivity were 10 mm and 90%, respectively. The hemispherical microlens cavity coupled semiconductor laser operates on the same principle as that of the confocal FP-cavity-coupled semiconductor laser. The phase of the feedback light from the hemispherical microlens cavity can be controlled stably by the voltage control applied to the PZT-θ in this figure. Because the free spectral range of this cavity (about 10 GHz) is larger than the relaxation oscillation frequency of the solitary laser, the resonant peak of FM noise at the resonant oscillation frequency is reduced.

Since the size of the external cavity is small, miniaturized packaging inside a small external package is possible. The miniaturized microcavity coupled semiconductor laser modules are shown in Fig. 3.17. The semiconductor laser used in these modules was a 1.3 μm InGaAsP DFB laser. The size of this model was $39(W) \times 88(L) \times 30(H)$ (mm^3). The linewidth maintained a value of less than 40 kHz. The output power was 2.3 mW. By miniaturization and the resultant isolation of the system from ambient temperature variation, long-term stability was drastically improved without use of a control loop.

The linewidth was sufficiently narrow for practical applications such as a passive ring cavity-type optical fiber gyroscope.[60,61] The 1.3 μm hemispherical microlens cavity coupled semiconductor laser modules are now used for experimental work on the passive ring cavity type optical fiber gyroscope.[61]

FIGURE 3.17. A miniature-packaged microcavity coupled semiconductor laser module. The laser used in this module was 1.3 μm InGaAsP DFB laser. The size was 88 mm(L)×39 mm(W)×30 mm(H).

3.3.4. Optical-Electrical Double Feedback

Because the confocal FP-cavity-coupled semiconductor laser shows wide-band FM noise suppression characteristics with a narrow linewidth of less than 10 kHz,[46,47,62] it could be expected that the linewidth of the semiconductor laser can be more efficiently narrowed by using an optical-electrical double feedback method that simultaneously uses the optical feedback from the external cavity and the electrical negative feedback from the reflection mode of a high-finesse stable optical cavity. Semiconductor lasers of narrowest linewidth are realized by using this optical-electrical double feedback method. Figure 3.18 shows the experimental setup for the optical-electrical double feedback semiconductor laser. The semiconductor laser used in this experiment was a channeled-substrate-planar-type AlGaAs semiconductor laser with a wavelength at 830 nm. The linewidth of this laser was 15 MHz. Before the electrical negative feedback scheme was used, the semiconductor was optically stabilized with the optical feedback scheme shown in Fig. 3.18. The free spectral range and the finesse of the external confocal FP cavity which were used for the optical feedback were 1.5 GHz and 75, respectively. The reduced linewidth with the optical feedback was narrower than 10 kHz. By using this semiconductor laser coupled to the external cavity, for the negative feedback, the control bandwidth of the order of 1 MHz could be secured by using two control routes, the external cavity length and the injection current for lower-frequency (< 5 kHz) and higher-frequency components, respectively. This bandwidth was sufficient to reduce the linewidth. For the negative electrical feedback, the frequency fluctuation of the confocal FP cavity coupled semiconductor laser was demodulated by the slope of the reflection-type resonance fringe of the supercavity (i.e., the slope of the reflected

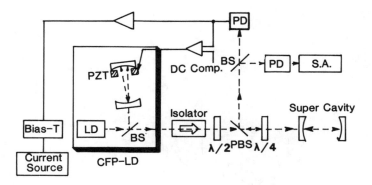

FIGURE 3.18. Experimental setup for the optical-electrical double feedback semiconductor laser. PBS: polarization beam splitter. BS: beam splitter. DC Comp.: dc compensator. PZT: piezoelectric transducer. S.A.: spectrum analyzer. PD's: photodetectors. CFP-LD: confocal FP-cavity-coupled semiconductor laser (after Ref. 52).

resonance fringe is used) and then the demodulated signal was negatively fed back to the laser through the injection current and the confocal FP cavity length control.

The most important factor for FM noise suppression in achieving a narrow linewidth is a stable frequency reference. As the frequency reference for this purpose, a hermetically sealed and temperature-controlled commercial supercavity (Newport SR-140) with a finesse of $\cong 34{,}500$ and a free spectral range of 6 GHz were used. The optical power for noise measurement was 0.62 mW. Figures 3.19(a) and 3.19(b) show the FM noise power spectral density. In comparing curve B (under electrical and optical feedback conditions) with A (under optical feedback condition) in Fig. 3.19(a), the FM noise power spectral density S_ν of the confocal FP-cavity-coupled semiconductor laser is suppressed by $\cong 50$ dB by the electrical negative feedback scheme, in which curve A was obtained with the lower gain and narrower bandwidth control. This means that the linewidth of the confocal FP-cavity-coupled semiconductor laser was narrowed to the order of 100 mHz. The measured power spectral density of curve A for f greater than several tens of kilohertz shows approximately -24 dB/decade mainly because of the FM noise discrimination characteristics of the reflection mode of FP interferometer, which is -20 dB/decade for $f < 1/4\pi\tau_{\mathrm{FP}}$. To confirm the FM noise suppression in the range of $f < 100$ Hz, the FM noise was measured as shown in Fig. 3.19(b). From Fig. 3.19 it can be seen that the FM noise was suppressed to a level of less than 0.3 Hz^2/Hz in the range of 10 Hz $< f <$ 1.5 MHz, except for the 50 Hz power supply line frequency and its higher harmonics. This noise level corresponds to a linewidth of the Lorentzian field power spectral profile of 1 Hz. The short-term frequency stability of the controlled laser was therefore considerably high. The calculated field spectral shape of the stabilized confocal FP-cavity-coupled semiconductor laser is shown in Fig. 3.20. The calculation procedure of the field spectral profile from the

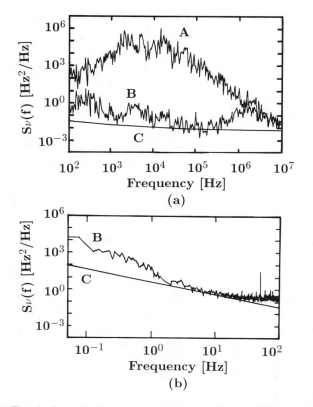

FIGURE 3.19. Electrically controlled power spectral densities of FM noise $[S_\nu(f)]$ for the confocal FP-cavity-coupled semiconductor laser. The field spectral linewidth of the noncontrolled confocal FP-cavity-coupled semiconductor laser was less than 10 kHz. Solid curves C indicate the FM noise detection limit imposed by the intensity modulated noise level of the confocal FP-cavity-coupled semiconductor laser. (a) results for a lower gain and narrower bandwidth (curve A) and a higher gain and wider bandwidth (curve B) controls measured by an rf spectrum analyzer. (b) Extension of curves B and C in (a) for a lower Fourier frequency range measured by a fast Fourier transform spectrum analyzed (after Ref. 52).

measured FM noise has been described in detail.[20] The linewidth of the calculated field spectral profile was 7 Hz with a resolution bandwidth of 2 Hz, which is to our knowledge the lowest published value. However, the field spectrum is frequency modulated by the 50 Hz power supply line frequency and its harmonic frequency. The modulation index was ~ 0.6, which was estimated from the field spectral shape, as can be seen in Fig. 3.20. The modulation index of 0.6 means that the concentrated power ratio within the carrier was 83%.

To summarize the discussion given above, Fig. 3.21 shows the progress of the authors' experiments in linewidth reduction.

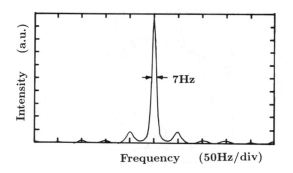

FIGURE 3.20. Calculated field spectral profile of the stabilized confocal FP-cavity-coupled semiconductor laser, in which curve B of Figs. 3.19(a) and 3.19(b) were used (after Ref. 52).

3.3.5. Optical Phase Locking using an Optical Frequency Comb Generator

The reliability of the optical heterodyne phase-locking technique has been sufficiently high for a stable frequency sweep of semiconductor lasers to satisfy requirement (2) described in the Introduction. However, as the heterodyne frequency is limited by the bandwidth of the photodetector, wide range sweeping is not possible. The range in which the frequency difference of the laser source can be precisely controlled is at most several tens of GHz. In order to increase the range in which the frequency difference of the laser source can be precisely controlled, we have recently proposed a new optical phase-locking loop using an optical frequency

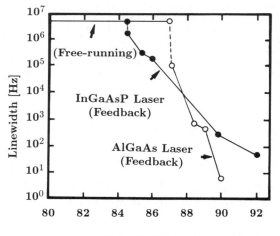

FIGURE 3.21. Progress of the authors' linewidth reduction experiments on semiconductor lasers.

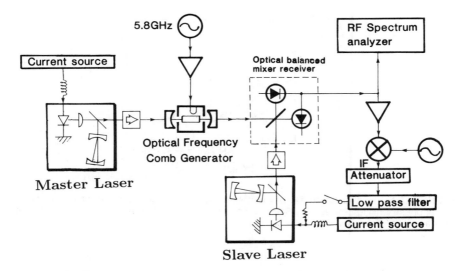

FIGURE 3.22. Experimental setup for phase locking using an optical frequency comb generator. Optical frequency comb generator is composed of an electro-optic phase modulator installed in FP cavity. The slave laser is heterodyne phase locked to one sideband of the optical frequency comb generated (after Ref. 63 © 1993 IEEE).

comb generator. We review here an experiment of the new method of optical phase locking using an optical frequency comb generator.

The experimental setup of optical phase locking using an optical frequency comb generator is shown in Fig. 3.22. The optical frequency comb generator is a system for generating high-order modulation sidebands, i.e., an optical frequency comb, by transmitting the semiconductor laser beam through an electro-optic modulator driven by a cw microwave with a frequency of f_m. The main component of the optical frequency comb generator which is shown in Fig. 3.22 is an electro-optic phase modulator which is installed in a FP cavity to realize a high modulation efficiency. In such a case, even if a heterodyne signal between the two lasers (master and slave lasers) cannot be observed because the frequency difference of the two lasers is too high, the heterodyne signal between the slave laser and the nearest modulation sideband which is generated by the optical frequency comb generator from the master laser can be measured. The frequency represented by this heterodyne signal f_h is $|\nu_1 - \nu_2 - Mf_m|$ and heterodyne phase locking can be performed with it, where M is a number of the sideband between the two lasers, and ν_1 and ν_2 are the frequencies of the two lasers, respectively.

In this experiment, a phase modulation as high as 0.2π radian has been achieved at the modulation frequency f_m of 5.8 GHz even with a bulky $LiNbO_3$ crystal for electro-optic phase modulation, where f_m was fixed to three times that of the free spectral range of the FP cavity. A finesse of this crystal-installed FP cavity as high as 200 was obtained. The optical frequency comb was generated by feeding an

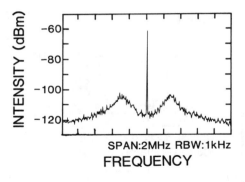

SPAN:2MHz RBW:1kHz
FREQUENCY

FIGURE 3.23. The spectral profile of the heterodyne signal observed under the condition of optical phase locking between the 84th sideband of the optical frequency comb generator and the slave laser (from Ref. 63).

FP-cavity-coupled 1.5 μm DFB laser beam into the cavity, and the heterodyne signal between one of the modulation sidebands in the optical frequency comb and another external cavity coupled laser was detected. Figure 3.23 represents the spectral profile of the heterodyne signal observed under the conditions of optical phase locking between the 84th sideband of the master laser generated by the optical frequency comb generator and the slave laser. Although the beat frequency f_h was 200 MHz, the separation between the carrier frequency of the modulated light, i.e., the master laser frequency and the slave laser frequency was as large as 487 GHz, i.e., the slave laser is optically phase locked to the laser modulated by the optical frequency comb generator with such a large frequency separation. The control bandwidth of this phase-locking loop was 250 kHz, and the phase error variance was estimated to be 0.01 rad^2.

The signal-to-noise ratio S/N (resolution band width = 10 kHz) of this detection was high, i.e., only 5 dB lower than that of the shot-noise-limited theoretical value, which is expressed as S/N (in units of dB) = $75 - 0.11 |M|$ where M is the order of the sideband in the optical frequency comb. Such a large value of the S/N ratio was attributed to using narrow linewidth lasers and an optical balanced mixer receiver. The maximum of the shot-noise-limited value of M, corresponding to S/N = 0 dB, is estimated to be 682, which gives the $M f_m = 4$ THz.

Based on these experimental results, it is expected that the frequency span of the optical frequency comb generator can be expanded to the limit set by the bandwidth of the mirror of the FP cavity, i.e., as wide as 20 THz, by using a higher efficiency phase modulator and by fabricating a monolithic FP cavity to reduce the loss. Such a wide frequency-span optical frequency comb generator can work as a local oscillator for optical frequency measurements of InGaAsP lasers in frequency division multiplexing coherent optical transmission in the 1.5 μm wavelength region.

3.4. FUTURE OF PHASE/FREQUENCY CONTROL IN SEMICONDUCTOR LASERS

As shown in Fig. 3.21, the realization of ultralow FM noise (ultranarrow linewidth < 1 Hz) is expected in the near future. And even in an optical phase-locked loop, 10 THz of heterodyne frequency will be seen in the near future by utilizing the optical frequency comb generator. Such expectation will meet requirements (1) and (2), which are described in the Introduction. However, in order to realize a light source similar to a frequency synthesizer used at microwave frequencies, requirement (3), which has not been discussed, must be satisfied for a precise setting of the frequency of the light source. As one method of fulfilling requirement (3) absorption lines of atoms or molecules can be used as optical frequency standard, but absorption lines accurately measured frequencies do not exist all optical frequency ranges. In this subsection, new technological ideas on the optical frequency synthesizer whose frequency can be accurately set with a wide range of optical frequencies is described. If such an optical frequency synthesizer can be realized, laser light can be handled like low-frequency electromagnetic waves and this synthesizer will become a very useful technique in many fields of basic physics, optical communication, and all high precision measurement.

3.4.1. Optical Frequency Synthesizer

What device is the ultimate optical frequency synthesizer? It is like the microwave frequency synthesizer, i.e., the device which generates a lightwave whose frequency has been accurately set. As one requirement for realizing such an optical synthesizer whose frequency can be tuned in the wide band, the laser medium must have a wide gain bandwidth. Typical lasers satisfying this requirement are dye lasers, which include liquid lasers, solid lasers, and semiconductor lasers. Among these, the semiconductor laser is compact, inexpensive, and the frequency can be controlled easily. The oscillation frequency range of the AlGaAs laser or InGaAs laser is from 0.6 to 2 μm, so it can serve as the light source which will be the core device for realizing an optical frequency synthesizer. Even in the 1.1 μm wavelength region where semiconductor lasers are lacking, solid lasers excited by semiconductor lasers or parametric oscillator[64,65] can be used. The next requirement is to measure the frequency of the laser and to control it precisely. If a conventional Michelson interferometer is used, frequency can be measured with an accuracy to about 10^{-6}, but more accurate measurement, which directly compares the laser frequency with microwave frequency is required. In order to compare the laser frequency with a microwave frequency, because the optical frequency is 2×10^{14} Hz or more, the optical frequency is reduced to the microwave frequency by the frequency chain using many lasers and optical frequencies mixers.[66] Thus, the frequency of some frequency standard lasers in the visible region have been measured, and the accuracy reaches 10^{-11}.[67] However, the optical frequency measurement system which has been used before is very complicated, and the wide band frequency measurement which covers a wide range of frequencies has not been realized. Recently,

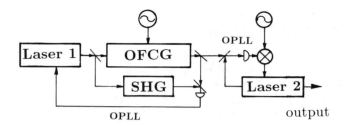

FIGURE 3.24. Construction of an optical frequency synthesizer. OFCG: optical frequency comb generator. SHG: crystal for generating the second harmonic wave. OPLL: optical phase-locking loop.

some new methods to solve the problems have been proposed. One of the proposed systems employs the frequency division of the difference in two laser frequencies by using the optical phase-locking technique and nonlinear optical crystals.[68,69,65] This system can achieve a very wide band frequency measurement because there are no limitations in setting the maximum frequency difference for division as long as the phase matching of nonlinear optical crystals is ensured. As in the other method, we propose a new system in which the optical frequency comb generator is utilized.

In Fig. 3.24, an optical frequency synthesizer in which an optical frequency comb generator is actively utilized is given. The wavelength of Laser 1 is assumed to be 1.56 μm (192 THz). A part of the light of Laser 1 is converted to a secondary harmonic wave (the wavelength is 0.78 μm) with nonlinear optical crystal. The remaining light of Laser 1 is converted to an optical frequency comb by the optical frequency comb generator. An optical frequency comb generator can generate an optical frequency comb with a span of about 10 THz. If another laser is locked to one sideband of the optical frequency comb by the phase-locking technique, then another optical frequency comb is generated from this laser, the total span of the optical frequency comb is increased. If this process is repeated several times, a span of the optical frequency comb can cover the wavelength range between the fundamental wavelength of Laser 1, i.e., 1.56 μm, and the secondary harmonic wavelength, i.e., 0.78 μm. As the differential frequency between the secondary harmonics and Laser 1 is coincident with the frequency of Laser 1, if the differential frequency of the secondary harmonics and Laser 1 is measured by a generated optical frequency comb, not only the frequency of Laser 1 but the frequency of all the sidebands of the optical frequency comb are determined and can be stabilized with an accuracy as high as the accuracy of the modulation frequency of the optical frequency comb generator. The generated optical frequency comb thus becomes a stable and highly accurate frequency standard. When a frequency tunable laser (Laser 2) is locked to the one sideband of the optical frequency comb with a phase-locking loop and a cesium microwave frequency standard with an accuracy of 10^{-14} for the modulation signal of the optical frequency comb generator, the frequency of Laser 2 can be locked to the desired frequency with an accuracy as high as the accuracy of the cesium microwave frequency standard. And if in the future, a better frequency standard than the cesium microwave frequency standard is established in the optical region, the standard of the optical region can be used for the standard of the microwave bandwidth.

To realize such a frequency synthesizer, many frequency tunable lasers must be provided in a wide frequency range, but in this proposed system, all can be realized with the semiconductor lasers. If such a technique for the optical frequency synthesizer is realized, not only the desired frequency but also the desired waveform can be produced by the optical frequency synthesizer. For example, it will become possible to generate ultrashort light pulses by matching the frequency and phase of plural cw lasers in the wide frequency range.[31]

Thus, methods of realizing an optical frequency synthesizer are introduced in this section, and this is related to the present progress in semiconductor laser and peripheral technologies. Specifically, conventional laser development is based mainly upon a technique of "searching for new materials" based on spectral data such as atoms and molecules, but the recent development is based on a strategy of "creating new materials" for the design semiconductor devices having new optical characteristics, solid-state laser crystals, or nonlinear wavelength crystals. Especially in the semiconductor laser, the quantum characteristics of electrons or light can be universally controlled. Thus, it can be said that the future optical frequency synthesizer will be realized from all solid-state light sources by solid-state crystals or semiconductors. Surely laser technology is at present analogous to electronic technology at the time between the vacuum tube radio and the transistor radio.

3.5. SUMMARY

This chapter reviewed the recent results of the authors' studies on highly sensitive FM noise detection and optical phase locking in order to suppress the laser FM noise originating from quantum fluctuations. The possibilities for an optical frequency synthesizer (a highly accurate wide band frequency tunable light source) based on optical frequency comb generators were also presented. It is expected that this optical frequency synthesizer will be built in the near future by increasing the width of the optical frequency comb generator, making use of a lot of semiconductor lasers, the frequency division of the difference between two laser frequencies by using nonlinear optical crystals and so on.

In addition to the frequency control methods of compensating for the effect of quantum FM noise described in this chapter, there may be other methods of realizing frequency stabilized lasers. One of these methods is to control directly the magnitude of the quantum noise, i.e., the probability of spontaneous emission, which can be a promising method even though it has not yet been applied to semiconductor lasers. There are at least two proposals for modifying the magnitudes of FM or phase noises: (1) It has been pointed out that laser oscillation without population inversion is possible using three-level atoms with A-shaped transition due to the quantum interference between the levels. The FM noise magnitude of this lasing light would be lower than the minimum value limited by conventional spontaneous emission fluctuations.[70] (2) The laser oscillation with correlated spontaneous emission is realized by using three-level atoms with V-shaped transition, which is also due to the quantum interference between the levels. The phase noise magnitude of this lasing light would be lower than the minimum value limited by quantum fluctuations (more precisely, vacuum fluctuations). It means that phase squeezing would be possible.[71]

Experiments on these two proposals have not yet been carried out for semiconductor lasers except for preliminary experiments on correlated spontaneous emission.[60] However, since it has been experimentally proved that electrical feedback to semiconductor lasers can generate squeezed light,[72] progress on these proposals can be expected in the near future. The future problems of using these quantum optical methods could be to find the application to frequency control and related experimental studies. When these problems are solved, novel light sources with low FM noise are expected in which the quantum noise magnitude is directly controlled.

REFERENCES

1. P. Vettiger, M. K. Benedict, G.-L. Bona, P. Buchmann, E. C. Cahoon, K. Datwyler, H.-P. Dietrich, A. Moser, H. K. Seitz, O. Voegeli, D. J. Webb, and P. Wolf, IEEE J. Quantum Electron. **27**, 1319 (1991).
2. P. J. A. Thijs, L. F. Tiemeijer, P. I. Kuindersma, J. J. M. Binsma, and T. V. Dongen, IEEE J. Quantum Electron. 1426 (1991).
3. M. Ohtsu and Y. Teramachi, IEEE J. Quantum Electron., **25**, 31 (1989).
4. M. Okai, S. Tsuji, and N. Chinone, IEEE J. Quantum Electron. **25**, 1314 (1989).
5. M. Ohtsu, *Highly Coherent Semiconductor Lasers* (Artech House, Norwood, 1992).
6. M. Ohtsu, *Coherent Quantum optics and Technology* (KTK Scientific, Tokyo, 1992).
7. M. Ohtsu and K. Nakagawa, (Wiley, New York, 1991).
8. G. Rempe, R. J. Thompson, and H. J. Kimble, Opt. Lett. **17**, 363 (1991).
9. A. L. Schawlow and C. H. Townes, Phys. Rev. **112**, 1940 (1958).
10. A. Blaquiere, Compt. Rend. **255**, 3141 (1962).
11. C. H. Henry, IEEE J. Quantum Electron. **QE-18**, 259 (1982).
12. Y. Yamamoto, O. Nilsson, and S. Saito, IEEE J. Quantum Electron. **QE-21**, 1919 (1985).
13. Y. Yamamoto and N. Imoto, IEEE J. Quantum Electron. **QE-22**, 2032 (1986).
14. J. Helmcke, S. A. Lee, and J. L. Hall, Appl. Opt. **21**, 1686 (1982).
15. H. Toba, K. Inoue, K. Nosu, and G. Motosugi, J. Opt. Commun. **9**, 50 (1988).
16. M. Ohtsu and S. Kotajima, IEEE J. Quantum Electron. **QE-21**, 1905 (1985).
17. W. Vassen, C. Zimmermann, R. Kallenbach, and T. W. Hänsch, Opt. Commun. **75**, 435 (1990).
18. H. Yasaka, Y. Yoshiuni, Y. Nakano, and K. Oe, Electron. Lett. **23**, 1161 (1987).
19. F. Favre and D. LeGuen, Electron. Lett. **16**, 709 (1980).
20. M. Ohtsu, M. Murata, and M. Kourogi, IEEE J. Quantum Electron. **26**, 231 (1990).
21. S. Matsumoto and M. Fujise, Electron. Lett. **25**, 814 (1989).
22. D. Shoemaker, A. Brillet, C. N. Man, O. Crégut, and G. Kerr, Opt. Lett. **14**, 609 (1989).
23. Ch. Salomon, D. Hils, and J. L. Hall, J. Opt. Soc. Am. B **5**, 1576 (1988).
24. Y. C. Chung and T. M. Shay, Opt. Eng. **27**, 424 (1988).
25. R. W. P. Drever, J. L. Hall, F. V. Kowalski, J. Hough, G. M. Ford, A. J. Munley, and H. Ward, Appl. Phys. B **31**, 97 (1983).
26. M. Houssin, M. Jardino, B. Gely, and M. Desaintfuscien, Opt. Lett. **13**, 823 (1988).
27. T. Day, E. K. Gustafson, and R. L. Byer, Opt. Lett. **15**, 221 (1990).
28. C. S. Adams and A. I. Ferguson, Opt. Commun. **75**, 419 (1990).

29. A. Sollberger, A. Heinämäki, and H. Melchior, J. Lightwave Technol. **LT-5**, 485 (1987).
30. R. V. Pound, Rev. Sci. Instrum. **17**, 490 (1946).
31. T. W. Hänsch and B. Couillaud, Opt. Commun. **35**, 441 (1980).
32. M. Kourogi and M. Ohtsu, Opt. Commun. **81**, 204 (1991).
33. A. Hemmerich, D. H. McIntyre, C. Zimmermann, and T. W. Hänsch, Opt. Lett. **15**, 372 (1990).
34. C. E. Tanner, B. P. Masterson, and C. E. Wieman, Opt. Lett. **13**, 357 (1988).
35. J. Harrison and A. Mooradian, IEEE J. Quantum Electron. **QE-25**, 1152 (1989).
36. D. Lenstra, B. H. Verbeek, and A. J. Bon Boef, IEEE J. Quantum Electron. **QE-21**, 674 (1985).
37. C. J. Nielsen and J. H. Osmundsen, J. Opt. Commun. **5**, 42 (1984).
38. J. M. Kahn, C. A. Burrus, and G. Raybon, IEEE Photon. Technol. Lett. **1**, 159 (1989).
39. G. Wenke, R. Gross, P. Meissner, and E. Pazak, J. Lightwave Technol. **LT-5**, 608 (1987).
40. N. A. Olsson and J. P. Van Der Ziel, J. Lightwave Technol. **LT-5**, 510 (1987).
41. M. Ohtsu, H. Suzuki, K. Nemoto, and Y. Teramachi, Jpn. J. Appl. Phys. **29**, L1463 (1990).
42. A. Schremer and C. L. Tang, Appl. Phys. Lett. **55**, 19 (1989).
43. F. Favre, D. LeGuen, and J. C. Simon, IEEE J. Quantum Electron. **QE-18**, 1712 (1982).
44. F. Favre and D. LeGuen, IEEE J. Quantum Electron. **QE-21**, 1937 (1985).
45. A. G. Bulushev, Y. V. Gurov, E. MN. Dianov, A. V. Kuznetsov, and V. M. Paramonov, Sov. J. Quantum Electron. **18**, 698 (1988).
46. B. Dahmani, L. Hollberg, and R. Drullinger, Opt. Lett. **12**, 876 (1987).
47. Ph. Laurent, A. Clairon, and Ch. Bréant, IEEE J. Quantum Electron. **25**, 1131 (1989).
48. M. Segev, S. Weiss, and B. Fischer, Appl. Phys. Lett. **50**, 1397 (1987).
49. C. H. Shin and M. Ohtsu, IEEE J. Quantum Electron. **29**, 374 (1993).
50. C. H. Shin, M. Teshima, M. Ohtsu, T. Imai, J. Yoshida, and K. Nishide, IEEE Photon. Technol. Lett. **2**, 297 (1990).
51. M. Ohtsu, H. Fukada, T. Tako, and H. Thuchida, Jpn. J. Appl. Phys. **22**, 1157 (1983).
52. C. H. Shin and M. Ohtsu, Opt. Lett. **15**, 1455 (1990).
53. K. Kuboki and M. Ohtsu, IEEE J. Quantum Electron. **25**, 2084 (1989).
54. C. H. Shin and M. Ohtsu, IEEE Photon. Technol. Lett. **2**, 297 (1990).
55. M. Kourogi, C. H. Shin, and M. Ohtsu, IEEE Photon. Technol. Lett. **3**, 270 (1991).
56. M. Okai, T. Tsuchiya, K. Uomi, N. Chinone, and T. Harada, IEEE Photon. Technol. Lett. **2**, 529 (1990).
57. M. Okai, T. Tsuchiya, K. Uomi, N. Chinone, and T. Harada, IEEE Photon. Technol. Lett. **3**, 427 (1991).
58. M. Kourogi and M. Ohtsu, IEEE Photon. Technol. Lett. **3**, 496 (1991).
59. K. Nakagawa, M. Kourogi, and M. Ohtsu, Opt. Lett. **17**, 934 (1992).
60. M. Ohtsu and S. Araki, Appl. Opt. **26**, 464 (1987).
61. T. Imai, N. Ken-ichi, H. Ochi, and M. Ohtsu, SPIE 1585, 153 (1991).
62. H. Li and H. R. Telle, IEEE J. Quantum Electron. **25**, 257 (1989).
63. M. Kourogi, K. Nakagawa, and M. Ohtsu, IEEE J. Quantum Electron. **29**, 2693 (1993).
64. R. C. Eckardt, C. D. Nabors, W. J. Kozlovsky, and Robert L. Byer, J. Opt. Soc. Am. B **8**, 646 (1991).
65. N. C. Wong, Opt. Lett. **15**, 1129 (1990).
66. D. A. Jennings, K. M. Evenson, and D. J. E. Knight, Proc. IEEE **74**, 168 (1986).
67. O. Acef, A. Clairon, M. Abed, P. Laurent, and Y. Millerioux, SPIE **1837**, 405 (1992).
68. H. R. Telle, D. Meschede, and T. W. Hänsch, Opt. Lett. **15**, 532 (1990).
69. T. Sato, S. Singh, S. Swartz, and J. L. Hall, 1990. IQEC'90 Technical Digest, paper QThD3.
70. O. Scully and S.-Y. Zhu, Phys. Rev. Lett. **62**, 2813 (1989).
71. M. Fleischauser, U. Rathe, and M. O. Scully, in International Conference on Quantum Electronics, Vienna, Austria, June 1992, pages 100–102.
72. Y. Yamamoto, S. Machida, and O. Nilsson, *Coherence, Amplification, and Quantum Effects in Semiconductor Lasers*, ed. by Y. Yamamoto (Wiley, New York, 1991), Chap. 11, pages 461–536.

CHAPTER 4

Mode-Locked Semiconductor Lasers

Roger Helkey* and John Bowers[†]

*MIT Lincoln Laboratory, Lexington, Massachusetts
[†]Department of Electrical and Computer Engineering,
University of California, Santa Barbara, California

4.1. INTRODUCTION

4.1.1. Mode-Locking Overview

Short optical pulses can be generated from a laser by mode-locking. Some advantages of semiconductor diode lasers as mode-locked sources are that they are compact, available over a wide range of wavelengths using bandgap engineering, allow integration with other optoelectronic devices, and are electrically pumped. They can be used for telecommunications systems, high speed A/D converters, electro-optic sampling systems, optical computing, phased array radar systems, optical microwave frequency sources, and multiple-wavelength high-speed physics experiments.

Mode-locking techniques utilize a time-dependent gain function, as illustrated in Fig. 4.1(a). The optical signal at the peak of the amplitude modulation receives the most gain, with the loss to the optical signal increasing away from this point. This causes the formation of optical pulses [Fig. 4.1(b)]. Each time a pulse travels through the cavity, the pulse tail receives less gain than the pulse peak, leading to pulse-width shortening. The optical pulsewidth decreases until the pulse shortening per pass is balanced by the pulse broadening per pass due to other mechanisms.

The optical spectrum of a mode-locked laser is a series of modes corresponding to the Fabry–Perot cavity modes of the laser [Fig. 4.1(c)]. However, short optical pulses only result from this optical spectrum if the optical modes are all in phase. The term mode-locking is a frequency domain description in which optical cavity modes are coupled in phase to produce a short pulse.

In this chapter we describe the different types of mode-locking. The theoretical

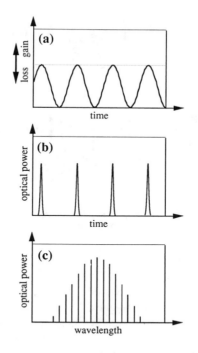

FIGURE 4.1. (a) Time dependence of net gain in the laser cavity. (b) Resulting optical output. (c) Optical spectrum of the mode-locked laser.

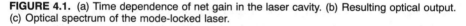

basis for semiconductor mode-locking is given with design curves for the dependence of pulse shortening on the laser parameters. The important process of colliding-pulse mode-locking is described theoretically and experimentally. The chapter concludes with a discussion of phase noise and techniques for reducing timing jitter.

4.1.2.. Mode-Locking Configurations

There are three classes of mode-locking, depending on the way the time-dependent gain function in Fig. 4.1 was generated. Active mode-locking is a technique in which the gain modulation is externally applied.[1] This can easily be done in semiconductor lasers by modulating the electrical current, which is possible up to very high frequencies [Fig. 4.2(a)].

Passive mode-locking uses a saturable absorber in order to operate under the same gain modulation principle [Fig. 4.2(b)], but the gain modulation is supplied by the optical pulse itself. The absorber attenuates the beginning of the optical pulse [Fig. 4.3], which leads to overall pulse shortening. After the absorber saturates, the center of the pulse experiences net gain. As the pulse continues to propagate, saturation of the gain medium reduces the gain to below threshold and shuts off

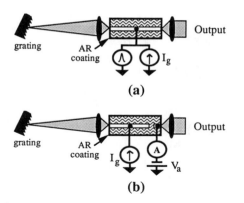

FIGURE 4.2. (a) Active mode-locking of a semiconductor laser diode. Time varying gain is applied by external current modulation. (b) Passive mode-locking using a reversed biased section as a saturable absorber.

lasing. The third technique is hybrid mode-locking, which uses a combination of active and passive modulation techniques.

There are several techniques to achieve passive mode-locking using semiconductor diode lasers. One method is ion implantation to introduce recombination centers at one facet, which decreases the carrier recombination time and forms a saturable absorber.[2] Another approach is to split the gain contact [Fig. 4.2(b)] and reverse bias one segment to form an integrated waveguide saturable absorber.[3] A third technique is to use a semiconductor medium in the external cavity which operates on an excitonic absorption transition.

The primary pulse-width shortening mechanisms in the cavity are gain modulation (active mode-locking) and saturable absorption (passive mode-locking). The primary pulse-width broadening mechanisms are self-phase modulation, dispersion, and bandwidth effects in the gain medium. The pulse width is determined by the balance between these mechanisms.

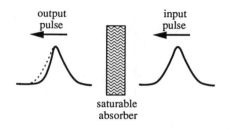

FIGURE 4.3. Passive mode-locking with a slow saturable absorber. Time varying loss is internally generated by the interaction of the pulse with the saturable absorber.

4.2. SUBPICOSECOND PULSE GENERATION

4.2.1. Gain and Absorber Region Models

For a short optical pulse, saturation of the gain region can be given by the following rate equations (see Ch. 8 for additional discussion of optical amplification):

$$P_{out}(\tau) = P_{in}(\tau)\exp[h(\tau)] \qquad (4.1)$$

$$\frac{dh(\tau)}{d\tau} = -\frac{P_{in}(\tau)}{E_{sat}}\{\exp[h(\tau)]-1\}, \qquad (4.2)$$

where the saturation energy E_{sat} and the amplifier power gain $G(\tau)$ are given by

$$E_{sat} \equiv \frac{h\nu A}{\Gamma\dfrac{dg(N_g)}{dN}} \qquad (4.3)$$

$$G(\tau) = \exp[h(\tau)], \qquad (4.4)$$

where $h\nu$ is the photon energy, A is the cross sectional area of the waveguide, Γ is the confinement factor, and dg/dN is the differential gain of the material at the gain region carrier density (N_g). The power gain G_f after passage of a pulse is lower due to gain saturation. For large unsaturated gain G_0, the saturated gain is[4]

$$G_f \approx \frac{G_0}{1+G_0 E_{in}/E_{sat}}. \qquad (4.5)$$

The power gain begins to saturate for an input pulse energy E_{in} such that $G_0 E_{in} \sim E_{sat}$ (the output energy in the absence of gain saturation is approximately equal to the saturation energy). For passive mode-locking with a gain and absorber region, this same model also can be applied to an absorber region, which in general will have a different saturation energy:

$$E_{sat}^{abs} \equiv \frac{h\nu A}{\Gamma\dfrac{dg(N_a)}{dN}} = \frac{E_{sat}}{\sigma} \qquad (4.6)$$

$$\frac{dh(\tau)}{d\tau} = -\frac{P_{in}(\tau)}{\sigma E_{sat}}\{\exp[h(\tau)]-1\}, \qquad (4.7)$$

where σ is the ratio of the saturation energy in the gain region to the saturation energy in the absorber region, and E_{sat} is the saturation energy of the gain region. Passive mode-locking requires that the absorber saturate more easily than the gain ($\sigma > 1$).[5]

When using an integrated saturable absorber and a uniform waveguide, the optical mode area is constant and the saturation energy ratio is given by the ratio of the differential gains. The saturable absorber can be formed from the same material

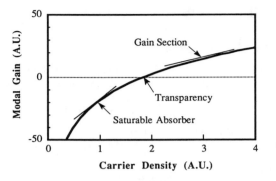

FIGURE 4.4. The differential gain in a semiconductor laser has a sublinear dependence on carrier density. For a constant optical mode area, the saturation energy ratio σ is given by the ratio of the slopes.

as the gain region, due to the sublinear dependence of differential gain on the carrier density illustrated in Fig. 4.4.[6]

By tapering the waveguide to change the optical mode area along the device,[7] the saturation energy ratio σ can either be enhanced or degraded. When using an external saturable absorber, the optics should be chosen to give a smaller beam spot size in the absorber than in the gain medium. Under normal mode-locking conditions, $E_{in} \gg E_{sat}^{abs}$, and the energy gain in the absorber is[4]

$$G_E \approx 1 - \frac{\ln(\alpha_0)}{\sigma E_{in}/E_{sat}}, \tag{4.8}$$

where α_0 is the unsaturated loss of the absorber, and the energy gain G_E is negative for absorption.

Bleaching energy is a useful value that can be defined for the absorber, which is the energy lost under strong saturation of the absorber. This value must not exceed the energy which can be supplied by the gain region, which is on the order of E_{sat}. The bleaching energy E_{bl} is given by

$$E_{bl} \equiv E_{in} - E_0 = E_{in}(1 - G_E) \tag{4.9}$$

$$E_{bl} \approx \frac{\ln(\alpha_0)}{\sigma} E_{sat} \quad \text{or} \quad \ln(\alpha_0) \approx \sigma \frac{E_{bl}}{E_{sat}}, \tag{4.10}$$

where E_0 is the absorber output energy. The absorber bleaching energy E_{bl} is an important design parameter. For a given absorber bleaching energy, the unsaturated absorption increases as the differential gain increases.

Increasing the differential gain parameter σ improves pulse shaping by sharpening the leading edge of the pulse. A larger value of σ also satisfies the mode-locking condition over a wider range of parameters by increasing the unsaturated

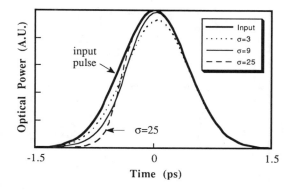

FIGURE 4.5. Gain shaping provided by the saturable absorber for several values of differential gain ratio between absorber and gain region. Input pulse energy $E_{in} = E_{sat}$ of the gain region.

absorption and suppressing CW lasing. Figure 4.5 shows the improvement in pulse shaping provided by the absorber as the differential gain ratio σ increases.

4.2.2. Pulse Shaping by Passive Mode-Locking

The pulse energy gain through the gain and absorber regions (the chip gain) must equal the round-trip loss of the cavity. The total cavity loss is given by the product of the absorber facet reflectivity, the external cavity mirror reflectivity, and the square of the coupling loss of the lens (since this loss occurs when coupling into and out of the waveguide). Figure 4.6 shows the calculated gain versus input pulse energy for a round trip through the gain and absorber, with an ideal mirror.

At low input energy, both the gain and absorber region are unsaturated and are not energy dependent. At an intermediate energy, the absorber is beginning to be bleached while the amplifier is not strongly saturated, giving the highest net gain for pulsed operation. As the input energy is increased further, the absorber becomes completely bleached, and increased amplifier saturation leads to less overall gain. Fig. 4.6 also shows that the gain region gives some pulse-width broadening.[27] However, the absorber gives stronger pulse shaping by removing the leading edge of the pulse, leading to overall pulse-width shortening.

There are several requirements for achieving mode-locking. There needs to be net cavity loss before and after the pulse, to prevent a CW mode from taking over. The peak energy gain needs to be higher than the energy gain for a low-energy input pulse. There also needs to be a net pulse-width reduction mechanism, to counterbalance the pulse-width broadening mechanisms of dispersion and bandwidth broadening.

For a given absorber value, the necessary unsaturated amplifier gain to give a pulse energy gain of 20 is shown in Fig. 4.7, along with the pulse-width shortening ratio at this peak energy gain. The required unsaturated amplifier gain is determined

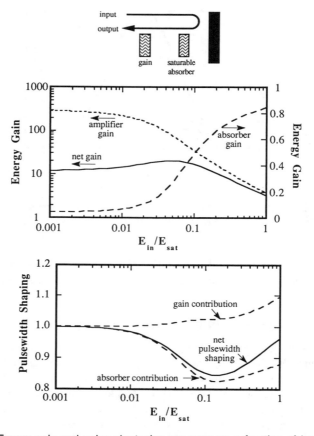

FIGURE 4.6. Energy gain and pulse shortening per pass as a function of input power for round-trip propagation with $\sigma = 3$, unsaturated amplifier gain = 17, and unsaturated absorber gain = 0.2 per pass. Peak energy gain ~ 20.

by the bleaching energy E_{bl}. The same gain curve applies for both values of σ. As the saturation energy ratio σ increases, the pulse-width shortening increases. A high saturation energy ratio is desirable not only for increasing the pulse-width shortening, but also for increasing the mode-locking parameter range. For the case of $\sigma = 3$, as the bleaching energy is reduced, the pulse-width shortening ratio goes to 1. At the same bleaching energy the mode-locking condition requiring net cavity loss before the pulse is no longer satisfied. In the other extreme, as the bleaching energy increases beyond E_{sat} the required unsaturated amplifier gain becomes extremely large. This is because the amplifier can only supply pulse energies on the order of E_{sat}.

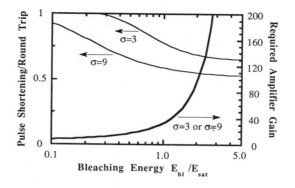

FIGURE 4.7. Pulse-width shortening and required unsaturated amplifier gain as a function of bleaching energy in order to give a round-trip energy gain of 20.

4.2.3. Experimental Results

A fundamental limitation in coupling a semiconductor diode laser into an external cavity is the finite facet reflectivity that remains after applying an antireflection coating. The minimum facet reflectivity is limited by the control of the thickness and index of the material being used to match the effective mode impedance of the laser to the impedance of air. Very tight tolerances are required in order to achieve a power reflectivity less than 10^{-4}.

In mode-locked semiconductor lasers, the finite facet reflectivity causes multiple pulses spaced at the round-trip time of the cavity. This degrades system performance, since the effective pulse width becomes that of the multiple pulse envelope, rather than that of each individual pulse. Using active mode-locking of a single section device, a facet reflectivity as low as 10^{-5} is necessary to prevent multiple pulsation.[8] This low facet reflectivity is impractical for most coating systems.[9] Techniques to reduce the multiple pulsation problem include using increased device length for active mode-locking,[10] and absorption recovery in passive mode-locking.[11]

One way to further reduce facet reflectivity is with angled facet devices.[12] Traveling wave amplifiers can be fabricated by placing the gain region at an angle to the crystallographic plane, so that the cleaved device has waveguides that are not perpendicular to the facet. This reduces the effective facet reflectivity, because most of the reflected light does not couple back within the acceptance angle of the guided mode of the waveguide.

To reduce multiple pulses, angled facet devices have been used for mode-locking,[13,14] using a traveling wave amplifier in a cavity between two external mirrors. However, this technique has the disadvantage that light must be coupled into and out of the semiconductor laser, which is more complex than the standard linear cavity geometry (Fig. 4.2), and has twice the coupling loss. In the linear

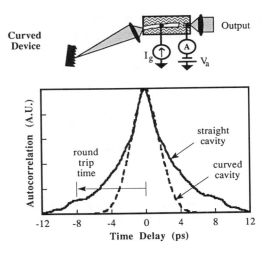

FIGURE 4.8. Noncollinear intensity autocorrelation using straight- and curved-waveguide devices.

cavity configuration, the gain region is at one end of the cavity, and one of the device facets acts as a cavity mirror.

When using an angled facet device, one of the laser facets is not available to act as a cavity mirror, so an external cavity mirror must be added. This means that the system will require a second cavity alignment, as well as increase the total coupling loss. The absorber should be placed at one end of the cavity. Since the angled-facet laser is in the middle of the cavity, the absorber cannot be integrated with the laser but must be added externally. The beam size in the absorber must be small to give a low saturation energy. In order to have a small optical spot size in the absorber, a second lens also must be added.

The angled-facet configuration eliminates multiple pulses, but the normal-facet configuration is simpler. Since only one low-reflectivity facet is desired for mode-locking, *only one facet should be angled with respect to the waveguide.* Two techniques for producing devices with a single-angled facet are an etched-facet/cleaved-facet device and a curved-waveguide device. The etched-angled-facet approach requires an additional alignment/etching step to form angled facets. The curved-waveguide technique has the advantage that both of the facets are formed by cleaving, without further processing steps.

Angled-facet InGaAs/AlGaAs devices were fabricated and compared to straight-cavity devices.[7] Both devices had an absorber length of ~ 40 μm and a cavity length of 300 μm. The curved device was curved to an angle of 5° over a distance of 220 μm, with a radius of curvature of 2.5 mm. The devices were fabricated by impurity-induced disordering.[15] Impurity-induced disordering[16,17] is a good fabrication technique for curved and angled gain regions, as it is insensitive to crystal orientation.

The two devices were antireflection coated using a quarter wave layer of non-stoichiometric SiN_x. The relative reduction in facet reflectivity can be found by measuring the amplitude ripple in the spontaneous emission spectrum due to the undesired Fabry–Perot cavity. In a straight waveguide device, the uncoated facet reflectivity is known. By biasing the device at its original threshold, the total material gain is equal to the uncoated mirror loss, and the coated mirror loss can be determined. The straight device had a gain ripple of 1 dB, corresponding to a facet power reflectivity of 10^{-3}. The curved-waveguide device had an estimated coated facet reflectivity of 5×10^{-5}. Passive mode-locking was initiated by reverse biasing the absorber to -0.5 V. The autocorrelation results are shown in Fig. 4.8. The straight device exhibits the small residual trailing pulse which is always seen in these types of structures. The curved device produces only a single pulse.

A major limitation of mode-locked semiconductor diode lasers is low output power. The average output power is typically a few mW for a single-stripe device. By increasing the modal cross section, the saturation energy can be increased.[18] Techniques to increase the mode-locked output power while maintaining a single lateral mode include external cavity spatial filtering of broad area and array structures[14,19] and resonant optical wave (ROW) coupled arrays.[20] External amplification can produce very high peak powers,[14] but requires an additional alignment. The simplest approach is to obtain more power from a basic single stripe laser.

In addition to curving the waveguide, the waveguide can also be flared to change the optical mode area. A tapered waveguide can be used which expands the optical mode from a wide multimode region for higher power, to a narrow region to force single lateral optical mode operation. Tapered waveguides also have been used as high-power CW sources.[21–23]

A tapered-waveguide device has been used to increase the mode-locked output power.[7] The waveguide active region has a linear taper from 2.5 to 7.5 μm over a 150-μm distance. This device is compared to a uniform-waveguide device with a similar absorber length, so the pulse energy should be proportional to the saturation energy and scale with optical mode area. However, the increase in the beam cross section is smaller than the increase in the waveguide width. The mode-locked output pulse energy was increased from 1.8 to 4.1 pJ, a factor of 2.3. At the same time, the tapered device pulse width widened by a factor of 1.2. Due to device heating, the absorber bias had to be increased from -0.5 to $+0.5$ V. This allowed the gain region to operate at lower bias current, but increased the absorber recovery time,[24] which slightly degraded mode-locking performance.

By mounting the device junction-side down to reduce heating, this technique could be extended to even higher output energy by tapering to an even broader absorber region. One potential problem with this flared-waveguide configuration is that the average beam diameter in the gain region is smaller than in the absorber region. The figure of merit for mode-locking is σ, the ratio of the saturation energy of the absorber to the gain defined earlier in the chapter. This saturation ratio is proportional to the volume of the optical mode. The value of E_{sat} for the amplifier is effectively a weighted value, since the mode area is changing as the pulse propagates. As a result, σ decreases due to the mode expansion in the waveguide, which can degrade the mode-locked pulse shaping for large tapering ratios. To achieve higher output power, other configurations could be used, such as tapering the waveguide in the other direction and extracting the output power from the wider external cavity side.

4.2.4. Self-Phase Modulation Effects on Optical Spectrum

In a semiconductor diode laser, gain saturation produces large refractive index changes. These are caused by the dependence of both the index and gain on the carrier density. This mutual dependence of gain and index is described by the parameter α, known as the linewidth enhancement factor:[25] Gain saturation which is induced by the optical pulse causes a corresponding index shift, which modulates the optical frequency:

$$\Delta \nu(\tau) \equiv \Delta \nu_{out}(\tau) - \Delta \nu_{in}(\tau) = \frac{\alpha}{4\pi} \frac{\partial h}{\partial t}, \tag{4.11}$$

where $\Delta \nu_{in}(\tau)$ and $\Delta \nu_{out}(\tau)$ are the instantaneous optical offset frequencies at the input and output of the semiconductor region, and $\Delta \nu(\tau)$ is the change in optical frequency caused by the self-phase modulation. This frequency shift is additive, and accumulates on each round trip through the optical cavity.

Substituting in the expression for the gain coefficient h gives[26]

$$\Delta \nu(\tau) = -\frac{\alpha}{4\pi} \frac{P_{in}}{E_{sat}} [G(\tau) - 1] \approx -\frac{\alpha}{4\pi} \frac{P_{out}}{E_{sat}} \exp\left(-\frac{U_{in}(\tau)}{E_{sat}}\right). \tag{4.12}$$

For small gain saturation $U_{in}(t) \ll E_{sat}$, so in the absence of gain saturation the chirp is proportional to the output power of the optical pulse.

For a Gaussian input pulse, P_{in} is given by

$$P(\tau) = \frac{E_{in}}{\pi^{1/2} \tau_0} \exp\left[-\left(\frac{\tau}{\tau_0}\right)^2\right], \tag{4.13}$$

where τ_0 is a parameter which determines the pulse width. For this input pulse, the relative optical frequency shift is given by

$$\Delta \nu(\tau) = -\frac{\alpha}{4\pi^{3/2} \tau_0} \frac{E_{in}}{E_{sat}} \exp\left[-\left(\frac{\tau}{\tau_0}\right)^2\right] [G(\tau) - 1]. \tag{4.14}$$

An important feature of the spectral broadening is the term for τ_0 in the denominator, which arises because the self-phase modulation is proportional to the optical instantaneous power. As the pulse width gets shorter, the spectral broadening due to self-phase modulation increases. This will be shown later to be a very strong pulse-width limiting mechanism.

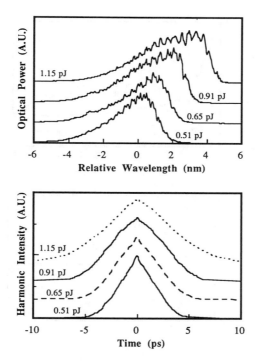

FIGURE 4.9. Measured (a) optical spectrum and (b) autocorrelation as a function of pulse energy.

The self-phase modulation modifies the phase of the optical signal which leads to spectral broadening, but does not change the shape of the pulse. Consequently, one might expect that the mode-locked pulse width would be independent of optical power. Figure 4.9 shows the effect of increasing pulse energy on a mode-locked GaAs/AlGaAs device. As the pulse energy increases, the optical spectrum widens, as expected from the equation for self-phase modulation. At the same time, however, the optical pulse gets wider. In order to understand this, it is necessary to also consider the finite gain-bandwidth and phase function of the material.

4.2.5. Bandwidth Shaping Effects on Pulsewidth

Unlike self-phase modulation, the amplifier spectral gain function does act as a pulse-width broadening mechanism. As discussed in Ch. 2, the spectral gain function leads to pulse-width broadening, which increases as the optical bandwidth increases. Increased optical bandwidth results from decreasing pulse width and chirp induced by self-phase modulation. The self-phase modulation induced by gain saturation widens the optical spectrum without changing the optical pulse width. However, this increased optical bandwidth then interacts with the amplifier gain and

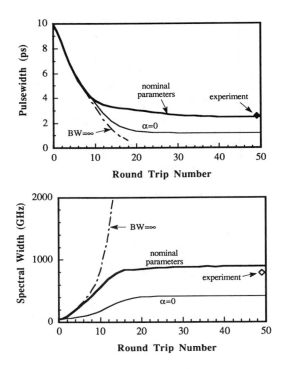

FIGURE 4.10. Calculated pulse width and spectral width after each round trip of the optical cavity, compared to experimental result using a 3-GHz external cavity.

phase response to cause pulse-width broadening. As a result, the pulse width is set by the interaction of self-phase modulation and the amplifier response.[27]

Figure 4.10 shows the calculated pulse width and spectral width for a device with a single pass absorber transmission $T_{abs} = 0.25$, saturation energy ratio $\sigma = 3$, linewidth enhancement factor $\alpha = 4$, Lorentzian linewidth = 5 THz, and absorber recovery time $\tau_{abs} = 5$ ps. The calculations only assume a phase response associated with the gain function, although phase dispersion would produce a similar result. The results are compared to experimental values for a 3-QW InGaAs device with a 60-μm absorber.

In the absence of self-phase modulation ($\alpha = 0$), the calculated pulse width is 1.2 ps, and the optical bandwidth is 420 GHz. With $\alpha = 4$, the pulsewidth increases to 2.4 ps, but the optical bandwidth broadens to 890 GHz. The primary goal in ultrashort pulse generation using external pulse compressors is generating a wide optical spectrum for compression. The width of the optical spectrum sets the ultimate pulse-width limit for semiconductor lasers. Even though small α can be used to directly generate shorter pulses, large α can be used to generate a broader spectrum leading to shorter pulses after compression.

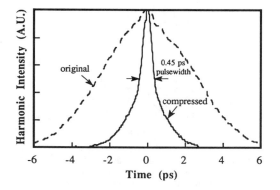

FIGURE 4.11. Pulse width before and after pulse compression to remove chirp caused by self-phase modulation. Compressed pulse width = 0.45 ps.

In the absence of bandwidth filtering (but still with $\alpha = 4$), there is no strong pulse-width broadening mechanism and the pulse becomes short. At the same time, there is no bandwidth limiting mechanism, so the optical bandwidth gets very large (giving short compressed pulses). This demonstrates that it *is not* important to design lasers for small α to give a smaller compressed pulse width. It *is* important to maximize the spectral flatness and phase flatness of the amplifier in order to generate a large optical spectrum for compression.

4.2.6. Dispersion-Based Pulse Compression

Self-phase modulation and the gain-bandwidth limit combine to give highly chirped 2–3- ps pulses. The solution to generating subpicosecond pulses with semiconductor diode lasers is to use dispersion to remove this chirp. The simplest way to provide the necessary dispersion is with a simple two-grating pulse compressor. Constructing a compact source requires a relatively high repetition rate (> 1 GHz) and a grating pair compressor without the telescope. Figure 4.11 shows the result using a 10-cm external cavity (1.5 GHz repetition rate). The compressor grating separation was ~ 12 cm, giving a total source area of 11.4×35.6 cm. The pulse width after compression was 0.45 ps, assuming a hyperbolic secant pulse. For further discussion of short pulse generation utilizing self-phase modulation to generate chirped pulses, see Ref. 28.

4.3. COLLIDING PULSE EFFECTS

4.3.1. Optical Pulse Collision

The mode-locking configuration discussed in the last section had the absorber placed at the output facet [Fig. 4.12(a)]. As a consequence, the forward pulse and

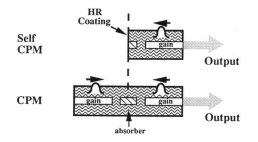

FIGURE 4.12. (a) Self-colliding pulse mode-locking configuration (SCPM). A single pulse collides with itself. (b) Semiconductor colliding pulse mode-locking configuration (CPM), where two separate pulses collide in an absorber of twice the length.

the reflected pulse form an optical standing wave in the absorber. This standing wave causes an additional coherent effect, which was not considered in the last section. In this section, the colliding pulse effect is shown to have two important components: a coherent effect due to grating formation, and an incoherent geometrical effect. These effects are examined separately, both theoretically and experimentally. It is shown that the carrier grating formation is not an important pulse shaping mechanism. However, the geometrical placement of the absorber is important.

Interference between the forward and reverse traveling pulses will cause the absorber to be bleached faster near the standing-wave peak than near the standing-wave null. This spatially dependent carrier generation causes an absorption grating with the same period as the optical standing wave. This absorption grating reduces the bleaching energy of the absorber, because less total volume of the absorber needs to be saturated for a given change in absorption. This effectively increases the differential gain in the absorber, which was shown in the last section to improve pulse shaping. A similar coherent effect also occurs in the colliding pulse mode-locking (CPM) configuration. A colliding pulse mode-locking example is shown in Fig. 4.12(b), where two counter-propagating pulses meet in the absorber.

Semiconductor lasers have been mode-locked in a linear cavity colliding pulse mode-locked (CPM) configuration [Fig. 4.12(a)], which has an absorber in the center of the cavity. If the counter-propagating pulses are identical, then by symmetry the CPM configuration is identical to the configuration shown in Fig. 4.12(b), which has an absorber of half the length at the end of the cavity.

The linear cavity configuration with the absorber at the end of the cavity can be called a self-colliding pulse mode-locked (SCPM) configuration,[29] because the same pulse collision pulse shaping is occurring due to the pulse colliding with itself.[30,31] Both configurations have been used for passive mode-locking of semiconductor diode lasers.[32–34] The SCPM configuration has the advantage of inherent pulse symmetry since there is only one pulse in the cavity. The SCPM configuration also has the advantage of smaller size, which is important for fabricating low-repetition-rate monolithic structures.

In dye lasers, pulse collision in mode-locking has been extensively studied and found to play an important pulse shaping role.[30,31,35-39] However, semiconductor lasers have a higher gain and loss per pass than dye lasers, and are physically much smaller. This can lead to differences in the importance of colliding pulse effects.

4.3.2. Carrier Grating Equations

The effects of pulse collision in saturable absorbers can be studied using a simple model in which the absorption grating is assumed to be sinusoidal:[37,39]

$$h = h_a + h_b \cos(2kx). \tag{4.15}$$

To be consistent with previous gain definitions, the grating is written in terms of the logarithmic gain parameter h, so the absorption term $\alpha = -h$. The DC absorption component is $-h_a$ and the sinusoidal absorption component due to the carrier grating is $-h_b$.

The equations for coherent pulse interaction[39] can be found from the coupling coefficients of a conductivity grating. The coupled equations for a short segment of length L are

$$A_{\text{out}}^+ = \left(1 + \frac{h_a}{2}\right) A_{\text{in}}^+ - \frac{h_b}{4} A_{\text{in}}^- \tag{4.16}$$

$$A_{\text{out}}^- = \left(1 + \frac{h_a}{2}\right) A_{\text{in}}^- - \frac{h_b}{4} A_{\text{in}}^+ \tag{4.17}$$

$$\frac{dh_a}{dt} = \frac{-\alpha_0 L - h_a}{\tau_c} - \frac{|A_{\text{in}}^+|^2 + |A_{\text{in}}^-|^2}{E_{\text{sat}}} h_a + \frac{|A_{\text{in}}^+||A_{\text{in}}^-|}{E_{\text{sat}}} h_a \tag{4.18}$$

$$\frac{dh_b}{dt} = -\frac{h_b}{\tau_c} - \frac{|A_{\text{in}}^+|^2 + |A_{\text{in}}^-|^2}{E_{\text{sat}}} h_b + \frac{2|A_{\text{in}}^+||A_{\text{in}}^-|}{E_{\text{sat}}} h_a, \tag{4.19}$$

where A^+ and A^- are the forward and reverse traveling field amplitudes normalized to the square root of optical power. The unsaturated power transmission of the segment is $\exp(-\alpha_0 L)$. E_{sat} is the previously defined saturation energy of the semiconductor material when used as an amplifier. In the absence of a carrier grating ($h_b = 0$) these equations reduce to the uniform carrier density case studied earlier.

4.3.3. CPM Analysis

In order to illustrate the effect of the carrier grating on the mode-locking process, the three SCPM geometry structures shown in Fig. 4.13 are analyzed. The "coherent CPM" case has an absorber at a high reflection mirror, and analyzes the coherent field interaction that causes grating formation. The "incoherent CPM" case assumes that the forward and reverse traveling waves add incoherently, so no grating is

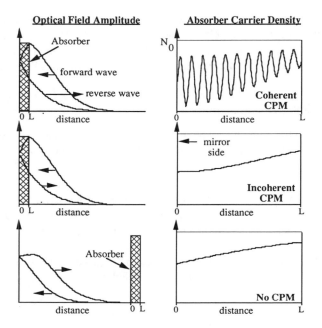

FIGURE 4.13. Three configurations used for calculations of colliding pulse effects. The shaded region in the optical field amplitude plot shows the region over which the absorber carrier density is plotted. Unsaturated round-trip power transmission = 0.04. $E_{in}/E_{sat} = 2$.

formed. This can be done by setting the sinusoidal component of the grating gain h_b to zero. The "no CPM" case has an absorber which is offset from the mirror by greater than the carrier recombination time, so that the absorber has to be bleached by the pulse on both the forward and reverse pass.

For high-energy pulses, the absorbed energy is a factor of two smaller for the "incoherent CPM" configuration than for the "no CPM" configuration. Half as many carriers are needed to reach transparency, reducing the energy needed to bleach the saturable absorber by a factor of two. This is because the absorber is bleached twice, on both the forward and reverse pass of the optical pulse. This reduced energy enhances the effectiveness of the saturable absorber in the "incoherent CPM" case, lowering the mode-locking pulse energy, which reduces the level of amplifier gain saturation as compared to the "no CPM" configuration.

In passive mode-locking, pulse-width narrowing is due to the saturable absorber's removing the front edge of the pulse, and provides a constant pulse shortening ratio as the pulse width is reduced. Pulse-width narrowing on each pass is a fixed ratio which is limited by amplifier gain saturation. Pulse-width broadening is dominated by self-phase modulation,[27] and the pulse width broadening ratio increases as the pulse width is reduced. The steady-state pulse width occurs when the pulse-width narrowing and broadening are equal for each round trip in the cavity. Due to

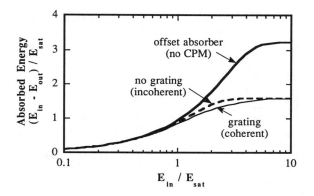

FIGURE 4.14. Calculated absorbed energy in the saturable absorber for large unsaturated absorption used with a semiconductor laser.

the "incoherent CPM" effect, the CPM mode-locked laser can be operated at a lower power than a non-CPM laser, which reduces the pulse broadening mechanisms of amplifier gain saturation and self-phase modulation. In order to take advantage of the "incoherent CPM" improvement, the absorber does not have to be at the facet, where the pulses collide in the absorber. The absorber must just be close enough to the facet so that the absorber does not have time to recover. This makes it a geometrical effect, rather than a coherent effect.

In the "coherent CPM" configuration which includes the optical standing wave, there is an additional bleaching energy reduction due to the grating formation. Since the carrier density does not have to be raised to transparency everywhere in the saturable absorber to reach a low-loss state, the saturable absorber bleaches with a lower pulse energy than would be necessary if the standing waves did not exist. This allows mode-locking with lower energy pulses, which reduces the pulse broadening mechanisms of amplifier gain saturation and self-phase modulation.

The calculated absorbed energy is shown in Fig. 4.14 for a semiconductor saturable absorber. The round-trip unsaturated power transmission is 0.04. The unsaturated gain per pass in the amplifier section of a semiconductor laser is much larger than in a dye laser. This means that the unsaturated absorption must also be higher than in a dye laser, due to the requirement that net gain in the cavity must be less than unity except during the passage of the pulse. The large initial absorption in the semiconductor saturable absorber attenuates the reverse traveling optical wave, causing a reduced standing wave. Only the section nearest the mirror receives the full benefit of coherent CPM, as this is the only section with a large standing wave. Thus the coherent CPM improvement due to the carrier grating is smaller for the semiconductor laser case than for the dye laser case.

The formation of a carrier grating in a semiconductor absorber is illustrated in Fig. 4.15. The carrier density and optical field amplitude are shown at three points in time. Initially the absorption is high [Fig. 4.15(a)], so the optical field amplitude

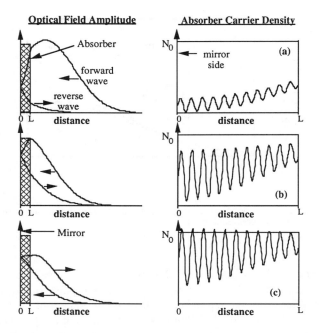

FIGURE 4.15. Calculated grating buildup in a semiconductor laser absorber at three points in time. The shaded region in the optical field amplitude plot shows the region over which the absorber carrier density is plotted. Unsaturated round-trip power transmission = 0.04. $E_{in}/E_{sat} = 2$. The peak carrier density is higher than N_0 due to the sinusoidal grating approximation.

at the mirror side of the absorber is small, and the carrier generation rate is low. The reflected wave is also small, so the optical standing wave at the input side of the absorber is small, and carriers are generated uniformly with only a small modulation depth grating on this side of the absorber.

As the input side of the absorber begins to saturate [Fig. 4.15(b)], carriers begin to be generated at the mirror side. A grating is formed on the mirror side because the optical standing-wave ratio is high. The optical standing-wave ratio is now also high on the input side of the absorber, but a large grating is not formed because this region has already been saturated.

As the optical pulse continues to propagate through the absorber [Fig. 4.15(c)], the relative modulation depth of the grating decreases as carriers are generated at the minimum points of the standing wave. Carrier diffusion will wash out the carrier grating after passage of the optical pulse. The computed carrier density has a maximum greater than the transparency value N_0 because of the numerical approximation of computing only the DC and sinusoidal grating components.

In the "incoherent CPM" case of no standing-wave formation, absorption recovery between forward and reverse optical passes will increase the energy required for mode-locking. The absorber will have to be saturated again on the

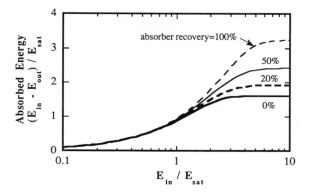

FIGURE 4.16. Calculated absorbed energy for different amounts of absorption recovery. Mirror power reflectivity = 100%. Unsaturated round-trip power transmission = 0.04.

return trip of the optical pulse. If the absorber is placed close to the output facet, the bleaching energy will depend on the amount of absorber recovery between the forward and reverse pass. Figure 4.16 shows the effect of going from no recovery (incoherent CPM) to full recovery (no CPM).

Reducing the output facet reflectivity will increase the effect of absorption recovery. As a smaller fraction of the optical energy is reflected, less energy is available to saturate the absorber on the return pass. Figure 4.17 shows the absorbed energy for 50% absorber recovery between passes. For a mirror power reflectivity much below 30%, the bleaching energy increases substantially. This result is independent of the material or wavelength.

The CPM grating analysis has shown that the grating does not play an important pulse shaping role for semiconductor lasers, so the pulse width should not depend strongly on the optical standing-wave ratio at the facet. However, the geometrical

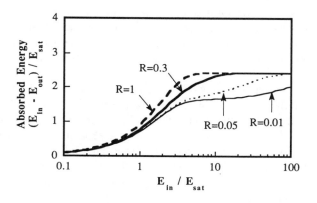

FIGURE 4.17. Calculated absorbed energy for different output mirror power reflectivities. Carrier recovery = 50% between passes. Unsaturated round-trip power transmission = 0.04.

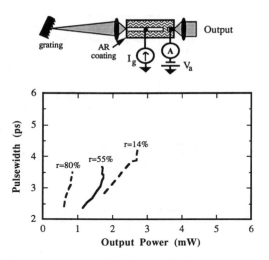

FIGURE 4.18. (a) Experimental configuration for passive mode-locking of a GaAs/AlGaAs semiconductor laser. (b) Measured dependence of the optical pulse width on the absorber mirror reflectivity. Lower mirror reflectivity gives higher output power, with somewhat wider output pulses.

position of the absorber is important, and the absorber should be close to the facet so that the optical round-trip time is much less than the absorber recovery time. The absorber offset becomes more important as the output facet reflectivity is reduced.

4.3.4. Coherent CPM Experimental Comparison

The effect of coherent pulse collision on passive mode-locking of semiconductor laser diodes was experimentally examined.[40] The SCPM configuration was used because the standing-wave ratio can be varied through the mirror reflectivity, and the collision overlap can be reduced by offsetting the absorber region from the mirror. Unlike dye lasers which require a high-Q cavity, semiconductor laser diodes can be passively mode-locked for a wide range of reflectivities of absorber side mirrors.

The multisection GaAs-AlGaAs lasers used in this experiment were fabricated using impurity-induced disordering from silicon diffusion.[16] The bulk active region thickness was 82 nm, and the typical threshold current was 15 mA for a 500-μm device. The gain section and absorber section were separated by a 4-μm gap which was proton bombarded to achieve high contact isolation.

A GaAs device was mode-locked using the linear cavity configuration shown in Fig. 4.18(a). The laser gain length was 500 μm and the absorber length was 16 μm. The device was antireflection (AR) coated and aligned in a 2-GHz external cavity. Passive mode-locking was achieved by reverse biasing the 16-μm segment to act as a saturable absorber.

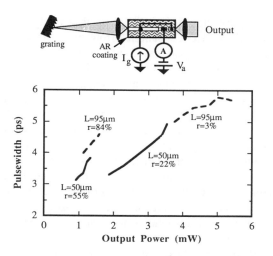

FIGURE 4.19. (a) Experimental configuration for offset absorber investigation. (b) Measured dependence of optical pulse width on the absorber offset. A larger absorber offset gives wider output pulses, with the effect more enhanced for lower facet reflectivity.

The experimental results for an absorber placed at the facet are shown in Fig. 4.18(b). All three results were obtained from the same device by using the device with a cleaved output facet, adding a SiN quarter wave layer to make a low-reflectivity output coating, and then adding a Si quarter wave layer to make a high-reflectivity output coating. At a given pulse width, the average optical power increases as the facet reflectivity is reduced. As the field reflectivity is decreased from $r = 0.8$ (high optical field standing-wave ratio) to $r = 0.55$ (medium optical field standing-wave ratio), the output power increases due to the greater mirror transmittance, while the pulse-width range remains the same. As the reflectivity is reduced to $r = 0.14$ (low standing-wave ratio), the output power for stable mode-locking increases even further, leading to a small increase in minimum pulse width due to increased gain saturation and self-phase modulation broadening. This small dependence of pulse width on the optical field standing-wave ratio indicates that the coherent CPM effect is *not* an important pulse shaping mechanism.

4.3.5. Incoherent CPM Experimental Comparison

Mode-locking with an absorber offset from the facet [Fig. 4.19(a)] was investigated to determine the effect of incoherent pulse collision. The absorber offset reduces the "incoherent CPM" effect by allowing partial absorption recovery before the reverse pass through the absorber. The offset distance required to allow the absorber to recover is determined by the absorber recovery time. Devices were measured with absorber offset distances of 50 μm (1.3 ps round-trip time) and 95 μm (2.5 ps round-trip time). Both devices were from the same GaAs/AlGaAs wafer as in the previous coherent SCPM experiment shown in Fig. 4.18. The length of each device was also 500 μm.

The absorber recovery time measurement using a pump-probe experiment has been reported using GaAs/AlGaAs samples from the same wafer.[24] The measured recovery time constant for reversed biased samples was ~ 5 ps. The rapid recovery time is due to carriers being swept out of the active region by the applied electric field. This rapid recovery time allows an absorber offset of as little as 50 μm to give a significant amount of absorber recovery during the two passes of the optical pulse.

The devices with absorbers offset from the facet were mode-locked in a 2-GHz external cavity. The experimental results are shown in Fig. 4.19(b). For a 50-μm absorber offset, the minimum pulse width has broadened from 2.4 ps in the previous no-offset experiment to 3 ps. For a larger absorber offset of 95 μm, the minimum pulse width has broadened even further to 4 ps. At the same time, the output power is increasing due to the higher bleaching energy of the absorber. This shows that the "incoherent CPM" effect is providing an important pulse shaping mechanism even for a small absorber offset corresponding to a short absorber recovery time.

The offset absorber experiment shows the same trend as before, that of a low facet reflectivity producing high output power for stable mode-locking, but at a wider pulse width. With an absorber offset of 95 μm and a mirror reflectivity of $r = 0.03$, an average power of 5.5 mW was achieved with a pulse width of 5.5 ps. This gives a peak power of 500 mW, and a pulse energy of 2.75 pJ. This technique produced a much larger pulse energy than using an absorber next to a cleaved facet. Because of the low facet reflectivity, a large pulse energy is needed in order to reflect enough energy to saturate the partially recovered absorber.

4.4. STABILIZATION

4.4.1. Stabilization Theory

Pulses produced by passive mode-locking are not synchronized to an external electrical reference. However, most mode-locking applications require such pulse synchronization, including time-division multiplexed optical transmission, millimeter-wave reference signal transmission, and optical computing. Depending on the implementation, applications such as nonlinear optical switching, photodetector testing, and electro-optic sampling may also require externally synchronized pulses.

The optical output power $P(t)$ of a passively mode-locked laser can be described as a series of pulses produced by an ideal oscillator with AM and PM modulation:

$$P(t) = \frac{\bar{P}T}{\sqrt{2\pi}\sigma_t} [1 + g_{am}(t)] \sum_{m=-\infty}^{\infty} e^{-[t-mT-J_{pm}(t)]^2/2\sigma_t^2}, \qquad (4.20)$$

where $\bar{P}$ is the average intensity, $T \equiv 1/f_{mod}$ is the pulse repetition period, σ_t is the rms pulse duration, $g_{am}(t)$ is the random amplitude modulation, and $J_{pm}(t)$ is the random timing fluctuation of the pulse train.

For most applications the optical pulse is used to extract frequency and timing information. Therefore, the result of the amplitude modulation noise expressed by

$g_{am}(t)$ is not as important as the frequency modulation expressed by $J_{pm}(t)$.

The phase modulation noise expressed by $J_{pm}(t)$, on the other hand, is usually more detrimental. Phase noise causes time-varying changes in the arrival time of the optical pulse, called timing jitter. Timing jitter degrades the timing resolution, giving an effective pulse width larger than the actual pulsewidth. For many applications, there is little advantage in having the actual pulse width shorter than the root-mean-square (rms) timing jitter.

Phase noise is more commonly described in the frequency domain. The phase noise is proportional to the Fourier transform of the timing jitter autocorrelation function:[41]

$$L(f) = (2\pi m f_{mod})^2 \int_{-\infty}^{\infty} \langle J_{pm}(t)J_{pm}(t+\tau)\rangle e^{-j2\pi f\tau} d\tau, \tag{4.21}$$

where $L(f)$ is defined as the ratio of the single sideband power in a 1-Hz bandwidth to the total signal power, at a frequency offset f away from the carrier frequency, f_{mod} is the fundamental pulse repetition frequency, m is the harmonic number at which the phase noise is being measured, f is the frequency offset from the carrier frequency $m*f_{mod}$, and $\langle ... \rangle$ is the mean expectation value function.

It is usually more convenient to compute jitter by frequency domain measurements of the phase noise. The root-mean-squared (rms) jitter can be found by integrating the phase noise:[42]

$$\sigma_{\text{timing jitter}} \equiv \sqrt{\langle J_{pm}(t)^2 \rangle} = \frac{1}{2\pi m f_{mod}} \sqrt{2 \int_{f_{low}}^{f_{high}} L(f) df}, \tag{4.22}$$

where f_{low} and f_{high} are the integration limits for the offset frequency range over which the timing jitter is defined. This equation gives the relation of the frequency domain description of the noise to the time domain description of the noise. The timing jitter performance can be completely determined by either jitter measurements in the time domain or by phase noise measurements in the frequency domain.

In many instances, the source phase noise can be described as a simple polynomial $L(f) = L_0 f^{-n}$. This gives a timing jitter of

$$\sigma_{\text{timing jitter}} = \frac{1}{2\pi m f_{mod}} \sqrt{\frac{2L_0}{1-n} \left[(f_{high})^{1-n} - (f_{low})^{1-n} \right]}. \tag{4.23}$$

Unstabilized oscillators such as passively mode-locked semiconductor lasers have a phase noise slope of 20 dB/decade corresponding to $n = 2$.[43,44] This is a consequence of the noise being generated from phase modulation by a white noise source. In this case, the timing jitter increases without limit as the lower integration frequency is reduced. Gain-switched lasers have a flat phase noise slope ($n = 0$), which is a consequence of the random pulse-to-pulse timing jitter, as each pulse starts from noise. In this case, timing jitter increases without limit as the upper integration limit is increased.

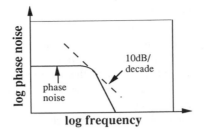

FIGURE 4.20. Phase noise of a mode-locked laser with low-frequency stabilization. The total timing jitter has an upper bound which can be found by integrating the noise contribution near the corner frequency.

Passively mode-locked lasers have infinite timing jitter when integrating the complete phase noise contribution down to $f_{low} = 0$. However, because any measurement takes place over a finite amount of time, the timing jitter integral has an effective lower integration limit. The effective lower integration frequency limit is[42]

$$f_{low} \approx \frac{1}{2*(\text{measurement time})}. \tag{4.24}$$

The timing jitter integral can also become unbounded as the upper integration limit is increased. If the system phase noise has a slope of less than 10 dB/decade, the timing jitter integral grows without limit as the upper integration frequency limit is increased. However, a flat noise floor does not contribute infinite timing jitter because the upper integration frequency has a maximum limit. An upper integration frequency limit is the comb frequency spacing. The electrical spectrum of repetitive pulses is a comb of frequencies spaced at the repetition frequency f_{mod}. The maximum carrier offset frequency from the nearest harmonic is $f_{mod}/2$, before the offset frequency is smaller relative to a different harmonic. The level of broadband phase noise is only significant in systems with large uncorrelated pulse-to-pulse jitter. The resonant external cavity used in mode-locking filters the broadband noise. The theory for computing timing jitter due to broadband noise has been developed for gain-switched lasers.[45]

In order for the total timing jitter to be bounded, the phase noise should have a slope < 10 dB/decade at low offset frequencies, and > 10 dB/decade at high offset frequencies. Because phase modulation causes symmetric sidebands on both sides of a carrier, only the spectral power of one sideband is plotted. Figure 4.20 shows the phase noise of a passively mode-locked laser with a phase noise slope of 20 dB/decade, which has been stabilized by a mechanism (described later) to an external reference oscillator at low offset frequencies. The peak timing jitter contribution comes from the transition corner where the slope is equal to 10 dB/decade. The total timing jitter has a bounded value which is found by integrating the noise contribution near the transition corner frequency.

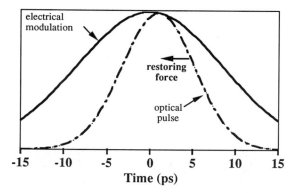

FIGURE 4.21. Timing stabilization of an optical pulse is provided by a pulsed electrical modulation signal. The modulation signal acts as a restoring force on the optical signal.

4.4.2. Modulation Stabilization of Repetition Rate

Previously, pulse stabilization of passively mode-locked semiconductor diode lasers has been done using gain modulation,[46] which by itself produces active mode-locking. The combination of active and passive mode-locking is called hybrid mode-locking.

The addition of active mode-locking acts as a pulse stabilizing mechanism. The electrical modulation signal shown in Fig. 4.21 acts as a restoring force on the optical pulse. If the optical pulse is not in the center of the modulation, the side of the pulse that is closer to the modulation center receives more gain. This nonsymmetric gain shifts the center of the optical pulse closer to the peak of the modulation.

The drawback of the gain modulation technique for repetition rate stabilization is that it is limited by the parasitics of the electrical contacts. For the millimeter-wave repetition rates that are possible for monolithic structures,[47] another form of stabilization is desirable.

4.4.3. Feedback Stabilization of Repetition Rate

Besides amplitude modulation, another stabilization technique is the feedback method shown in Fig. 4.22(a). This is a common method for stabilizing electrical oscillators.[48,49] The timing of the oscillator is compared to an external frequency reference. An error signal is generated based on the timing difference between the oscillator and reference frequency. $K_f(s)$ is the frequency dependent transfer function used to determine the loop characteristics. This error signal is used as feedback to a voltage control which changes the timing of the oscillator.

Ideally, the feedback should drive the error voltage to zero, which means the oscillator and reference signal are perfectly synchronized in phase. In practice, oscillator noise causes a time-varying error voltage which is only partly canceled by

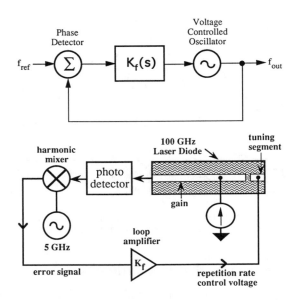

FIGURE 4.22. (a) Amplifier feedback stabilization. (b) Oscillator feedback stabilization. (c) Possible stabilization of a 100-GHz monolithic mode-locked device. The harmonic mixer performs frequency multiplication of the microwave source.

the feedback loop. The loop is considered locked when the average phase error is reduced to zero. The amount of oscillator noise that is not canceled by the feedback loop depends on the feedback loop parameters.

The timing stabilization of optical sources has been demonstrated on an actively mode-locked Nd:YAG laser,[41] where the pulse shaping and primary timing stabilization was provided by an optical modulator. The feedback mechanism was a phase shifter before the optical modulator, which shifted the modulation timing window to reduce timing jitter.

Repetition rate stabilization of passively mode-locked dye and color center lasers has also been demonstrated using a piezoelectric element to control the cavity length.[50] This has the advantage of replacing a more complex optical modulator with a piezoelectric element for repetition rate tuning, but limits the loop bandwidth to a few kHz, which is too low for stabilizing semiconductor diode lasers.

Feedback has been demonstrated as a means to stabilize passively mode-locked semiconductor lasers.[51] Figure 4.22(b) shows a potential compact millimeter-wave source which includes the functions of optical modulation and frequency multiplication from a microwave reference. This type of source would be useful for signal distribution of millimeter-wave reference signals over optical fiber, in such systems as phased array radar.[52]

The operation of the loop can be analyzed when the loop is locked, so that $f_{out} = f_{ref}$. If the oscillator has an internal phase noise perturbation θ_n, the stabilized phase noise reduction function $N(s)$ is given by[42]

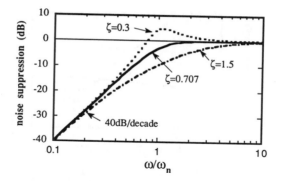

FIGURE 4.23. Noise suppression function $N(s)$ for different values of damping factor ζ.

$$N(s) \equiv \frac{\theta_{\text{out}}}{\theta_n} = \frac{1}{1 + K_\phi K_f K_v / s}, \quad (4.25)$$

where θ_{out} is the stabilized phase perturbation, K_ϕ is the phase detector transfer function, K_f is the feedback amplifier transfer function, and K_v is the voltage-controlled oscillator transfer function. The variable $s = j2\pi f$ is the complex frequency variable used in Laplace-transform notation.

If $|1 + (K_\phi^* K_f^* K_v / s)| > 1$, the phase noise of the source is reduced by the negative feedback of the loop. If $|1 + (K_\phi^* K_f^* K_v / s)| < 1$, the feedback is positive, and the control loop enhances the oscillator phase noise, rather than reducing it.

A common feedback amplifier has the transfer function K_f given by

$$K_f(s) = \frac{1 + s/\omega_2}{s/\omega_1}, \quad (4.26)$$

where $\omega_1 = 1/R_1 C$ and $\omega_2 = 1/R_2 C$. With this feedback amplifier, the resulting phase noise reduction function $N(s)$ is

$$N(s) \equiv \frac{\theta_{\text{out}}}{\theta_n} = \frac{(s/\omega_n)^2}{1 + 2\zeta(s/\omega_n) + (s/\omega_n)^2}, \quad (4.27)$$

where the closed loop resonance frequency is given by ω_n, and the loop damping is given by ζ:

$$\omega_n \equiv \sqrt{K_\phi K_v \omega_1} \quad \zeta \equiv \frac{\omega_n}{2\omega_2}. \quad (4.28)$$

The noise suppression function $N(s)$ is shown in Fig. 4.23 for several values of damping factor ζ. Critical damping occurs when $\zeta = 0.707$. For the under-damped case of $\zeta < 0.707$, the noise with feedback will be higher than the open-loop noise

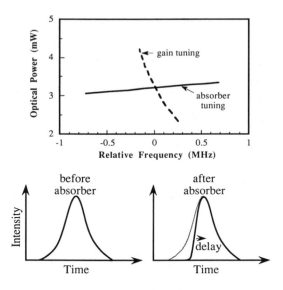

FIGURE 4.24. (a) Measured optical power dependence on repetition frequency for gain tuning and absorber tuning. (b) Delay due to absorption saturation.

near the resonance frequency.[41] For the over-damped case of $\zeta > 0.707$, the noise reduction is lower than for a critically damped loop with the same 3-dB noise suppression frequency.

4.4.4. Repetition Rate Bias Tuning

The bias tuning range was measured using a 5-GHz external cavity mode-locked semiconductor laser. The active device was a 360-μm-long GaAs/AlGaAs bulk active region laser. Passive mode-locking was initiated using an 8-μm integrated-waveguide saturable absorber, which was placed at one facet. The optical pulses had a pulse width of 3 ps and a spectral width of 3 nm.

For a monolithic multisection device with a uniform gain region (without using passive waveguide), the two parameters that can be varied are gain-region current and absorber-region voltage. The gain and absorber tuning characteristics are plotted against the relative repetition frequency for gain and absorber tuning in Fig. 4.24(a). The important figures of merit for bias tuning mechanisms are Δf (the total tuning range) and dP/df (the rate of change of power with respect to tuning range). The total tuning range Δf is much larger for the absorber tuning. In addition, the ratio dP/df is smaller (0.2 mW/MHz for the absorber vs -5 mW/MHz for the gain region). This parameter is important, as a constant power output prevents amplitude noise being generated as the bias voltage is changed to cancel out phase noise. The absorber bias is therefore a more appropriate control element for feedback stabilization.

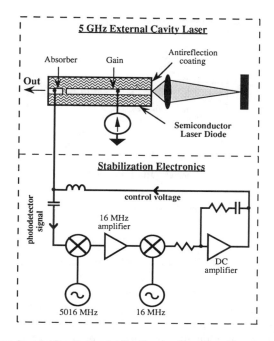

FIGURE 4.25. Experimental feedback stabilization configuration. The short segment acts as a saturable absorber, a photodetector, and a repetition rate tuning element.

Gain and absorber saturation cause effective time delays by shifting the pulse center. The time delay caused by absorber saturation is shown in Fig. 4.24(b). Changes in bias conditions alter the unsaturated gain, and the effective time delay Δt_{gain}, which changes the repetition rate. A time delay shift Δt_{index} also results from the change in index associated with any gain change. The same mechanism takes place for external bias changes as for noise-induced bias changes. For timing jitter induced by spontaneous emission in these devices and $\alpha = 4$, the saturation-induced time delay caused by spontaneous emission-induced noise has been calculated to be $\sim$ ten times the index-induced time delay.[53] This means that gain saturation is the dominant repetition rate tuning mechanism.

4.4.5. Experimental Stabilization Results

The 5-GHz mode-locked laser was stabilized using the experimental configuration shown in Fig. 4.25.[54] The saturable absorber performed the functions of pulse shaping, photodetection, and repetition rate tuning. The reversed-biased saturable absorber was used as a photodetector to monitor the pulse repetition rate. The pulse timing output was compared to both a microwave synthesizer and then a low-frequency oscillator using a two-step down conversion. The second comparison generated a DC error signal which passed through a type-II control loop, and was

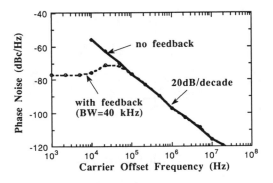

FIGURE 4.26. Measured single-sideband phase noise with and without feedback stabilization. Calculated stabilized timing jitter = 4 ps.

applied back to the saturable absorber to control the repetition rate. The 16-MHz IF frequency was chosen to be high enough so that the tuning range of the laser did not extend to the other sideband (5032 MHz would also mix down to a 16-MHz IF). This implementation has an advantage over previous feedback stabilization experiments in that the pulse-shaping mechanism, the photodetector, and the tuning mechanism are all monolithically integrated along with the gain element into a single device.

In this experimental demonstration, the 16-MHz oscillator for the second-stage down conversion is a crystal oscillator which is not phase locked to the 5016-MHz microwave reference oscillator. Phase locking the two oscillators is not important, as the phase noise is completely dominated by the microwave oscillator. However, most microwave oscillators are phase locked to a low-frequency reference which is externally available, so both of these references are usually already phase locked together.

The measured phase noise is shown in Fig. 4.26 for both open-loop (no feedback) and closed-loop (with feedback) configurations. The unstabilized system has the characteristic 20-dB/decade slope of the phase noise, which produces unbounded timing jitter for arbitrarily long measurement times. For a one-second measurement interval, the unstabilized timing jitter was 1 ns. For longer measurement times, the timing jitter would be even higher. The pulse interval is 200 ps for the 5-GHz repetition rate. A timing jitter of 1 ns implies that during a one-second measurement, the number of pulses is 5×10^9, with an rms error of ± 5 pulses.

The loop feedback bandwidth for the closed-loop configuration was 40 kHz. With repetition rate stabilization, the resulting timing jitter found from integrating all of the phase noise was 4 ps. Unlike the unstabilized case, the timing jitter does not continue to increase as the lower integration limit is reduced. The majority of the timing jitter contribution came from the phase noise near the control loop corner frequency of 40 kHz.

In principle, increasing the loop bandwidth would increase the corner frequency

TABLE 1. A comparison of mode-locked laser characteristics. Included are elements required for a free-standing source, such as the pump laser.

Characteristics (including pump)	Semiconductor laser	Dye laser (argon)	Ti:sapphire laser (argon)	Fiber laser (Semiconductor)
Size (cm^3)	1	10^6	10^6	100
Power consumption (W)	1	100	100	10
Water cooling	No	Yes	Yes	No
Weight (kg)	10	200	200	30
Planar processing	Yes	No	No	No
Cost ($k)	10	100	50	20
Pulse width (fs)	1000	20	20	40
Pulse energy (pJ)	50	200	1000	200
Jitter (100 Hz to 1 MHz) (ps)	0.1	10	100	10
Wavelengths available (μm)	0.5–1.7	0.3–1.7	0.7–1.0	0.7–1.6
Tunability (nm)	30	30	30	30
Lifetime (h)	10^6	10^3	10^3	10^6

and reduce the phase noise. In this case, however, the loop bandwidth could not be further increased because of instability caused by excess phase shift in the control loop. Timing jitter could be reduced even further by using higher frequency components and increasing the feedback loop bandwidth.

4.5. OUTLOOK

This chapter has summarized the properties of mode-locked semiconductor lasers with an emphasis on the factors that determine mode-locking pulse width, pulse shape, and spectral characteristics. The physics involved in the colliding pulse mode-locking process has been examined and design considerations for minimum pulse width and maximum power have been given. Finally, techniques to stabilize the timing of output pulses and lock it to an external reference have been discussed. A summary of the characteristics that have been achieved is given in Table I. The purpose of this section is to examine potential advances in these characteristics and consider new applications that these devices could address.

We can see from Table I that the primary advantages of mode-locked semiconductor sources are in the area of manufacturability. The small size, weight, and low power consumption provide for simple packaging in a compact electronic package suitable for inclusion on a printed circuit board. The long lifetimes of semiconductor sources are a major asset for large-scale implementation of short pulse sources. The only source which is comparable in these characteristics is the semiconductor laser-pumped fiber laser.

The primary advantages of a mode-locked semiconductor laser are its smaller size and its lower cost, since the planar processing techniques allow for large volume production at low cost with all of the laser elements integrated onto one chip. The primary limitations of semiconductor lasers are the relatively long pulses

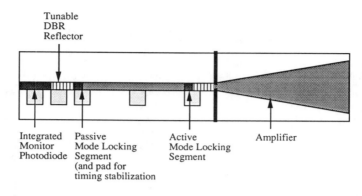

Tunable
DBR
Reflector

Integrated
Monitor
Photodiode

Passive
Mode Locking
Segment
(and pad for
timing stabilization

Active
Mode Locking
Segment

Amplifier

FIGURE 4.27. Schematic diagram of an integrated amplified mode-locked source.

(> 100 fs), and relatively low pulse energies (10 pJ). The solution to these problems lies in the rapidly advancing fields of semiconductor and fiber amplifiers. Recently, sources with excellent characteristics have been demonstrated by amplifying the output of a small source with good lateral mode characteristics and good spectral characteristics.[21-23,55] The same approach will be important for short pulse sources. The output of a mode-locked semiconductor laser can be amplified,[28,56] with an integrated or external amplifier to peak powers in excess of 1 kW. The analogy with electronics is clear. Early microwave sources were typically klystron tubes which delivered adequate power at microwave and millimeter wave frequencies, but were bulky, expensive, noisy, and had poor spectral characteristics. Now, for most applications, the output of a well-controlled, crystal or YIG oscillator is amplified in a solid-state or tube amplifier up to the powers needed for the microwave application. If extreme stability is needed, or locking to an external reference is required, then the output is locked to a highly stable crystal-controlled oscillator. This is exactly the approach taken in the last section of this chapter, where the output of a passively mode-locked laser is locked to a crystal source with very little additional timing jitter over that of the crystal source.

Using the techniques of flared waveguides described earlier, this amplifier can easily be integrated with the mode-locked source itself. Thus, the ideal mode-locked source would look something like the one shown schematically in Fig. 4.27. The source is hybrid-mode-locked, wavelength tunable via a tunable DBR mirror, and amplified in an integrated, flared waveguide amplifier. An external fiber or grating could be used for additional compression below 100 fs. The primary disadvantage of an integrated source such as this is the lack of an optical isolator between the source and the amplifier. The spontaneous emission coming back into the oscillator degrades the mode-locking performance. This is easily solved with hybrid mounted components and a YIG isolator. One solution is to limit any spontaneous emission from outside the flared input of the amplifier from being absorbed in the waveguide. Absorbing regions or reflecting channels, as indicated in Fig. 4.27, should help this problem. Another approach is to combine a mode-locked semiconductor laser with a fiber amplifier. This is a particularly attractive approach for soliton generation where higher repetition rates and lower jitter are needed than can generally be supplied by mode-locked fiber lasers.

With these techniques, compressed pulse widths under 50 fs should be possible over a variety of wavelengths. Pulse energies in excess of 1 nJ with time bandwidth limited spectral output should be possible. The capabilities of such sources in integrated form need to be investigated. Waveguide isolators of careful design will be needed to achieve such high energies in an integrated form.

Another approach is to use the mode-locked semiconductor laser as a low-intensity noise, low-timing jitter source to injection lock and stabilize other sources such as Argon or Nd:YAG lasers. The ability to adjust the composition of the semiconductor material to match the lasing line of other sources is an important advantage of semiconductor lasers.

The authors would like to thank the Office of Naval Research and the National Science Foundation for their financial support, and Al Goodman, Al Harvey, Tom Reynolds, Dennis Derickson, Wenbin Jiang, Alan Mar, Wei-Xiong Zou, John Wasserbauer, Radha Nagarajan, Ernie Caine, and Robert Thornton for their suggestions and assistance.

REFERENCES

1. P.-T. Ho, L. A. Glasser, E. P. Ippen, and H. A. Haus, Appl. Phys. Lett. **33**, 241 (1978).
2. A. G. Weber, M. Schell, G. Fischbeck, and D. Bimberg, IEEE J. Quantum Electron. **28**, 2220 (1992).
3. C. Harder, J. S. Smith, K. Y. Lau, and A. Yariv, Appl. Phys. Lett. **42**, 772 (1983).
4. R. J. Helkey, Ph.D. thesis, University of California, Santa Barbara, California (1993).
5. H. A. Haus, IEEE J. Quantum Electron. **QE-11**, 736 (1975).
6. G. P. Agrawal and N. K. Dutta, *Semiconductor Lasers*, 2nd ed. (Van Nostrand Reinhold, New York, 1993).
7. R. J. Helkey, W. X. Zou, A. Mar, D. B. Young, and J. E. Bowers, Device Research Conference, Santa Barbara CA, June 21–23, 1993.
8. M. Schell, A. Weber, E. Schol, and D. Bimberg, IEEE J. Quantum Electron. **27**, 1661 (1991).
9. T. Saitoh, T. Mukai, and O. Mikami, J. Lightwave Technol. **LT-3**, 288 (1985).
10. A. Mar, D. Derickson, R. Helkey, J. E. Bowers, R. T. Huang, and D. Wolf, Optics Lett. **17**, 868 (1992).
11. D. J. Derickson, R. J. Helkey, A. Mar, and J. E. Bowers, IEEE Photon. Technol. Lett. **4**, 333 (1992).
12. B. L. Frescura, C. J. Hwang, H. Luechinger, and J. E. Ripper, Appl. Phys. Lett. **31**, 770 (1977).
13. D. J. Bradley, M. B. Holbrook, and W. E. Sleat, IEEE J. Quantum Electron. **17**, 658 (1981).
14. P. J. Delfyett, C.-H. Lee, G. A. Alphonse, and J. C. Connolly, Appl. Phys. Lett. **57**, 971 (1990).
15. W. X. Zou, T. Bowen, K.-K. Law, D. B. Young, and J. L. Merz, IEEE Photon. Technol. Lett. **5**, 591 (1993).
16. R. L. Thornton, R. D. Burnham, T. L. Paoli, N. Holonyak, and D. G. Deppe, Appl. Phys. Lett. **47**, 1239 (1985).
17. D. G. Deppe and N. Holonyak, J. Appl. Phys. **64**, R94 (1988).
18. J. C. Simon, J. Lightwave Technol. **5**, 1286 (1987).
19. L. Pang and J. G. Fujimoto, Opt. Lett. **17**, 1599–1601 (1992).
20. A. Mar, R. J. Helkey, J. E. Bowers, D. Botez, C. Zmudzenski, C. Tu, and L. Mawst, IEEE Photon. Technol. Lett. **5**, 1335 (1993).
21. G. Bendelli, K. Komori, S. Arai, and Y. Suematsu, IEEE Photon. Technol. Lett. **3**, 42 (1991).
22. P. A. Yazaki, K. Komori, G. Bendelli, S. Arai, and Y. Suematsu, IEEE Photon. Technol. Lett. **3**, 1060 (1991).
23. E. S. Kintzer, J. N. Walpole, S. R. Chinn, C. A. Wang, and J. L. Missaggia, Digest of Optical Fiber

Communication Conference, vol. 5 (Optical Society of America, Washington, D.C., 1992), paper TuH5.

24. J. R. Karin, R. J. Helkey, D. J. Derickson, R. Nagarajan, D. S. Allin, J. E. Bowers, and R. L. Thornton, Appl. Phys. Lett. **64**, 676 (1994).
25. C. H. Henry, IEEE J. Quantum Electron. **18**, 259 (1982).
26. G. P. Agrawal and N. A. Olsson, IEEE J. Quantum Electron. **25**, 2297 (1989).
27. D. J. Derickson, R. J. Helkey, A. Mar, J. R. Karin, J. G. Wasserbauer, and J. E. Bowers, IEEE J. Quantum Electron. **28**, 2186 (1992).
28. P. J. Delfyett, L. T. Florez, N. Stoffel, T. Gmitter, N. C. Andreadakis, Y. Silberberg, J. P. Heritage, and G. Alphonse, IEEE J. Quantum Electron. **28**, 2203 (1992).
29. D. J. Derickson, J. G. Wasserbauer, R. J. Helkey, A. Mar, J. E. Bowers, D. Coblentz, R. Logan, and T. Tanbun-Ek, Digest of Optical Fiber Communications Conference (Optical Society of America, Washington, D.C., 1992), paper ThB3.
30. D. Kuhlke, W. Rudolph, and B. Wilhelmi, Appl. Phys. Lett. **42**, 325 (1983).
31. A. E. Siegman, *Lasers* (University Science Books, Mill Valley CA, 1986).
32. Y. K. Chen and M. C. Wu, IEEE J. Quantum Electron. **28**, 2176 (1992).
33. S. Sanders, L. Eng, J. Paslaski, and A. Yariv, Appl. Phys. Lett. **56**, 310 (1990).
34. J. Paslawski and K. Y. Lau, IEEE Photon. Technol. Lett. **59**, 7 (1991).
35. G. H. C. New, IEEE J. Quantum Electron. **QE-10**, 115 (1974).
36. I. S. Ruddock and B. J. Bradley, Appl. Phys. Lett. **29**, 296 (1976).
37. J. Herrmann, F. Weidner, and B. Wilhelmi, Appl. Phys. B **26**, 197 (1981).
38. W. Dietel, Opt. Commun. **43**, 69 (1982).
39. M. S. Stix and E. P. Ippen, IEEE J. Quantum Electron. **QE-19**, 520 (1983).
40. R. J. Helkey, D. J. Derickson, A. Mar, J. G. Wasserbauer, and J. E. Bowers, Digest of Conference on Lasers and Electro-Optics (Optical Society of America, Washington, D.C., 1992), paper JThB2.
41. M. Rodwell, D. Bloom, and K. Weingarten, IEEE J. Quantum Electron. **25**, 817 (1989).
42. W. P. Robins, *Phase Noise in Signal Sources* (Peter Peregrinus, London, 1982).
43. D. B. Leeson, Proc. IEEE **54**, 329 (1966).
44. W. F. Egan, *Frequency Synthesis by Phase Lock* (Wiley, New York, 1981).
45. D. A. Leep and D. A. Holm, Appl. Phys. Lett. **60**, 2451 (1992).
46. D. J. Derickson, P. A. Morton, and J. E. Bowers, Appl. Phys. Lett. **59**, 3372 (1991).
47. K. Y. Lau and J. Paslaski, IEEE Photon. Technol. Lett. **3**, 974 (1991).
48. E. Labin, Philips Res. Rept. **4**, 291 (1949).
49. J. Noordanus, IEEE Trans. Commun. Technol. **COM-17**, 257 (1969).
50. D. Walker, D. Crust, W. Sleat, and W. Sibbet, IEEE J. Quantum Electron. **28**, 289 (1992).
51. R. J. Helkey, D. J. Derickson, A. Mar, J. G. Wasserbauer, J. E. Bowers, and R. L. Thornton, Electron. Lett. **28**, 1920 (1992).
52. A. S. Daryoush, IEEE Microwave Theory Techniques **38**, 467 (1990).
53. D. Derickson, Ph.D. thesis, University of California, Santa Barbara, CA (1992).
54. R. J. Helkey, A. Mar, W. X. Zou, D. B. Young, and J. E. Bowers, SPIE Optoelectronic Packaging and Interconnects Proceedings, vol. 1861 (SPIE, Bellingham, WA, 1993).
55. D. Mehuys, D. F. Welch, and L. Goldberg, Electron. Lett. **28**, 1944 (1992).
56. A. Mar, R. Helkey, J. Bowers, D. Mehuys, and D. Welch, IEEE Photon. Technol. Lett. **6**, 1067 (1994).

CHAPTER 5

Vertical-Cavity Surface-Emitting Lasers

Connie J. Chang-Hasnain

Department of Electrical Engineering, Stanford University,
Stanford, California

5.1. INTRODUCTION

Semiconductor diode lasers emitting normal to the substrate plane, known as surface emitting lasers, are highly promising for a range of applications in optical interconnects, optical communications, optical recording, optical signal processing, etc. The most attractive aspect perhaps lies in the prospect of eliminating low yield and labor intensive fabrication steps such as wafer lapping (to $\leqslant 100$ μm thick), cleaving and dicing, facet coatings, and diode bonding. The possibility of being able to fabricate and test lasers on a wafer scale and perform nonintrusive testing are very important factors. In addition, the size of each laser can be made very small to facilitate the fabrication of a large two-dimensional (2D) array.

At present there are two approaches aimed at realizing surface-emitting lasers. The first represents an extension of the existing technology for semiconductor edge-emitting lasers that uses a 45° slanted mirror[1] or a second-order grating[2] to vertically couple the light out [Figs. 5.1(a) and 5.1(b)]. The second[3] uses highly reflective mirrors to clad the active region resulting in a vertical cavity that produces an output beam propagating normal to the substrate surface [Fig. 5.1(c)]. The focus of this chapter will be on the vertical-cavity surface-emitting lasers (VCSELs).

The vertical-cavity design offers important advantages over other surface-emitting laser designs. The unique topology of a vertical cavity facilitates a circular low divergence beam, which can be coupled easily and efficiently with an optical fiber and bulk optics. The emission wavelengths are determined by epitaxial growth rather than by photolithography, and thus can potentially be made with higher accuracy. Large, monolithic arrays of single-wavelength lasers with distinct, equally spaced wavelengths can be fabricated. Preprocess screening of laser samples to determine laser wavelengths or other characteristics are available. The small lateral

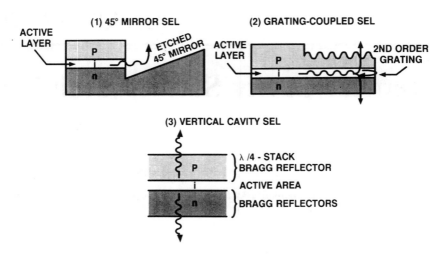

FIGURE 5.1. Three major approaches used to obtain laser emission in the direction normal to the substrate: (a) a 45° slanted mirror (b) a second-order grating, (c) a vertical-cavity configuration.

dimensions allow for fabrication of large 2D arrays with high packing density and integration with other optical and electronic devices. In addition, the active volume can potentially be scaled down to very small size (10^{-13}–10^{-14} cm^{-3}) such that extremely low threshold currents (μA range) are attainable.

5.1.1. Historic Overview

The first VCSEL was pioneered in 1979 by Soda, Iga, and co-workers[3] as shown in Fig. 5.2(a). It had a double heterostructure (DH) design with an InGaAsP active layer emitting at 1.3 μm. The mirrors were metal and had fairly substantial absorption, and thus comparitively low reflectivity. It lased at 77 K with high threshold currents. In 1982, Burnham, Scifres, and Streifer filed a patent[4] on various epitaxial designs for a VCSEL. One of their designs is shown in Fig. 5.2(b). Some of these designs show remarkable resemblance to the structures used today. The epitaxial growth technology was, however, not yet precise enough for fabricating such a structure. Epitaxially grown mirrors were first demonstrated in 1983.[5] Most of the later work focused on structures with a bulk active layer and a combination of dielectric mirrors and epitaxially grown semiconductor mirrors,[6,7] as shown in Fig. 5.2(c). The thick active region was considered necessary for a high single-pass gain because of the low mirror reflectivity. The threshold current density of these lasers was quite high.

Recently, with the advances in epitaxial technologies, particularly in the control of thickness and composition of epitaxial layers, semiconductor Bragg reflectors with extremely high reflectivity can be made. This led to the realization and the first demonstration in 1989 of a cw, room-temperature operated, low threshold current,

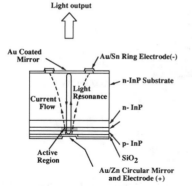

(a) Metallic Mirror Strucutre

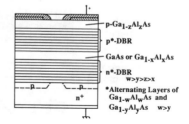

(b) Epitaxially Bragg Reflecors Structure

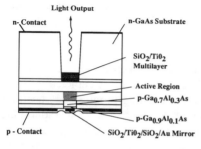

(c) Dielectric Mirror Structure

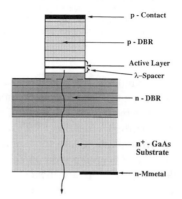

(d) Air-post Index-Guided Structure

(e) Ion-Implanted Gain-Guided Structure

(f) Passive Antiguide Region (PAR) Structure

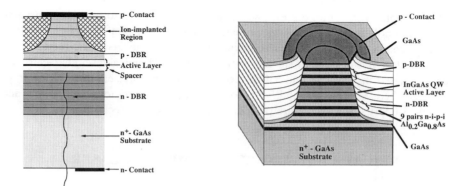

FIGURE 5.2. Various VCSEL structures.

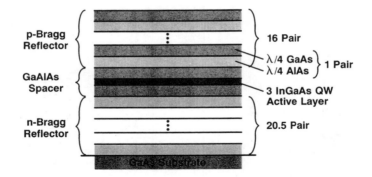

FIGURE 5.3. Schematic of a bottom emitting VCSEL design consisting of a 20.5-pair n-doped quarter-wave GaAs/AlAs stack as the lower DBR, three 80 Å $In_{0.2}Ga_{0.8}As$/GaAs strained quantum wells in the center of a one-wave $Ga_{0.5}Al_{0.5}As$ spacer, and a 14-pair p-doped upper GaAs/AlAs DBR, grown on an n^+-GaAs substrate. The top-most GaAs layer is half-wave thick to provide phase matching for the metal contact (after Ref. 39).

VCSEL.[8] The laser structure consists of a quantum-well active region sandwiched between n- and p-doped semiconductor Bragg reflectors, as depicted in Fig. 5.2(d).

Triggered by this demonstration, rapid and exciting progress has been made in threshold current reduction,[9,9a,9b] wavelength tuning,[10,10a,10b] and monolithic integration with a photodetector[11] and with a thyristor.[12] In the meantime, many advances have been reported on 2D VCSEL arrays, including novel multiple-wavelength arrays,[13,14] phase-locked arrays,[15,16] and large addressable arrays.[17–19] Demonstrations in systems experiments using VCSEL arrays have also been made for wavelength-division-multiplexed (WDM) optical communications[20,21] and board-to-board optical interconnections.[19] Further improvements are continuously being made on the basic laser structure to provide ease of fabrication with the use of ion implants[22,23] [Fig. 5.2(e)] and natural oxide,[9a,9b] and to provide transverse mode and polarization control with a passive antiguide region[24,25] [Fig. 5.2(f)].

This chapter is organized as follows. In Sec. 5.2, the general structure of a VCSEL design is discussed. The fabrication process and typical laser characteristics are discussed in Secs. 5.3 and 5.4. In Sec. 5.5 multiple-wavelength VCSEL arrays are described. Sections 5.6 and 5.7 report recent progress on long-wavelength and visible VCSELs. Finally, future prospects are discussed in Sec. 5.8.

5.2. VERTICAL-CAVITY SURFACE-EMITTING LASER DESIGN

A generic VCSEL heterostructure is illustrated in Fig. 5.3. The structure can be divided into two components: the top and bottom distributed Bragg reflectors (DBRs), and the center active/spacer region. The round-trip gain-path length is very short for a VCSEL, simply twice the active-layer thickness. Thus, the reflectivities

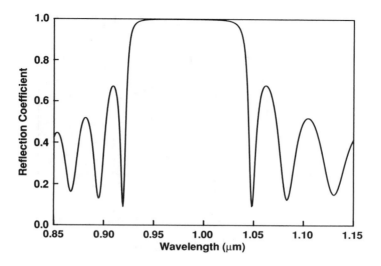

FIGURE 5.4. Calculated reflectivity spectrum for a plane wave ($\lambda = 0.98$ μm) entering from semi-infinite GaAs material onto a 20-pair AlAs/GaAs DBR and exiting into air.

of the DBRs need to be very high. In this section, general design considerations and estimates of threshold current and efficiency are discussed.

5.2.1. Bragg Reflectors

The DBRs are typically made with alternating layers of quarterwave thick high- and low-refractive index epitaxial materials, e.g., $Al_xGa_{1-x}As$ and $Al_yGa_{1-y}As$. The net reflectivity depends on the number of pairs, the boundary conditions, and the index difference. The only difference between the upper and lower DBRs of a VCSEL is in the last layer, because of the different output boundary conditions. A VCSEL emitting from the epitaxy side (top emitting) has air and GaAs as the boundaries for its upper and lower DBRs, respectively, whereas a VCSEL with its emission through the back of the substrate (bottom emitting) may have metal and GaAs as the boundaries, respectively. An appropriate phase-matched half-wave layer may be used to match the boundary condition with metal. The bottom DBRs for both types of VCSELs usually have an integer number of quarter-wave pairs plus half a pair.

Figure 5.4 shows the calculated reflectivity spectrum for a plane wave ($\lambda = 0.98$ μm) entering from a semi-infinite GaAs material onto a twenty-pair AlAs/GaAs DBR and exiting into air. The reflectivity at the designed wavelength of 0.98 μm is as high as 0.99 with a broad bandwidth of 100 nm due to a relatively large refractive-index difference, 3.6 and 2.9 for GaAs and AlAs respectively. The peak reflectivity as a function of the number of DBR pairs is shown in Fig. 5.5 with air and GaAs as exiting media for various Al contents. For the same reflectivity, the number

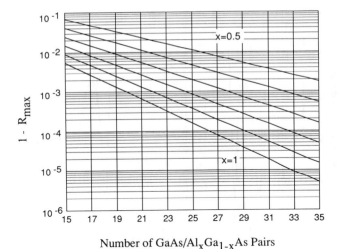

Number of GaAs/Al$_x$Ga$_{1-x}$As Pairs

FIGURE 5.5. Peak reflectivity as a function of the number of GaAs/Al$_x$Ga$_{1-x}$As DBR pairs with air and GaAs as exiting media for various Al compositions.

of pairs required drastically increases with the reduction of Al content due to a smaller refractive-index difference.

The calculation of the reflectivity spectra of a DBR can be found in several text books.[26–28] The following formula summarizes one of the approaches. Assume a plane wave incident into a DBR from the layer N, as shown in Fig. 5.6. The net field reflectivity at layer i is represented by Γ_i, whereas the local reflectivity between layers i and i-1 is represented by r_i:

$$\Gamma_{i+1}(\lambda) = \frac{r_{i+1}(\lambda) + \Gamma_i(\lambda)e^{-j2\beta_i(\lambda)l_i^{\text{eff}}}}{1 + r_{i+1}(\lambda)\Gamma_i(\lambda)e^{-j2\beta_i(\lambda)l_i^{\text{eff}}}}, \quad \Gamma_1(\lambda) = r_1(\lambda),$$

$$r_i = \frac{n_i - n_{i-1}}{n_i + n_{i-1}}, \quad \beta_i(\lambda) = \frac{2\pi n_i(\lambda)}{\lambda}$$

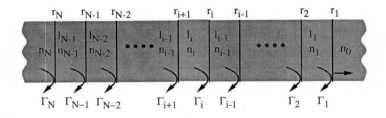

FIGURE 5.6. Schematic of a periodic layered media.

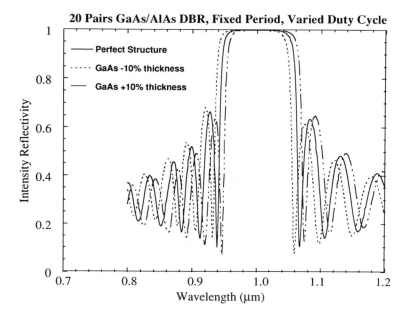

FIGURE 5.7. Reflectivity spectra for a 20-pair DBR with fixed periods and varied duty cycles.

where $\beta_i(\lambda)$ = propagation constant of the ith layer, $l_i^{\text{eff}} = l_i \cos(\theta_i)$ = effective thickness of the ith layer, $r_i(\lambda)$ = field reflectivity of the ith interface only (looking to the right), $\Gamma_i(\lambda)$ = net field reflectivity of the DBR stack looking to the right from beyond the ith interface, n_i is the refractive index of the ith layer, and θ_i is the propagation angle in ith layer.

Using this formula, a simple program can be written on a personal computer to calculate the reflectivity spectra for various combination of indices, input beam angles, and thicknesses.

One of the most common reasons for a VCSEL wafer failing to lase is the lack of precision in the growth of the DBR layers. With poor DBR layer growth, the reflectivity of the mirrors is not high enough for lasers to reach threshold. There are a number of possible growth miscalibrations. Two probable scenarios are: (1) Fixed period with a varied duty cycle and (2) fixed duty cycle but with a walk off on thickness during the growth.

In the first case, each period of the GaAs/AlAs layers is an equivalent half-wave in thickness; however, each individual layer is not exactly quarterwave. Thus, when the GaAs layer is w% more than a designed value of $\lambda/4n_{\text{GaAs}}(\lambda)$, the AlAs is w% thinner than $\lambda/4n_{\text{AlAs}}(\lambda)$. This condition arises sometimes when the growth rate of each material is not accurately calibrated, whereas the period is calibrated. (It is typically easier to measure the period of a periodic structure than the thickness of each individual layer.) Figure 5.7 shows the calculated results. A small amount of spectral shift, 10 nm for 10% variations, is observed. The reflectivity does not

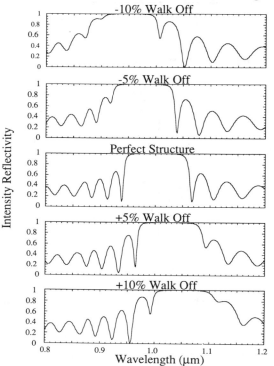

FIGURE 5.8. Reflectivity spectra for a 20-pair DBR with a walk off on the growth rate during growth.

deteriorate much for such miscalibration, and the amount of spectral shift is tolerable.

The second case arises when the layers are calibrated at the beginning of the growth and the growth rate changes linearly during the growth (walk off). Walk off is likely irrespective of whether the growth technique is molecular beam epitaxy (MBE) or metalorganic chemical vapor deposition (MOCVD). Figure 5.8 shows the spectra for the cases for which the last layer is −10%, −5%, 0%, 5%, and 10% off from its optimum value. Significant degradation of peak reflectivity and spectral bandwidth can occur in some cases. This can be a serious problem in the epitaxial growth of VCSEL if a large percentage of walk off is obtained with a particular growth system.

5.2.2. Active Gain Region

In a many-pair DBR, if the thickness of one of the layers in the middle region deviates from the quarter-wave thickness, a standing wave can form in the structure.

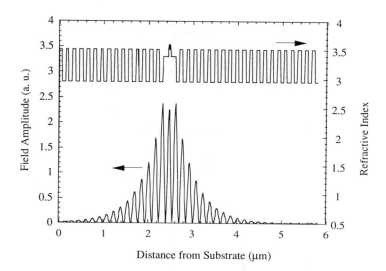

FIGURE 5.9. Optical intensity of the Fabry–Perot mode and the refractive index as a function of distance in a typical VCSEL.

The simplest case to visualize is when this layer is a half-wave thick. In this case, the wave at the center of this layer sees high reflectivity mirrors in both forward and backward directions. Thus, a cavity is formed, and Fabry–Perot modes can be calculated. Depending on whether this layer has a low or high refractive index, the optical field can have a maximum at the center or at the edges of this layer, respectively. For better carrier-transport characteristics, this center layer for a VCSEL typically has a lower bandgap energy and, thus, a high refractive index, typically either GaAs or $Al_xGa_{1-x}As$ with a small value of x. In order to have the standing wave peak at the center where the quantum-well active region is, the center spacer is designed to be one wavelength (or an integer multiple of wavelengths) thick. The maximum of the optical field of the Fabry–Perot mode is thus located at the center to have the maximum overlap with active layers so that the gain in the quantum wells can be used efficiently. Figure 5.9 shows the optical intensity of the Fabry–Perot mode and the refractive index as a function of distance in a typical VCSEL.

Although the parameters of the VCSELs are at an extreme, e.g., ultrashort cavity length and ultrahigh mirror reflectivity, the design criteria are no different from those for edge-emitting lasers. Hence, the threshold conditions and general characteristics are the same as those for edge-emitting lasers and are determined by the well-understood wave equation and rate equations.

The effective cavity length is calculated by examining the phase of a DBR as a function of the wavelength of a DBR:

$$\phi(\beta) = 2\beta L \Rightarrow L = \frac{1}{2}\frac{d\phi}{d\beta} = \frac{-\lambda^2}{4\pi n_c}\frac{d\phi}{d\lambda},$$

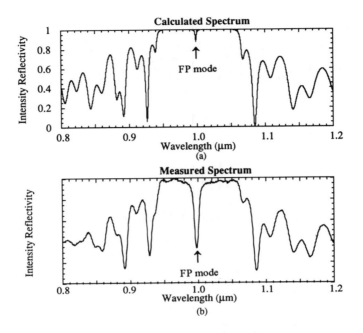

FIGURE 5.10. Calculated and measured reflectivity spectra of a typical VCSEL.

where L = penetration depth of the DBR stack, $\phi(\beta)$ = phase change in the complex reflectivity coefficient, and n_c = refractive index of the cladding layer.

The effective cavity length varies with the exact design; however, it is typically around 1–2 μm. Hence, the Fabry–Perot (FP) mode spacing is very large—in the order of 100 nm. As a result, there is only one FP mode in the DBR mirror bandwidth, as shown in Fig. 5.10. Thus, it is this FP mode and not the gain spectrum that determines the lasing wavelength. In the case when the gain spectrum is totally off from this FP mode wavelength, the laser will simply not lase. This is actually one of the most common reasons for a VCSEL wafer not to yield any working lasers. And if the laser does lase, it must do so at the FP wavelength. As with any other Fabry–Perot cavity, there are transverse modes at slightly different wavelengths around each FP mode.[28] Therefore, to obtain a true single-wavelength VCSEL, a design that controls transverse modes is essential. This point will be discussed later in the chapter. A measured spectrum of the same design is shown in Fig. 5.10(b). Excellent agreement with theory is obtained.

The effective cavity-length approximation allows an equivalent model of a VCSEL to be made, as shown in Fig. 5.11. The threshold condition is given by

$$R_1 R_2 e^{2g_{th}L_{QW} - 2\alpha_i L_{cav}} = 1,$$

and the threshold gain g_{th} is

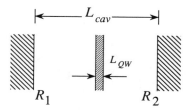

FIGURE 5.11. Equivalent VCSEL model.

$$g_{th} = \frac{1}{2L_{QW}} \left(\ln \frac{1}{R_1 R_2} + 2\alpha_i L_{cav} \right).$$

If the optical gain as a function of carrier density is written as

$$g_{th} = \frac{\partial g}{\partial n} (n_{th} - n_0),$$

the threshold current is given by

$$I_{th} = q \frac{n_{th}}{n_i} (L_{QW} A) \frac{1}{\tau_e},$$

where τ_e = electron lifetime, A = active-region area, and η_i = internal quantum efficiency.

The differential quantum efficiency can also be estimated with the same equation used for edge-emitting lasers:

$$\eta_D = \eta_i \frac{\ln \dfrac{1}{R_1 R_2}}{\left(\ln \dfrac{1}{R_1 R_2} \right) + 2\alpha_i L_{cav}}.$$

For example, for a VCSEL with $\alpha_i = 20$ cm^{-1}, $L_{cav} = 1.5$ μm, $\tau_e = 2$ ns, $L_{QW} = 24$ nm, $\eta_i = 1$, and $R_1 R_2 = 0.99$, the threshold current density is estimated to be 1.2 kA/cm^2 and $\eta_D = 0.63$. Increasing the reflectivity to 0.995 reduces the threshold current density to 770 A/cm^2.

5.2.3 Graded Heterojunctions

A graded refractive index, separate-confinement heterostructure (GRIN-SCH) around the quantum-well region is often used for better carrier confinement similar to that in a typical edge-emitting laser design. However, this approach is accompanied by a lower optical confinement factor (overlap integral of the optical field and the active region) in a VCSEL. As a result of the index profile in the GRIN-SCH

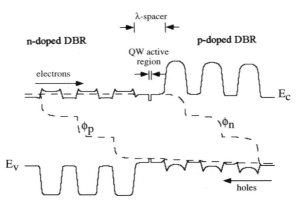

FIGURE 5.12. Energy band diagram of a VCSEL, where E_c and E_v are conduction and valence bond edges, and ϕ_n and ϕ_p are the Fermi levels of the electrons and holes, respectively.

region, the calculated optical intensity versus distance curve (Fig. 5.9) often has its peak intensity lowered slightly in the quantum well region.

The barriers at the heterojunctions of the Bragg reflectors contribute to significant increases in drive voltage and series resistance, which together create a serious heating problem. The resistance and voltage drop at the heterobarriers can be calculated using a self-consistent comprehensive model including drift-diffusion and thermionic currents.[28a] A schematic of calculated energy-band diagram of a VCSEL is depicted by Fig. 5.12. Graded layers are typically used at the heterojunctions to reduce the voltage and resistance.[9] Although effective, this measure has not completely eliminated the heating problem. In addition, greater numbers of pairs are needed in the DBRs to compensate for the reduction of reflectivity due to the graded layers. Recently, there have been many studies on this issue. Results obtained using stepped doping profiles[29,30] in conjunction with the graded heterostructure appear to be promising.

5.3. VCSEL FABRICATION

5.3.1. VCSEL Structures

There are a number of variations on the techniques used to fabricate VCSELs. A brief summary is presented here on proton-implanted lasers, air-post-type lasers, and passive-antiguide-region (PAR) lasers.

The proton-implanted VCSELs are fabricated by photolithographically masking off the lasers such that they are protected from the proton implantation. Since the implant creates damages in the region surrounding each laser, it provides the desirable current confinement as well as some unintentional optical losses. A careful balance between the two effects is necessary.[23] Since the laser is defined by the current/gain path, this type of laser is called a gain-guided VCSEL. With a deeper implant, the damages are closer to the active region. It is not very clear at this time whether such damage may adversely affect the long-term stability and reliability of the lasers. The advantages of this method are (1) the yield is exceedingly high, and (2) the processing steps leave the wafer planar and it is thus easier to make metal contacts. Currently, most top emitting VCSELs are fabricated with this technique[31] whereas bottom emitting VCSELs are fabricated with several different methods.

Air-post VCSELs are fabricated by using various dry etching techniques, including reactive ion etching (RIE), chemically assisted ion-beam etching (CAIBE), and electron cyclotron resonance etching (ECRE). The advantage of using dry etching is that very small (submicron diameter) laser posts can be made. The electrical confinement is typically excellent as long as the damages on the side walls can be controlled or eliminated. To date, most very low threshold current lasers are achieved with such VCSELs. However, the index difference between the DBR and air is so high that these VCSELs tend to emit multiple transverse modes.[32] Moreover, the nonplanar nature of such lasers make them difficult to contact, especially in the form of arrays.

The passive-antiguide-region (PAR) VCSEL is a special case of buried heterostructure (BH) VCSEL that provides stable transverse mode control. The VCSEL wafer is first etched with a dry etching technique to form laser posts. The depth of etching may surpass the laser active region to obtain both carrier and optical confinements in a final device. The laser posts are then chemically etched such that sloping walls are formed around the posts to facilitate MOCVD regrowth. Complete removal of oxides on AlAs layers before the MOCVD regrowth is essential for obtaining crystallized regrowth. The regrown heterostructure, designed to form current-blocking claddings around the lasers, includes a 0.2 μm undoped GaAs layer, nine periods of 0.4-μm-thick n-i-p-i-doped $Al_{0.2}Ga_{0.8}As$ layers, and finally another 0.2 μm undoped GaAs layer. Excellent single-crystal regrowth at the Bragg reflector interfaces has been obtained,[24] as depicted by the scanning electron micrograph (SEM) of the cross section of a typical laser (Fig. 5.13). The polycrystal formed on top of the laser posts as well as the SiO_2 masks can be chemically removed subsequently. Finally, p-type metal contacts were evaporated on the laser posts. The laser output in this particular case was measured through the polished substrate.

5.3.2. Laser-Array Addressing Scheme

Independently addressable diode laser arrays have been fabricated with gain-guided VCSEL.[18] The processing steps include an additional mask evaporated on the wafer to provide contact pads for bonding. Figure 5.14 shows a photograph of a fabricated 8×8 independently addressable VCSEL array. For larger arrays, independent addressing becomes difficult. Using a matrix (also known as row column) addressing scheme, N^2 contacts can be reduced to 2N contacts for an N×N array. The fabrication of a matrix addressable array,[17] shown in Fig. 5.15, requires the

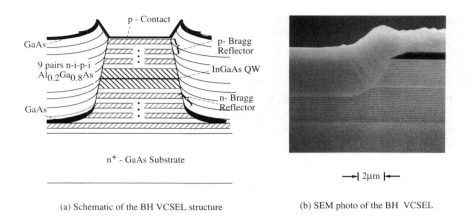

(a) Schematic of the BH VCSEL structure   (b) SEM photo of the BH VCSEL

FIGURE 5.13. Scanning electron micrograph (SEM) cross section of a PAR VCSEL.

epitaxial material to be grown on a semi-insulating substrate, along with some additional processing steps. The processing steps include etching groves between the columns of lasers to isolate the n-doped layers between columns, etching the edges of each column down to the n-doped layers, evaporating the n-contacts for the columns, planarizing the wafer by filling the etched grooves with polyimide, and evaporating p-metal to connect the p-contacts of the rows.

For high-speed or high-power operation, VCSELs can be mounted junction-side down onto a metalized BeO substrate with In solder bumps and put into an appropriate package. The flip-chip bonding requires much thicker p-side metal, typically obtained by Au plating, and better isolation between the devices, achieved by deep etching between the lasers. Figure 5.16 shows the schematic of a flip-chip mounted 2×8 VCSEL array and a photograph of such an array in a high-speed package.[33] The performance of these packaged lasers is discussed in the following section.

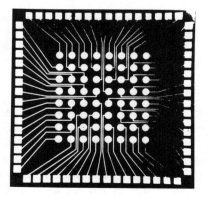

FIGURE 5.14. Photograph of an 8×8 independently addressable VCSEL array (after Ref. 17).

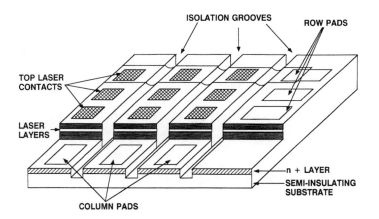

FIGURE 5.15. Schematic of a matrix-addressable VCSEL array. The *p*-contacts of all lasers in a row are connected, whereas the *n*-contacts of lasers in a column are connected. By applying bias to the *i*th row and *j*th column, the *ij*th laser is turned on [after Ref. 17(a)].

5.4. LASER CHARACTERISTICS

5.4.1. Transverse Modes

The transverse modes of typical VCSELs are nearly TEM modes, because the laser's transverse dimension is typically significantly larger than its effective cavity length. Although exact modal analyses may reveal that the transverse modes of a VCSEL are not exactly TEM modes, the longitudinal field components for typical VCSELs are likely to be very small. Thus, TEM modes can be used for all practical purposes, such as calculations on modal confinement factor and far field beam divergence. As a consequence, the complete set of possible transverse modes is the same for lasers irrespective of the shapes of their contacts. However, since the current/carrier distribution determines the gain profile, whose overlap with the modes determines the modal gain for each transverse mode, the specific onset of a particular higher-order mode may vary with the VCSEL shape.

The near-field patterns of a 15 μm square gain-guided VCSEL at various cw current levels are shown in Fig. 5.17. The laser starts to lase in the fundamental TEM_{00} mode, as shown in Fig. 5.17(a). At higher currents, higher-order modes successively appear due to gain spatial hole burning. The TEM_{01*} and TEM_{10} modes dominate for only a small range of drive currents [Fig. 5.17(b) and 5.17(c)]. At still higher cw current, the laser emits both TEM_{00} and TEM_{11} [Fig. 5.17(d)]. In addition, the larger the VCSEL, the more modes it emits. A 5 μm square VCSEL emits only the TEM_{00} mode, whereas a 10 μm×5 μm rectangular VCSEL emits two modes, TEM_{00} and TEM_{10}. On the other hand, a 20 μm square laser supports transverse modes up to TEM_{22}. These observations are consistent with those for well-behaved gain-guided edge-emitting lasers.[34]

The transverse modes of gain-guided VCSELs are linearly polarized in typically

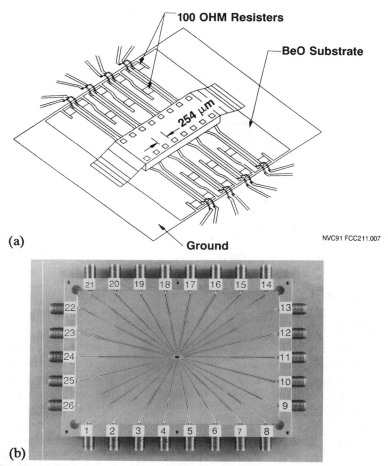

FIGURE 5.16. (a) Schematic of a 2×8 VCSEL array flip chip mounted onto a metalized BeO substrate. The metalized pattern on the BeO substrate is matched to the spacings of the VCSEL array. Additional metalized lines are for wire bonding. Resistors are fabricated on the BeO substrate to provide impedance matching for the VCSELs (after Ref. 33 © 1991 IEEE). (b) Photograph of a flip chip mounted 2×8 VCSEL array in a high-speed package. Laser spacings are 254 and 508 μm for the lasers in one row and one column, respectively.

uncorrelated directions.[35] When two modes are simultaneously excited, they often exhibit orthogonal polarization directions because of unintentional inhomogeneity in the gain or thermal distribution. Figure 5.18 shows the spectrally and polarization-resolved near-field patterns of a VCSEL at various cw current levels. Near threshold, the laser emits a single highly linearly polarized TEM_{00} mode. At higher current, when two transverse modes are simultaneously excited, the two modes tend to oscillate at two orthogonal polarizations. Not only has the polarization for a particular mode been observed to vary randomly from one laser to next, it also

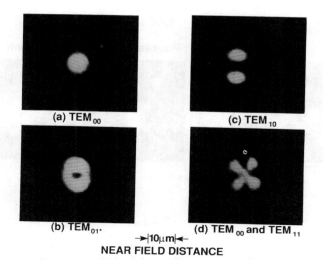

(a) TEM$_{00}$

(c) TEM$_{10}$

(b) TEM$_{01*}$

(d) TEM$_{00}$ and TEM$_{11}$

→|10μm|←

NEAR FIELD DISTANCE

FIGURE 5.17. cw near-field patterns of a typical 15 μm square gain-guided VCSEL emitting a single (a) TEM$_{00}$ mode, (b) TEM$_{01*}$ mode, (c) TEM$_{10}$ mode, and (d) TEM$_{00}$ and TEM$_{11}$ mode (after Ref. 25).

varies with drive current. At $1.8\,I_{th}$, the laser emits the donut-shaped mode TEM$_{01*}$ [Fig. 5.8(b)], whose spectrally resolved near-field shows a single-mode-like pattern [Fig. 5.18(e)] limited by spectral resolution. However, with the use of a polarizer, this mode can be clearly separated into the TEM$_{10}$ mode at 84° and the TE$_{01}$ mode at 4° [Fig. 5.18(f)], showing that the ring-shaped TEM$_{01*}$ mode is indeed a superposition of TEM$_{10}$ and TEM$_{01}$ modes.

The air-post VCSELs typically emit in multiple transverse modes even near threshold. Shown in Fig. 5.19 is a near-field intensity pattern and a spectrally resolved near field of a 10 μm air-post VCSEL at $1.1\,I_{th}$. The spectrally resolved near field patterns reveal simutaneous emission of multiple transverse modes. Nonetheless, single transverse mode air-post VCSELs have been reported. However, the repeatable single-mode operating range is small.

At low current levels, a gain-guided VCSEL emits in a single pure transverse mode. The modal evolution is well behaved. Both characteristics make gain-guided VCSELs preferable for systems prototype experiments. As a consequence, most systems demonstrations using VCSELs have thus far been made using gain-guided VCSELs.

Recent results on PAR VCSELs show high promise of achieving single transverse-mode operation over a large current range with a large laser aperture.[24,25] The light output versus current characteristics of an 8 and 16 μm PAR VCSEL are shown in Fig. 5.20. These devices exhibit 0.8 and 1.2 mA threshold current, respectively. Typical near-field patterns for an 8-μm-diam PAR VCSEL at various currents are shown in Fig. 5.21(a). The spontaneous emission is emitted from the entire laser

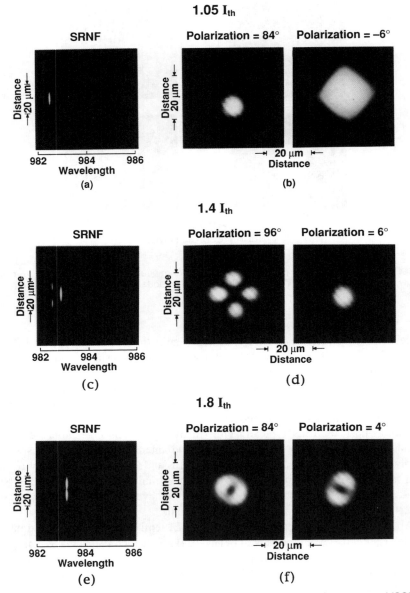

FIGURE 5.18. Spectrally and polarization-resolved near-field patterns for the same VCSEL at various cw current levels: (a) and (b) 1.05 I_{th}: The laser emits a single TEM_{00} mode. Near-field images filtered through a polarizer with its angle set at 84° and −6° with respect to the [011] crystal axis. The two pictures in (b) are not aligned with each other. The intensity of the image at −6° is about 1000 times weaker than that at 84°. (c) and (d) 1.4 I_{th}: Near-field images filtered through a polarizer with angles set at 96° and 6°. The two transverse modes exhibits orthogonal polarizations. The TEM_{00} mode changed its polarization orientation with current from (b) to (d). (e) and (f) 1.8 I_{th}: Near-field images filtered through a polarizer with angles set at 84° and 4° (after Ref. 35).

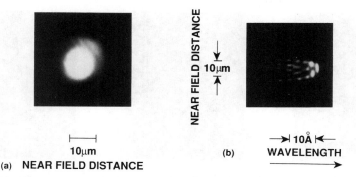

(a) **NEAR FIELD DISTANCE** (b) **WAVELENGTH**

FIGURE 5.19. Near field and spectrally resolved near-field patterns of an airpost VCSEL.

aperture. The laser emits a single fundamental TEM_{00} mode near threshold and at twice threshold with a nearly Gaussian intensity profile. Stable single TEM_{00} mode emission is achieved with currents exceeding $12\,I_{th}$.[25] The beam width is stable and unchanged with current, in contrast to the case with gain guided VCSELs, where emission sizes shrink as current increases, attributed to thermal lensing effect.

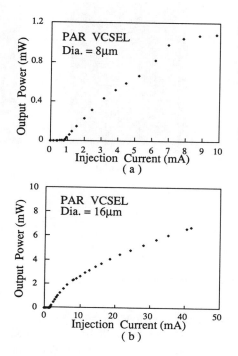

FIGURE 5.20. The L-I characteristics of an (a) 8-μm and (b) 16-μm-diam PAR VCSEL.

I=0.9I$_{th}$ I=1.2 I$_{th}$ I=2.0I$_{th}$

10μm

(a)

I=0.9I$_{th}$ I=1.3 I$_{th}$ I=2.0I$_{th}$

10μm

(b)

FIGURE 5.21. Near-field patterns of an (a) 8-μm diam and (b) 16-μm-diam PAR VCSEL.

Figure 5.21(b) shows typical near-field patterns for a 16 μm PAR VCSEL at various currents. Similar to the case for an 8 μm PAR VCSEL, the spontaneous emission is emitted from the entire laser aperture. This condition indicates that uniform current injection without filamentations is obtained for broad area

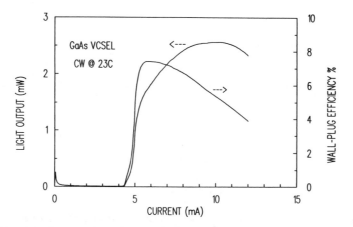

FIGURE 5.22. Light intensity vs current and voltage vs current characteristics for a top emitting VCSEL under cw operation at room temperature (after Ref. 30 © 1994 IEEE).

VCSELs. Figure 5.21(b) also shows the near-field patterns for a typical 16 μm BH VCSEL near and at twice threshold current. The 16 μm lasers stay single mode at current levels up to $\sim 4\,I_{\text{th}}$. At higher injection currents, the TEM_{10} mode begins to lase because of spatial hole burning.

All of the PAR VCSELs are linearly polarized in the [01̲1] direction when emitting a single transverse mode. The polarization is attributed to a crystal-orientation-dependent strain in the quantum-well region due to the anisotropic wet-chemical etching and MOCVD regrowth. This characteristic makes the PAR VCSELs highly important for applications requiring repeatable and reliable polarization control.

5.4.2. Power and Voltage versus Current Characteristics

The optical light output versus current (L-I) and voltage versus current (V-I) characteristics[36] of a top emitting 15 μm square VCSEL in cw operation is shown by Fig. 5.22. The drive current and voltage at the laser threshold for this particular laser are 3 mA and 3 V. This laser exhibits the highest electric power to optical power conversion efficiency (wall-plug efficiency), 7%, reported before 1994. It emits a single TEM_{00} mode with up to 0.5 mW output power at 10 mA bias current. Above that level the second transverse mode is excited, causing a kink to appear on the L-I curve. The L-I curve saturates due to heating.[37] Recent advances on native oxide defined VCSELs have pushed the single mode wall plug efficiency to a record of 50%.[9b]

The L-I and V-I characteristics for a bottom emitting VCSEL ideally are not very different from those of a top emitting laser. The best reported wall-plug efficiency for a 10 μm diameter is $\sim 8\%$ and for 20 μm diameter $\sim 17\%$.[38] However, highly nonlinear L-I curves are often observed with these lasers due to the external feedback provided by the polished substrate, which acts as a third mirror forming an

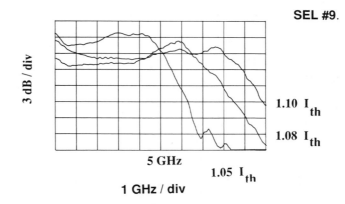

FIGURE 5.23. Typical small-signal intensity modulation response of the laser showing the increasing resonance frequency as the bias current level is raised (after Ref. 33 © 1991 IEEE).

additional optical cavity.[39] The additional feedback can be minimized using an antireflection coating on the back of the substrate. Conversely, the feedback can be enhanced to create a coupled-cavity tunable laser.[39]

5.4.3. Optical Spectra

The optical spectrum of a VCSEL emitting a single TEM_{00} mode typically exhibits a side-mode suppression greater than 45 dB and a linewidth of about 100 MHz.[40] However, the spectrum and linewidth deteriorates rapidly when the higher-order modes begin to lase. The wavelength separations between adjacent transverse modes are about 2–3 Å.[32]

5.4.4. Modulation Response

An example of the high-speed modulation response of a VCSEL is shown in Fig. 5.23. A typical 3 dB bandwidth of $\sim$ 8 GHz is observed by many groups.[14] Modulation bandwidth is currently limited by the highest single-mode output power, heating, and quantum well transport characteristics.[41] An extremely high relaxation oscillation frequency of 39 GHz was reported for a VCSEL fabricated with bulk active layers and driven by short pulses.[42] Although the relaxation oscillation is not the only parameter that determines the laser modulation bandwidth, this result indicates that a much higher modulation bandwidth can be obtained for VCSELs with proper design. Further study on this subject will no doubt lead to higher modulation bandwidth.

5.4.5. Optical Fiber Coupling

One of the major advantages provided by a VCSEL is its convenience and high efficiency for fiber coupling. The highest coupling efficiency reported is nearly 90%

for a top emitting VCSEL butt coupled into a single mode fiber without AR coating.[43] The excellent mode matching is considered to be the main reason for the high efficiency. For bottom emitting VCSELs, because of the thickness of the substrate and the laser beam divergence of 5°–7°, coupling into a single mode fiber is more difficult. Via holes can be etched from the back of the substrate to improve the coupling. The coupling of bottom emitting VCSELs into multimode fibers however, will not be an issue.

Coupling a VCSEL array into an array of single mode fibers is presently considered challenging. The main challenge lies in the packaging of fibers with a fixed separation. Si wafers with etched via holes have been used to hold the fibers.[19] The fiber array packaging technology, however, is expected to improve in the future.

5.5. VCSEL ARRAYS

5.5.1. Multiple Wavelength Array

Optical sources capable of Tb/s (10^{12} bits/s) data rates are essential for applications in multimedia optical fiber communications, interconnection of large computers, and real-time optical signal and image processing. A highly promising approach is to use a system that utilizes the laser wavelength as an additional parameter for multiplexing and coding, also known as wavelength-division multiplexing (WDM).

An important optical source for WDM systems is a monolithic array of single-wavelength lasers emitting distinct wavelengths with uniform wavelength spacings. Multiple-wavelength arrays have been fabricated with distributed feedback (DFB) lasers.[44,45] However, since the laser array is one dimensional (1D), the number of laser wavelengths and the wavelength spacing are both limited.

In this section, we describe a VCSEL array that, with proper control during the wafer epitaxial growth, can provide hundreds of independent wavelengths.[39] The wavelength spacing demonstrated is uniform and as small as 0.3 nm. The spacing can, in principle, be made smaller than 0.1 nm or as large as several nanometers. Using this novel design, not only can a larger number of controllable wavelengths be obtained, but the processing is as simple as that for a single VCSEL.

A WDM-based system experiment that exploited such a laser array is also described here. This was the first system experiment using a VCSEL array.[20,21] The results indicate negligible optical and electrical crosstalk between the lasers.

5.5.1.1. Laser Array Design

The Fabry–Perot mode wavelength of a VCSEL depends critically on the cavity thickness variation, as shown by the calculated reflectivity spectra for a complete VCSEL with ideally designed thickness and with two pairs of the Bragg reflectors being 10% thicker than designed in Fig. 5.24. Also shown in Fig. 5.25, the wavelength varies nearly linearly and monotonically with the layer thickness, the number of layers and their relative position in the cavity. Thus, the VCSEL lasing wave-

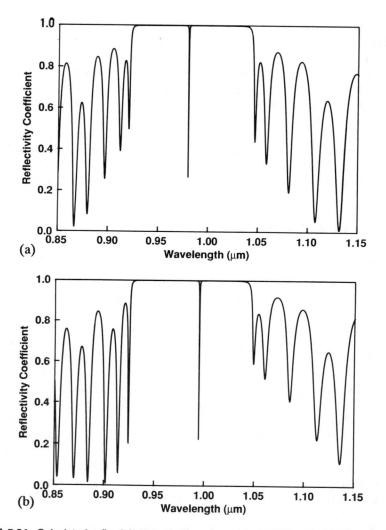

(a)

(b)

FIGURE 5.24. Calculated reflectivity spectra for a complete VCSEL with (a) ideal designed thickness and (b) two pairs of the Bragg reflectors being 10% thicker than designed. The calculation shows that there exists only one Fabry–Perot mode within the mirror bandwidth due to the VCSELs ultrashort cavity. This mode depends critically on the total cavity length.

length can be tailored to either longer or shorter wavelength with a small variation in cavity thickness without significant compromise in the lasing characteristics.

An intentional thickness gradient across the wafer translates into a wavelength gradient of the lasers fabricated across the wafer. Figure 5.26 shows the schematic of a three-element laser array based on this idea. This design to obtain multiple wavelength VCSEL array was first proposed and demonstrated using the natural

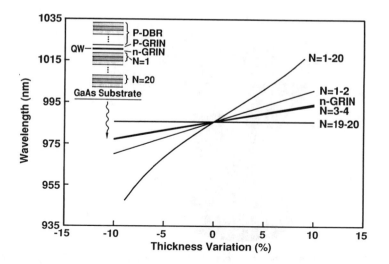

FIGURE 5.25. Fabry–Perot mode wavelength as a function of thickness for different combinations of layers being varied. It varies nearly linearly and monotonically with the layer thickness variation. The closer the variation is to the active layer the more its effect; similarly, the more layers are varied, the more wavelength shift can be obtained.

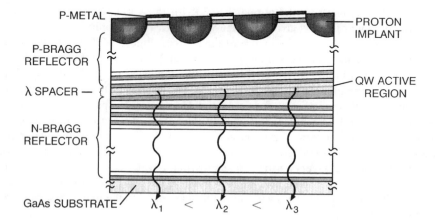

FIGURE 5.26. Schematic of an array consisting of three VCSELs as defined by proton implant. One pair of Bragg reflectors near the active region is shown to have a thickness gradient, which is used to alter the Fabry–Perot mode wavelength of the adjacent lasers. The amount of thickness variation in the spatially chirped layers is highly exaggerated.

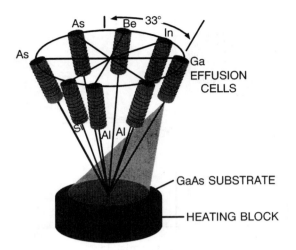

FIGURE 5.27. Schematic of the arrangement of atomic sources in an MBE system. All the sources are placed in a cone off-normal from the substrate (after Ref. 39 © 1991 IEEE).

nonuniform deposition in a molecular beam epitaxy (MBE) system.[13] Recently, there are a number of methods used to create the cavity thickness variation to attain multiple wavelength VCSELs.[45a,45b,45c] In this section, only the method used in Ref. 13 will be summarized.

One way to create such a thickness gradient is to simply keep the wafer stationary during part of the molecular beam epitaxy (MBE) growth. The thickness variation originates from the fact that the atomic sources in an MBE system are incident on the wafer at an angle from the normal ($\sim$ 33° for the Varian Gen II system) and hence the number of atoms arriving at the wafer varies monotonically in the direction parallel to the planes of incidence of the sources. Figure 5.27 shows schematically the arrangement of the atomic sources in an MBE system. Since the MBE material is grown in an As-rich environment, the thickness of the MBE growth is determined by the number of group III sources that arrives at the wafer. Thus, the directions of thickness variation for GaAs and AlAs layers are parallel to the directions of the Ga and Al sources, respectively. Therefore, we can obtain the desired small but definite thickness variation across the wafer by rotating the wafer for uniformity during the MBE growth of the VCSEL structure except for two pairs of AlAs and GaAs DBR layers during whose growth the wafer is kept stationary. If the Ga and Al sources are placed next to each other in the MBE system, the direction of the resulting cavity thickness variation will run parallel to the line intersecting the directions of the two sources.

A 2D laser array having no redundant wavelengths can be obtained if the direction of growth does not coincide with the array axes. The laser wavelength depends linearly on the cavity thickness and, in addition, the thickness gradient implemented by keeping the substrate stationary is linear over a small distance (small compared

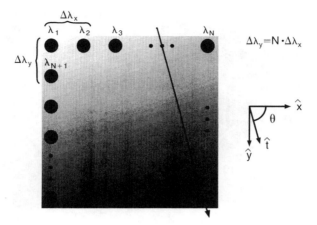

INCREASING DBR THICKNESS

FIGURE 5.28. Schematic of the 2D rastered multiple-wavelength (RMW) VCSEL array. By aligning the array obliquely with respect to the direction of thickness variation, a 2D laser array with no redundant wavelengths is obtained (after Ref. 39 © 1991 IEEE).

to the distance between the atomic sources and the wafer, which is 5–10 cm). Hence, the wavelength separation $\Delta\lambda$ between any two lasers is proportional to the distance between the lasers projected onto t. For a 2D array with its x axis making an oblique angle θ_x with t as shown in Fig. 5.28, we obtain

$$\Delta X \sim d \cos \theta_x \quad \text{and} \quad \Delta Y \sim d \sin \theta_x,$$

where d is the spacing between neighboring lasers, and $\Delta\lambda_x$ and $\Delta\lambda_y$ are the wavelength separation between neighboring lasers in the x and y direction, respectively. Therefore,

$$\Delta\lambda_y \cong \tan \theta_x \, \Delta\lambda_x = N\Delta\lambda_x,$$

$$N \equiv \tan \theta_x.$$

Hence, the wavelength of the $(N+1)$th laser on a row is the same as that of the first laser on the next row, provided $\tan \theta_x$ is close to an integer. Restricting the array to have N columns, a 2D array with no redundant wavelengths is obtained. To obtain a large N, θ_x must be large and close to 90°. The number of rows in such an array, M, can be made very large. Ideally, M is limited only by the total gain bandwidth ~ 100 nm, which has to be larger than or equal to the total wavelength span

$$\lambda_{\text{span}} = (NM-1)\Delta\lambda_x.$$

The actual physical dimension may impose a limit if uniform wavelength separation throughout the array is required. The direction of increasing wavelength rasters

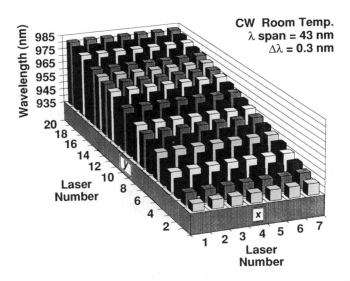

FIGURE 5.29. Experimentally measured cw wavelength distribution of the 7×20 RMW VCSEL array. Each laser emits a unique wavelength. The direction of increasing wavelength rasters through the 2D array (after Ref. 39 © 1991 IEEE).

through the elements of the array. Such an array is a rastered multiple-wavelength, or RMW, laser array.

In an experimental demonstration, the angle θ_x was chosen to be $\sim 82°$, and $\Delta\lambda_y$ is seven times $\Delta\lambda_x$. Although larger θ_x would give a larger N, alignment becomes progressively more difficult.

5.5.1.2. Laser Array Characteristics

The wavelength distribution of a 7×20 RMW array of 20 μm square lasers is shown in Fig. 5.29. The total wavelength range is as large as 43 nm, from 940 to 983 nm. The average wavelength separation between two neighboring lasers on a row and a column are 0.3 and 2.1 nm, respectively. The laser spectra are measured under cw operation with the lasers emitting a single TEM_{00} mode. The wavelength separation can be easily made smaller by reducing the laser spacing or the number of chirped layers. It can also easily be made larger by doing the opposite.

Figure 5.30 shows the cw threshold current distribution of the 2D RMW laser array. The average cw threshold current is 8.5 mA. The 2D laser array exhibits good uniformity with 85% of the lasers having threshold currents within ±2 mA of the average values. All lasers show comparable electrical and optical characteristics.

5.5.2. Independently Addressable Array

The yield and uniformity of independently addressable VCSEL arrays can be very high because of the simplicity of array fabrication. Typical threshold current variation for an 8×8 VCSEL array is about 2 mA. The actual amount is wafer dependent.

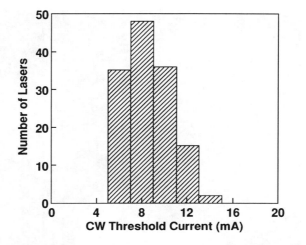

FIGURE 5.30. Distribution of measured cw threshold currents of the 7×20 RMW VCSEL array (after Ref. 39 © 1991 IEEE).

Systems experiments using individually addressable VCSEL arrays have been demonstrated for optical data link applications.[19] Figure 5.31 shows a photograph of a complete 36-channel optical interconnection scheme using a 2×18 top emitting VCSEL array, which is pigtailed into a 2×18 optical fiber bundle and a 2×18 GaAs p-i-n detector array. The schematic of one such connection is shown in Fig. 5.32. The VCSEL array exhibits high uniformity as shown by the L-I characteristics in Fig. 5.33. The data rate at higher than 622 Mb/s per channel was demonstrated, with a resultant aggregate bandwidth of 22 Gb/s.

5.5.3. High-Speed Multiple Wavelength VCSEL Arrays

Figure 5.16(a) shows the schematic of a VCSEL flip chip mounted onto a BeO substrate.[14,33] Such a BeO carrier is mounted into a high-speed package. Figure 5.16(b) shows a photograph of a flip chip mounted 2×8 VCSEL array mounted in a high-speed package. The laser row and column spacings are 254 and 508 μm, respectively. Though comparable with those used for edge emitting laser arrays, these spacings are designed to be fairly large in order to minimize thermal crosstalks and to allow most of the array to be operated simultaneously. Each laser is connected in parallel with a 100 Ω resistor fabricated on the BeO substrate to provide high-speed impedance matching. Each microstrip line from the BeO substrate terminates at an SMA connector. The lasers share the common ground contact, the ground termination being made via ribbon wires from the n-type substrate at the top of the bonded device. The light output from each laser was transmitted through the ~ 200-μm-thick substrate and butt coupled into a single-

FIGURE 5.31. Photograph of a complete 36-channel optical interconnection scheme using a 2×18 top-emitting VCSEL array, pigtailed into a 2×18 optical fiber bundle, and a 2×18 GaAs *p-i-n* detector array (after Ref. 19).

mode fiber. The output power from the laser was about 0.5 mW at 1.1 I_{th}. Up to eight lasers were biased simultaneously in the experiment. Optical isolators were not used in the measurements.

Figure 5.34 shows the output spectra of the 16 lasers. The sidemode suppression for each laser is greater than 45 dB; the chirp-broadened 20 dB spectral width of the lasers under 5 Gb/s modulation is under 0.3 nm. The wavelength spacing is approximately 0.9 nm between the row elements. There is a small overlap in the wavelength range covered by the lasers in the two rows; however, due to the limited spectral width, all 16 lasers can be used as independent wavelength sources. The use of wavelength-rastered arrays can also remove the wavelength duplication in larger arrays. Simultaneous operation was demonstrated on eight VCSELs of the same row, the number being limited solely by the availability of equipment.

Figure 5.23 shows the typical small-signal intensity modulation response of the laser showing the increasing resonance frequency as the bias current level is raised. The maximum speed of the present device is limited to be about 8 GHz by the appearance of multiple modes.

In order to verify that the lasers are well-behaved under large-signal modulation, sensitivity measurements were carried out at 5 and 2.5 Gb/s with a (2^7-1)-length

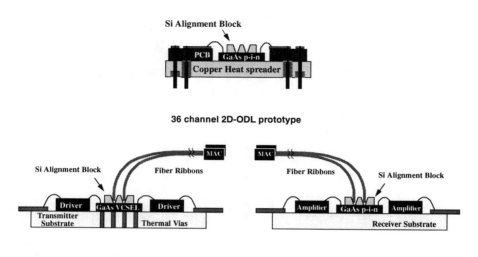

36 channel 2D-ODL concept

FIGURE 5.32. Schematic of one of the fiber connections (after Ref. 19).

pseudorandom patterns. Longer word lengths were not used due to the low-frequency cutoff in the receiver response. Figure 5.35 shows bit error ratio as a function of received power for VCSEL 12 modulated alone and with two nearest neighboring lasers (254 μm) being simultaneously modulated by two independent sources. At 10^{-9} error ratio, a sensitivity penalty of 0.7 dB was measured. No crosstalk effect was observed when the modulation was applied to the next nearest (508 μm spaced) neighbors. Thus, this 16-wavelength VCSEL array[14] promises a record high aggregate bandwidth of 80 Gb/s.

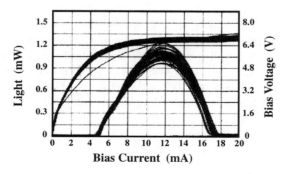

FIGURE 5.33. L-I and I-V characteristics of a 2×18 VCSEL array used in the 36-channel optical interconnection experiment (after Ref. 19).

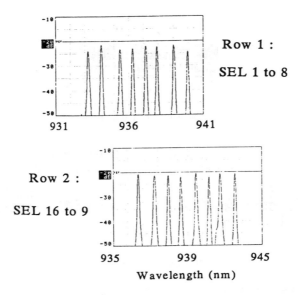

FIGURE 5.34. Output spectra of 16 lasers. The wavelength spacing is approximately 0.9 nm between the row elements. There is a small overlap in the wavelength range covered by the lasers in the two rows, however, all 16 lasers can be used as independent wavelength sources (after Ref. 33 © 1991 IEEE).

These measurements were repeated at 2.5 Gb/s, as shown in Fig. 5.36. An error ratio of 10^{-9} was attained for a receiver input power of -27.5 dBm. Negligible effect on the sensitivity curve (< 0.2 dB) was observed with the two nearest adjacent lasers being simultaneously modulated. Although the loss and dispersion in optical fibers are not optimum for 980 nm (~ 1 dB/km and ~ 40 ps/nm km, respectively), the feasibility of fiber transmission using a 8 km span of fiber for these VCSEL at 2.5 Gb/s is demonstrated as shown in Fig. 5.36. The 1.4 dB penalty is attributed to the optical reflections in the systems and can be largely eliminated with an optical isolator.

There are three crosstalk mechanisms for laser arrays: Optical, thermal, and electrical. With a laser spacing of 254 μm and a substrate thickness of about 200 μm, the optical cross coupling of adjacent lasers into the fiber is less than -50 dB. The thermal rf crosstalk, measured as a wavelength shift induced by the bias change of an adjacent laser, is between 0.02 and 0.05 Å/mA. The dominant component of crosstalk in this experiment is, thus, electrical crosstalk, which can be reduced with better packaging design.

5.6. LONG-WAVELENGTH VCSEL

Much progress has been made for VCSELs emitting in the 1.3–1.5 μm wavelength range. Since the wavelength of interest is longer and the index difference in quar-

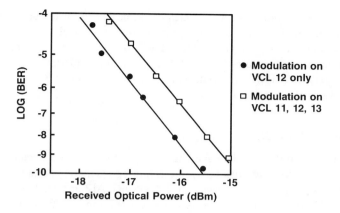

FIGURE 5.35. Bit-error rate as a function of received power for VCSEL 12 being modulated alone at 5 Gb/s and with two nearest neighboring lasers being simultaneously modulated by two independent sources also at 5 Gb/s (after Ref. 14).

ternary material systems is substantially smaller, a great many layer pairs are required for an epitaxial DBR. This requirement makes not only growth time and precision an issue, but more fundamentally the diffraction loss in such a thick DBR will be very significant.[46] Thus, the same design concept for GaAs-based VCSELs may not be the optimal solution for making long-wavelength VCSELs.

Currently, two approaches appear to be promising. The first one employs dielectric mirrors that use materials with high thermal conductivity. A VCSEL with

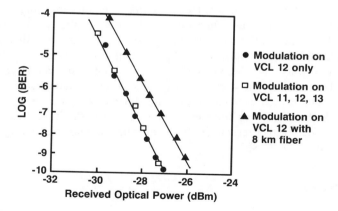

FIGURE 5.36. Bit-error rate as a function of received power for VCSEL 12 being modulated alone at 2.5 Gb/s and with two nearest neighboring lasers being simultaneously modulated by two independent sources also at 2.5 Gb/s. An error rate of 10^{-9} was attained at a receiver input power of -27.5 dBm (after Ref. 14).

TiO_2/ZnO_2 dielectric mirrors that operates cw at nearly room temperature (14 °C) was recently reported.[47] The second approach utilizes a novel technique of wafer bonding[48,49] and employs a GaAs/AlAs DBR bonded to the InP-based active layers.[50] A room-temperature pulsed 1.3 μm VCSEL fabricated this way was reported to exhibit a low threshold. This area will no doubt see a great many breakthroughs in the near future.

5.7. VISIBLE VCSEL

VCSELs emitting in the visible wavelength range are of interest because of potential applications in compact disk (CD) players, laser scanners, and optical interconnects. The 0.6–0.7 μm laser uses an InGaP/InGaAlP material system lattice matched to a GaAs substrate. The design of a 0.6–0.7 μm VCSEL can thus be very similar to that of an 0.8–1 μm VCSEL. A recent article reported the room-temperature cw operation of a 0.65 μm VCSEL with a design using InGaP strained quantum wells and AlGaAs/AlAs DBRs.[51] Although the material growth issues on the InGaP material system (lattice matched to a GaAs substrate) still exist, the understanding of this system will no doubt advance rapidly. Thus, it is anticipated that rapid progress will continue in visible VCSELs and VCSEL arrays.

5.8. FUTURE PROSPECTS

VCSELs offer a unique combination of easily fabricatable large 2D arrays and novel wavelength tailorable functionality. The low-divergence circular output beam and the potential for very low threshold current operation are also important advantages. It is expected that these devices will find important applications both as low-cost high-performance lasers and as highly functional laser arrays. One most important aspect of VCSELs that needs to be improved is stable single transverse and polarization mode operation. In addition, long term reliability study, which is a function of device fabrication and structure, needs to be conducted thoroughly.

It is clear that for certain applications highly multimode but low-cost lasers will be important. In this case, VCSELs are already suitable for commercialization. However, a laser emitting in more than one mode is highly undesirable in terms of noise[52,53] and modulation characteristics. Thus, for many applications, it is crucial to design a VCSEL emitting light in a single mode, both single transverse and polarization mode. Further research will no doubt be undertaken resulting in further improvement of such VCSELs.

REFERENCES

1. Z. L. Liau and J. N. Walpole, Appl. Phys. Lett. **50**, 528 (1987).
2. N. W. Carlson, G. A. Evans, D. P. Bour, and S. K. Liew, Appl. Phys. Lett. **56**, 16 (1990).
3. H. Soda, K. Iga, C. Kitahara, and Y. Suematsu, Jpn. J. Appl. Phys. **18**, 2329 (1979).
4. R. Burnham, D. R. Scifres, and W. Streifer, U.S. Patent No. 4309670, Jan. 1982.

5. A. Chailertvanitkul, S. Uchiyama, Y. Kotaki, Y. Kokubun, and K. Iga, Annual Meet. Jpn. Soc. Appl. Phys., Tokyo, Japan, April 1983.
6. K. Iga, F. Koyama, and S. Kinoshita, IEEE J. Quantum Electron. 24, 1845 (1988).
7. D. Botez, L. M. Zinkiewicz, T. J. Roth, L. J. Mawst, and G. Peterson, IEEE Photon. Technol. Lett. 1, 205 (1989).
8. J. L. Jewell, A. Scherer, S. L. McCall, Y. H. Lee, S. Walker, J. P. Harbison, and L. T. Florez, Electron. Lett. 25, 1123 (1989).
9. R. S. Geels, S. W. Corzine, J. W. Scott, D. B. Young, and L. A. Coldren, IEEE Photon. Technol. Lett. 2, 234 (1990).
9a. D. L. Huffaker, J. Shin, H. Deng, C. C. Lin, D. G. Deppe, and B. G. Streetman, App. Phys. Lett., 65, 2642–2644 (1994).
9b. K. L. Lear, K. D. Choquette, R. P. Schneider, S. P. Kilcoyne, and K. M. Geib. Electr. Lett., 31 (3), 208–209 (1995).
10. C. J. Chang-Hasnain, J. P. Harbison, C. E. Zah, L. T. Florez, and N. C. Andreadakis, Electron. Lett. 27, 1002 (1991).
10a. L. Fan, M. C. Wu, H. C. Lee, and P. Grodzinski, Electron. Lett., 30, 1409–1410 (1994).
10b. N. Yokouchi, T. Miyamoto, T. Uchida, Y. Inaba, F. Koyama, and K. Iga, IEEE Photonics Tech. Lett., 4, 701–703 (1992).
11. G. Hasnain, K. Tai, Y. H. Wang, J. D. Wynn, K. D. Choquette, B. E. Weir, N. K. Dutta, and A. Y Cho, Electron. Lett. 27, 1630 (1991).
12. Ping Zhou, Julian Cheng, C. F. Schaus, S. Z. Sun, C. Hains, K. Zheng, A. Torres, D. R. Myers, and G. A. Vawter, Appl. Phys. Lett. 59, 2504 (1991).
13. C. J. Chang-Hasnain, J. R. Wullert, J. P. Harbison, L. T. Florez, N. G. Stoffel, and M. W. Maeda, Appl. Phys. Lett. 58, 31 (1991).
14. C. J. Chang-Hasnain, M. W. Maeda, A. Von Lehmen, Chinlon Lin, and H. Izadpanah, LEOS Annual Meeting, San Jose, CA Nov. 1991; M. W. Maeda, A. Von Lehmen, J. R. Wullert, M. Allersma, H. Izadpanah, C. J. Chang-Hasnain, M. Z. Iqbal, and Chinlon Lin, Optical Fiber Communications Conference, San Jose, CA, Feb. 1992.
15. M. Orenstein, E. Kapon, J. P. Harbison, L. T. Florez, and N. G. Stoffel, Appl. Phys. Lett. 60, 1535 (1992).
16. M. E. Warren, P. L. Gourley, G. R. Hadley, G. A. Vawter, T. M. Brennan, B. E. Hammons, and K. L. Lear, Appl. Phys. Lett. 61, 1484 (1992).
17. M. Orenstein, A. Von Lehman, C. J. Chang-Hasnain, N. G. Stoffel, L. T. Florez, J. P. Harbison, J. Wullert, and A. Scherer, Electron. Lett. 27, 437 (1991).
18. A. Von Lehman, C. J. Chang-Hasnain, J. Wullert, L. Carrion, N. G. Stoffel, L. T. Florez, and J. P. Harbison, Electron. Lett. 27, 583 (1991).
19. G. Hasnain, R. A. Novotny, J. D. Wynn, and R. Leibenguth, Conference on Lasers and Electro-Optics, Anaheim CA, 1992.
20. M. W. Maeda, C. J. Chang-Hasnain, Chinlon Lin, J. S. Patel, H. A. Johnson, and J. A. Walker, IEEE Photon. Technol. Lett. 3, 268 (1991).
21. C. J. Chang-Hasnain, M. W. Maeda, J. P. Harbison, L. T. Florez, and Chinlon Lin, J. Lightwave Technol. 9, 1665 (1991).
22. K. Tai, R. J. Fischer, C. W. Seabury, N. A. Olsson, T.-C. D. Huo, Y. Ota, and A. Y. Cho, Appl. Phys. Lett. 55, 2473 (1989).
23. M. Orenstein, A. Von Lehman, C. J. Chang-Hasnain, N. G. Stoffel, J. P. Harbison, and L. T. Florez, Appl. Phys. Lett. 57, 2384 (1990).
24. C. J. Chang-Hasnain, Y. A. Wu, G. S. Li, G. Hasnain, K. D. Choquete, C. Caneau, and L. T. Flores, Appl. Phys. Lett. 63, 1307 (1993).
25. Y. A. Wu, C. J. Chang-Hasnain, and R. Nabiev, Electron. Lett. 29, 20 (1993).
26. M. A. Plonus, Applied Electromagnetics (McGraw Hill, New York, 1978).
27. P. Yeh, Optical Waves in Layered Media (Wiley, New York, 1988).
28. A. Yariv, Quantum Electronics, 2nd ed. (Wiley, New York, 1975).
28a. R. F. Nabiev and C. J. Chang-Hasnain, IEEE Photonics Tech. Lett., (to be published).
29. M. G. Peters, D. B. Young, F. H. Peters, J. W. Scott, B. J. Thibeault, and L. A. Coldren, IEEE Photon. Technol. Lett. 6, 31 (1994).
30. K. L. Lear, R. P. Schneider, K. D. Choquette, S. P. Kilcoyne, J. J. Figiel, and J. C. Zolper, IEEE Photon. Technol. Lett. 6, 1053 (1994).
31. Y. H. Lee, B. Tell, K. F. Brown-Goebeler, J. L. Jewell, and J. V. Hove, Electron. Lett. 26, 710 (1990).

32. C. J. Chang-Hasnain, M. Orenstein, A. Von Lehmen, L. T. Florez, J. P. Harbison, and N. G. Stoffel, Appl. Phys. Lett. **57**, 218 (1990).
33. M. W. Maeda, C. J. Chang-Hasnain, A. Von Lehmen, H. Izadpanah, C. Lin, M. Z. Iqbal, L. T. Florez, and J. P. Harbison, IEEE Photon. Technol. Lett. **3**, 863 (1991).
34. C. J. Chang-Hasnain, E. Kapon, and R. Bhat, Appl. Phys. Lett. **54**, 205 (1989).
35. C. J. Chang-Hasnain, J. P. Harbison, L. T. Florez, and N. G. Stoffel, Electron. Lett. **27**, 163 (1991).
36. G. Hasnain, J. D. Wynn, S. Gunapalo, and R. E. Leibengnth, paper JTHA2, Conference on Lasers and Electroptics, Anaheim, CA May 14, 1992.
37. G. Hasnain, K. Tai, L. Yang, Y. H. Wang, R. J. Fischer, J. D. Wynn, B. Weir, N. K. Dutta, and A. Y. Cho, IEEE J. Quantum Electron. **27**, 1377 (1991).
38. R. S. Geels, S. W. Corzine, B. Thibeault, and L. A. Coldren, Opt. Fiber Commun. Conf., San Jose, CA, February 2–7, 1992.
39. C. J. Chang-Hasnain, J. P. Harbison, C.-E. Zah, M. W. Maeda, L. T. Florez, N. G. Stoffel, and T.-P. Lee, IEEE J. Quantum Electron. **27**, 1368 (1991).
40. R. S. Geels, S. W. Corzine and L. A. Coldren, IEEE J. Quantum Electron. **27**, 1359 (1991).
41. K. Y. Lau and A. Yariv, IEEE J. Quantum. Electron. **QE-21**, 121 (1985).
42. J. Lin, J. K. Gamelin, K. Y. Lau, S. Wang, M. Hong, and J. P. Mannaerts, Appl. Phys. Lett. **60**, 15 (1992).
43. K. Tai, G. Hasnain, J. D. Wynn, R. J. Fischer, Y. H. Wang, B. E. Weir, J. Gamelin, and A. Y. Cho, Electron. Lett. **26**, 1628 (1990).
44. M. Nakao, K. Sato. T. Nishida, T. Tamamura, A. Ozawa, Y. Saito, I. Okada, and H. Yoshihara, Electron. Lett. **25**, 148 (1989); see also, M. Nakao, K. Sato, T. Nishida, and T. Tamamura, IEEE J. Select. Areas Commun. **8**, 1178 (1990).
45. C. E. Zah, K. W. Cheung, S. G. Menocal, R. Bhat, M. Z. Iqbal, F. Favire, N. C. Andreadakis, P. S. D. Lin, A. S. Gozdz, M. A. Koza, and T. P. Lee, Optical Fiber Commun. Conference, San Diego, CA, Feb. 1991, paper WB5.
45a. L. E. Eng, W. Yuen, K. Bacher, J. S. Harris, and C. J. Chang-Hasnain, IEEE J. Quantum Electron. Special Issue on Semiconductor Lasers, June 1995.
45b. F. Koyoma, T. Mukaihara, Y. Hayashi, N. Ohnoki, N. Hatori, and K. Iga, Electron. Lett., November 1994.
45c. T. Wipiejewski, M. G. Peters, and L. A. Coldren, LEOS Annual Meeting, Boston, November 1994.
46. D. I. Babic, Y. Chung, N. Dagli, and J. E. Bowers, IEEE J. Quantum Electron. **29**, 1950 (1993).
47. T. Baba, Y. Yogo, K. Suzuki, F. Koyama, and K. Iga, Quantum Optoelectronics Conference, paper PD2-2, Palm Springs, CA (1993).
48. Z. L. Liau and D. E. Mull, Appl. Phys. Lett. **56**, 737 (1990).
49. Y. H. Lo, R. Bhat, D. M. Hwang, M. A. Koza, and T. P. Lee, Appl. Phys. Lett. **58**, 1961 (1991).
50. J. J. Dudley, D. I. Babic, R. Mirin, L. Yang, B. I. Miller, R. J. Ram, T. Reynolds, E. L. Hu and J. E. Bowers, Appl. Phys. Lett. **64**, 1463 (1994).
51. R. P. Schneider and J. Lott, Appl. Phys. Lett. **63**, 917 (1993).
52. D. M. Kuchta, J. Gamelin, J. D. Walker, J. Lin, K. Y. Lau, J. Smith, M. Hong, and J. P. Mannaerts, Appl. Phys. Lett. **62**, 1194 (1993).
53. M. S. Wu, L. A. Buckman, G. S. Li, K. Y. Lau, and C. J. Chang-Hasnain, International Semiconductor Lasers Conference, Maui, Hawaii, September 1994.

CHAPTER 6

Visible Semiconductor Lasers

Gen-ichi Hatakoshi

Materials and Devices Research Laboratories, Research and Development Center, Toshiba Corporation, 1, Komukai Toshiba-cho, Saiwai-ku, Kawasaki 210, Japan

6.1. INTRODUCTION

Short-wavelength compact light sources are deemed crucial for the development of optical information processing systems such as bar-code scanners, laser printers, and high-density optical disk memory systems. For example, small spot size is required in order to increase the recording density in optical disk systems. The use of short-wavelength light enables us to focus the laser beam into a small spot, because the focused spot diameter is proportional to the wavelength. Another advantage of shortening the wavelength is in the visibility of the output beam. The adjustment of an optical system becomes remarkably easy using such visible light sources. Visible semiconductor lasers can also be used as a direct substitute for conventional gas lasers in a variety of applications, wherein semiconductor lasers have the characteristic advantages of small size, low cost, low power consumption, and high reliability.

Several materials have been intensively investigated for obtaining short-wavelength laser diodes oscillating in the red light region. Continuous-wave (cw) oscillations in the 0.6 μm wavelength range at room temperature have been achieved by GaAlAs,[1] InGaAsP,[2,3] and InGaAlP[4-6] material systems. The InGaAlP quaternary alloy is the most promising material for 0.6 μm-band laser diodes, because it offers a large direct band-gap energy and can also form double heterostructures with complete lattice matching to a GaAs substrate. The problem in realizing InGaAlP lasers was that it was very difficult to obtain high-quality crystals by the conventional liquid phase epitaxy (LPE) methods because of the large segregation coefficient of aluminum.[7] The development of molecular beam epitaxy (MBE)[8-10] and metalorganic chemical vapor deposition (MOCVD)[4-6,11-13] techniques has enabled the production of high-quality thin-film crystals which can be used to form double heterostructures.

Room-temperature cw operation of InGaAlP lasers was first achieved in 1985.[4-6] Since then, much effort has been expended on the improvement of the laser's performance. Transverse-mode stabilization, threshold-current reduction, high-temperature operation, and high reliability are required for the practical applications of InGaAlP lasers. The development of InGaAlP laser diodes has continued towards high-power and shorter wavelength operation. The possibility of high-power operation makes the laser commercially attractive especially in optical disk applications. Semiconductor lasers operating at shorter wavelengths have the advantages of producing a small focused spot size and having greater relative luminosity for the human eye.

The problem in realizing such a high-power or a short-wavelength InGaAlP laser diode is that it is difficult to obtain such laser operation at high temperatures. The deterioration in the temperature characteristics arises from electron leakage from the active layer to the p-type cladding layer. Several methods such as high doping of the p-type cladding layer, strained active layer, off-angle substrate, multiquantum well (MQW) active layer, and the multiquantum barrier (MQB) structure have been developed to improve the laser performance. Using these techniques, the maximum operating temperature has been significantly increased resulting in the realization of highly reliable high-power and short-wavelength lasers.

6.2. CHARACTERISTICS OF THE InGaAlP DOUBLE HETEROSTRUCTURES

6.2.1. Material Parameters for the InGaAlP System

Characterization of the material system is required for designing semiconductor laser devices. Many studies have been carried out to determine the InGaAlP material parameters such as band-gap energy, band offset, and refractive index. The InGaAlP material system has a characteristic feature wherein the band-gap energy value depends on the growth conditions, for example, on the growth temperature.[11,14] This is attributed to the dependence of atomic ordering on the growth temperature.[15,16] An ordered structure of an InGaAlP alloy is produced by the formation of a natural superlattice.[15-19]

Measurements of the band-gap energy have been reported for InGaAlP alloys lattice matched to GaAs as a function of the Al composition ratio.[8,13,15,20] The band structures have also been investigated using the capacitance-voltage profiling method.[21,22] Band offset measurements have been reported for InGaP/GaAs[23-25] and InGaP/InGaAlP.[26] The refractive index dependence on the photon energy has been measured for InGaP[27] and InGaAlP.[9,28] These parameters are fundamental for designing a double heterostructure.

6.2.2. Gain-Current Characteristics

A linear relation between the gain g and the injected current density J, as in other materials, gives a good approximation for double heterostructure lasers with the InGaAlP system. In this approximation, g is represented by[29]

$$g = \beta(\eta_i J/d - J_0), \tag{6.1}$$

where η_i is the internal quantum efficiency below the threshold, d is the active layer thickness, and β and J_0 are the gain coefficients. At the oscillation threshold, g is equal to the sum of the internal loss and the cavity mirror loss. Thus, the threshold current density J_{th} is represented as follows:

$$J_{th} = \frac{d}{\eta_i} \left\{ \frac{1}{\beta \Gamma} \left[\alpha + \frac{1}{2L} \ln \left(\frac{1}{R_1 R_2} \right) \right] + J_0 \right\}, \tag{6.2}$$

where Γ is the optical confinement factor, α is the total internal loss, L is the cavity length, and R_1 and R_2 are reflectivities for the front facet and rear facet, respectively.

In InGaAlP laser diodes with an InGaP active layer, the gain coefficients β and J_0 have been obtained experimentally by the measurement of the cavity length dependence of the light-output versus current characteristics.[30] The experimental results are shown in Fig. 6.1. It can be seen that both β and J_0 have an exponential dependence on the temperature, as represented by the following equations:

$$\beta(T) = \beta^{(0)} \exp(-[T-293]/T_0), \tag{6.3}$$

$$J_0(T) = J_0^{(0)} \exp([T-293]/T_0), \tag{6.4}$$

where $\beta^{(0)}$ and $J_0^{(0)}$ are the values of β and J_0, respectively, at 20 °C (293 K), and are given by

$$\beta^{(0)} = 2.27 \times 10^{-2} \text{ cm } \mu m/A, \tag{6.5}$$

$$J_0^{(0)} = 4470 \text{ A/(cm}^2 \text{ } \mu m). \tag{6.6}$$

These values agree well with the results obtained by theoretical calculations.[31] Note that β and J_0 have the same temperature dependence represented by the characteristic temperature of T_0, which is given by

$$T_0 = 200 \text{ K.} \tag{6.7}$$

Thus, the temperature dependence of the threshold current density can be represented by

$$J_{th} \propto \exp(T/T_0). \tag{6.8}$$

The above T_0 value of 200 K is consistent with the theoretically calculated results.[31,32]

6.2.3. Effect of Carrier Overflow

The apparent characteristic temperatures in actual devices are, in general, lower than the value given by Eq. (6.7). This is attributed to carrier overflow from the

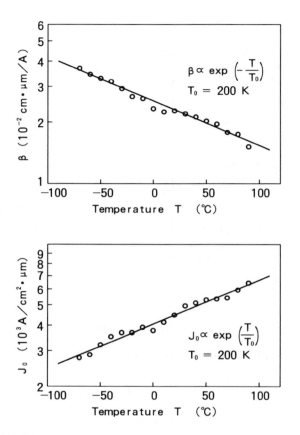

FIGURE 6.1. Measured gain coefficient as a function of temperature.

active layer to the p-cladding layer.[30,32-36] Carrier overflow is an important factor limiting high-temperature and/or short-wavelength oscillation of InGaAlP lasers.

In an InGaAlP double heterostructure, the small conduction-band heterobarrier causes a carrier overflow (see Fig. 6.2), which restricts both short-wavelength and high-power operation.[34,35] The barrier height, δ_P in Fig. 6.2, can be determined by the band-gap difference ΔE_g between the active layer and the cladding layer and the quasi-Fermi level Φ_P for the p-cladding layer, and is given as follows:

$$\delta_P = \Delta E_g - \Phi_P - (\phi_n + \phi_p), \tag{6.9}$$

$$\Delta E_g = E_g^{clad} - E_g^{act}. \tag{6.10}$$

The parameters ϕ_n and ϕ_p in Eq. (6.9) are the quasi-Fermi levels for electron and hole in the active layer, respectively, and are related to the injected carrier densities n and p by the following equations.

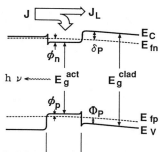

FIGURE 6.2. Band diagram for InGaAlP double heterostructure.

$$n = N_C F_{1/2}[\, \phi_n/(kT)\,], \tag{6.11}$$

$$p = N_V F_{1/2}[\, \phi_p/(kT)\,], \tag{6.12}$$

where N_C and N_V are the effective densities of states for the conduction band and valence band, respectively, and $F_{1/2}[x]$ represents the Fermi–Dirac integral.

The value of Φ_P in Eq. (6.9) is determined by the acceptor concentration p in the p-cladding layer, and can be decreased by increasing p. This means that carrier overflow can be reduced by using a highly doped p-cladding layer. Figure 6.3 shows calculated examples of band diagrams and carrier distributions representing the effect of p-cladding doping.[35] The results shown in this figure were obtained by a self-consistent device simulator for laser diodes.[37] The cases for low doping (a) and high doping (b) for the p-cladding layer are shown in the figure. In the lower figures, the solid and the dashed lines represent the carrier distributions for electrons and holes, respectively. As shown in Fig. 6.3(a), electron overflow from the active layer to the p-cladding layer is seen to occur for the case of low acceptor concentration. The carrier depletion at the interface of the active layer and the p-cladding layer is due to a valence-band discontinuity. A large heterobarrier height is obtained in the case of a highly doped p-cladding layer, as shown in Fig. 6.3(b). In this case, the electron overflow is seen to be significantly reduced.

6.2.4. Current-Voltage Characteristics

A characteristic feature of an InGaAlP/GaAs heterojunction system is that the valence-band discontinuity between InGaAlP and GaAs is very large.[21] This property should be taken into account because GaAs is usually used as a substrate or a contact layer in InGaAlP lasers, as shown in Fig. 6.4. The band discontinuity causes a large spike in the valence-band profile at the interface between p-InGaAlP and p-GaAs, which acts as a barrier for the hole carriers and has a significant effect on the current-voltage (I–V) characteristics of the heterojunction.[38] This effect depends on the Al composition and acceptor concentration of the InGaAlP layer. A relatively

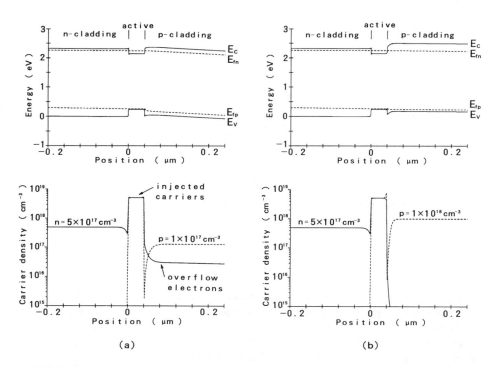

FIGURE 6.3. Calculated examples of the band diagrams and carrier distributions. (a) $p = 1\times10^{17}$ cm^{-3}, (b) $p = 1\times10^{18}$ cm^{-3}.

high voltage drop occurs at the p-InGaAlP/p-GaAs interface in the case of a high Al composition InGaAlP layer, such as that used for the cladding layers of InGaAlP lasers.

Figure 6.5 shows the dependence of the I–V characteristics on the acceptor

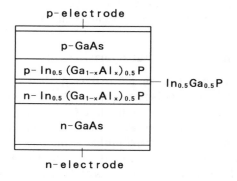

FIGURE 6.4. InGaAlP double heterostructure.

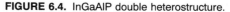

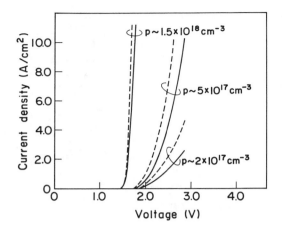

FIGURE 6.5. Dependence of the current density vs voltage characteristics on acceptor concentration in the InGaAlP layer for the structure shown in Fig. 6.4.

concentration of the $In_{0.5}(Ga_{0.3}Al_{0.7})_{0.5}P$ layer for the structure shown in Fig. 6.4.[38] The solid and dashed lines represent the experimental results and theoretical calculation, respectively. In the calculation, the band-gap energy values of $In_{0.5}(Ga_{1-x}Al_x)_{0.5}P$ for the Γ and X valleys are given, respectively, by[8,21,39]

$$E_g^{\Gamma} = 1.9 + 0.6x \quad (eV), \qquad (6.13)$$

$$E_g^{X} = 2.25 + 0.1x \quad (eV). \qquad (6.14)$$

The band offset values, used in the calculation, of $In_{0.5}(Ga_{1-x}Al_x)_{0.5}P$ relative to GaAs are given by[21]

$$\Delta E_C^{\Gamma} = 0.19 + 0.27x \quad (eV), \qquad (6.15)$$

$$\Delta E_V = 0.30 + 0.32x \quad (eV), \qquad (6.16)$$

for the conduction band and valence band, respectively.

The large magnitude of the applied voltage required for current injection, in the case of the low acceptor concentration shown in Fig. 6.5, is attributed to the large valence-band discontinuity ΔE_V, as given by Eq. (6.16). The voltage drop decreases with an increase in the acceptor concentration, as shown in Fig. 6.5. It can be seen that the experimental results (solid lines) agree well with the theoretically calculated results (dashed lines).

The heterobarrier height is reduced by introducing a p-InGaP layer between the p-InGaAlP and p-GaAs layers. InGaP has an intermediate band-gap energy between InGaAlP and GaAs, and thus both the band offset values for InGaAlP/ InGaP and InGaP/GaAs interfaces are smaller than that for InGaAlP/GaAs. There-

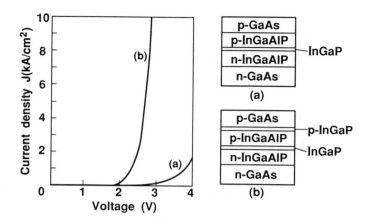

FIGURE 6.6. Comparison of the current density vs voltage characteristics for two kinds of structures.

fore, the voltage drop can be reduced by using a p-InGaAlP/p-InGaP/p-GaAs structure. Figure 6.6 shows a comparison of the calculated I–V characteristics for two kinds of structures shown on the right.[38,40,41] A remarkable reduction in the voltage drop is obtained by using structure (b), where a p-InGaP layer is introduced between the p-InGaAlP and p-GaAs layers.

The I–V characteristics of a $p–p$ isotype InGaAlP/GaAs heterojunction shows various features, which should be taken into account in device design. For example, the ohmic characteristics are markedly affected in the case of a low acceptor concentration. Figure 6.7 shows some calculated examples of the I–V characteristics and differential resistance characteristics for a p-InGaP/p-GaAs heterojunction. In the case of a relatively high acceptor concentration ($p = 5 \times 10^{17}$ cm^{-3}) in the p-InGaP layer, a normal ohmic characteristic is obtained. The I–V characteristics show a linear dependence, and the differential resistance is almost constant independent of the current density. On the other hand, the I–V relation is nonlinear and nonsymmetrical with respect to the bias direction in the case of a low acceptor concentration ($p = 1 \times 10^{17}$ cm^{-3}). The differential resistance also shows a nonsymmetrical relation. These results indicate the importance of p-type doping for the InGaAlP material system.

6.2.5. Thermal Effects on the Device Characteristics

InGaAlP materials have a relatively large thermal resistivity, which has a significant effect on the device performance. The large thermal resistivity of the InGaAlP cladding layer causes a temperature rise in the active layer resulting in an increase

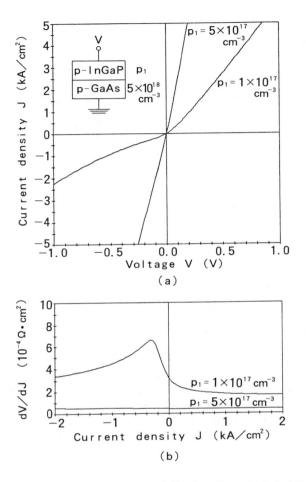

FIGURE 6.7. Calculated examples of the current density-voltage characteristics and the differential resistance characteristics for a p-InGaP/p-GaAs heterojunction.

in the threshold current. Therefore, it is necessary to optimize the cladding layer thickness, as well as the cavity length and heat sink material, in order to reduce the thermal resistance of the device.

Figure 6.8 shows a calculated example of the temperature rise in the active layer, at the oscillation threshold, as a function of the cladding layer thickness H[42] for gain-guided InGaAlP lasers with an inner stripe structure.[43,44] The temperature rise in the active layer causes an increase in the threshold current of the device. Therefore, it is important to minimize this temperature rise. In an InGaAlP laser, the cladding layer thickness H has a large effect on the temperature rise ΔT. As shown in Fig. 6.8, ΔT increases as H increases. This is due to the increase in the thermal resistance and the generation of heat in the p-cladding layer. The increase in ΔT for

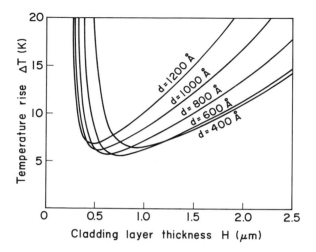

FIGURE 6.8. Calculated example of the temperature rise in the active layer, at the oscillation threshold, as a function of cladding layer thickness H.

a small H region is due to the absorption loss of the waveguide mode. Thus, it is clear that an optimum value exists for the cladding layer thickness, which minimizes the temperature rise in the active layer. The optimum thickness depends on the active layer thickness, as shown in Fig. 6.8. The cladding layer thickness should be designed to have an optimum value, especially for high power operation.

The thermal resistance also depends on the stripe width,[45,46] which is an important parameter in designing high-power InGaAlP lasers with broad-stripe structures. Figure 6.9 shows a calculated example of the thermal resistance R_{th} for InGaAlP lasers as a function of the stripe width W and the cavity length L for the case of a

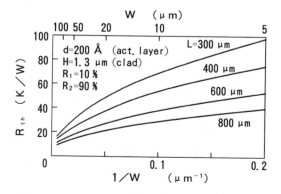

FIGURE 6.9. Calculated example of the thermal resistance R_{th} of InGaAlP lasers as a function of stripe width W and cavity length L.

junction-side-down configuration.[46] The thermal resistances were calculated by considering the heat generation in the active layer, cladding layers, substrate, and contact layer.[42] The thermal resistance of a broad-stripe laser is nearly proportional to the inverse of the stripe width and the inverse of the cavity length. However, its linearity is not complete, as shown in the figure. The slope of the curve for R_{th} versus $1/W$ decreases in the narrow stripe region, because the heat generated in the active layer can be conducted to the heat sink after spreading at laterally through the laser structure and the substrate before passing through the contact layer. On the other hand, such a heat spreading mechanism is more restricted in a broad-stripe laser, and the heat flow becomes almost one dimensional. This causes the nonlinearity of the curve shown in Fig. 6.9. Another factor affecting the deviation from $1/W$ dependence is the offset of the thermal resistance, i.e., the curve in Fig. 6.9 does not meet the origin at an infinite stripe width. This offset is due to the thermal resistance at regions far from the active layer, where the heat is spread widely and the effect of the stripe width is very small. This offset in the thermal resistance and the nonlinearity of the curve have significant effects on the temperature characteristics of broad-stripe lasers.

The temperature characteristics of a laser diode can be described by a simple model, where the output power P, injected current I, threshold current I_{th} at a junction temperature T_j are related by the following equations:

$$P = a \eta_d V_j [I - I_{th}(T_j)],$$
(6.17)

$$I_{th}(T_j) = I_{th\,0} \exp(T_j/T_0),$$
(6.18)

$$T_j = T + R_{th}(IV_j - P/a),$$
(6.19)

where V_j is the junction voltage, η_d is the external quantum efficiency, T is the ambient temperature, T_0 is the characteristic temperature, and R_{th} is the thermal resistance of the device. The factor a in Eq. (6.17) is a coefficient representing the ratio of the front facet power to the total output power. Using the above equations, the maximum operating temperature T_{max} and the maximum output power P_{max} are represented as follows:[46]

$$T_{max} = T_0 \left[\ln \left(\frac{T_0}{R_{th} I_{th\,0} V_j} \right) - 1 \right],$$
(6.20)

$$P_{max} = \frac{a \eta_d}{R_{th}(1 - \eta_d)} (T_{max} - T).$$
(6.21)

Figure 6.10(a) shows a calculated example of the maximum operating temperature as a function of the stripe width W.[46] As shown in the figure, T_{max} decreases with increasing W. This is due to the nonlinearity in the R_{th} vs $1/W$ relation shown in Fig. 6.9. This dependence of T_{max} on W has been confirmed by the experiment, as shown in Fig. 6.10(b).[46]

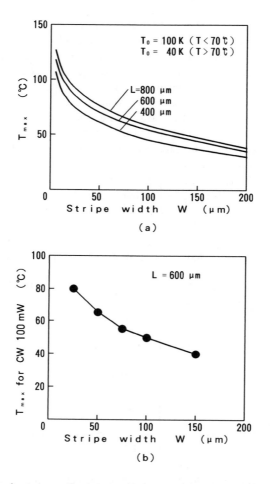

FIGURE 6.10. Maximum operating temperature as a function of stripe width *W*. (a) Calculated results. (b) Experimental results.

The above results show that a narrow stripe laser is preferable for high-temperature operation. However, its maximum output power is restricted by catastrophic optical damage (COD). In a broad-stripe laser, which is preferable for high-power operation, the maximum operation temperature is restricted by the thermal effect. Therefore, the stripe width should be optimized in order to obtain both high power and high temperature operation.

6.3. DEVICE STRUCTURES AND CHARACTERISTICS OF InGaAlP LASER DIODES

Examples of the device structures used in InGaAlP laser diodes are shown in Fig. 6.11. These structures are fabricated employing metalorganic chemical vapor depo-

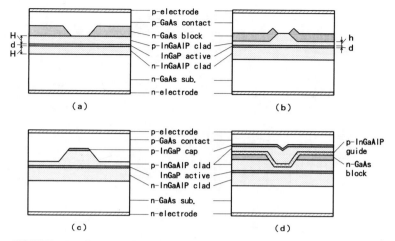

FIGURE 6.11. Examples of device structures used in InGaAlP laser diodes.

sition (MOCVD). Figure 6.11(a) shows a typical gain-guided InGaAlP laser and Figs. 6.11(b)–6.11(d) show transverse-mode stabilized lasers. The gain-guided and transverse-mode stabilized lasers are used in different application fields, depending on which particular laser characteristics are required.

6.3.1. Optical Properties of Gain-Guided and Transverse-Mode Stabilized InGaAlP Lasers

In the laser shown in Fig. 6.11(a),[43,44] the injected current is confined within the stripe region by the n-GaAs current-blocking layer, and an optical mode is formed by the gain distribution produced by the current. The laser shown in Fig. 6.11(b) is fabricated by a three-step MOCVD growth, which includes the formation of an n-GaAs current confining layer utilizing a selective growth on a ridge-shaped double heterostructure.[47] This structure is called a selectively buried-ridge waveguide (SBR) laser. In this laser, the n-GaAs current confining layer also acts as a light absorbing layer, which forms an effective refractive index step along the junction plane. Thus, the optical mode is confined within the ridge stripe region by this refractive index step.

The application field of the gain-guided laser is somewhat restricted because of its multimode oscillation and large astigmatism. However, this laser has the advantage of having a simple structure, which can be fabricated by a relatively simple process, and high reliability. On the other hand, the advantage of the transverse-mode stabilized laser is its small astigmatism, as compared with the gain-guided laser. The astigmatism of laser diodes depends on the optical mode profile, which in turn is determined by the laser structure. Astigmatism and other optical properties of semiconductor lasers are very important in practical applications such as light sources for laser printers and optical disk systems.

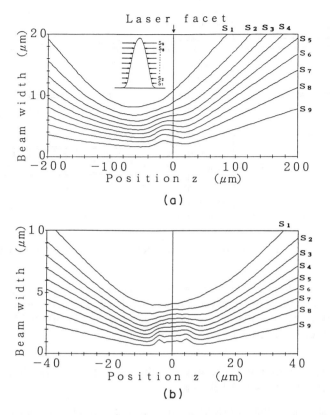

FIGURE 6.12. Calculated beam widths near the laser facet, for (a) gain-guided and (b) SBR lasers.

The optical properties of laser diodes are determined by the optical mode profiles at the laser facet. In gain-guided lasers, the gain distribution in the stripe region influences the phase profile of the optical mode and has a large effect on the astigmatism. On the other hand, the phase change in the stripe region of an SBR laser is smaller than that of a gain-guided laser. Therefore, the astigmatism of an SBR laser is expected to be smaller than that of a gain-guided laser. Figure 6.12 shows the calculated beam widths near the laser facet, for (a) gain-guided and (b) SBR lasers.[41,48] The beam widths inside the laser cavity, which are represented in the negative z region, are virtual width values calculated by the Fresnel–Kirchhoff diffraction formula, and correspond to those for the focused image of the output beam. The curves in this figure represent the approximate normal of the wavefront in the direction parallel to the junction plane. The beam waist position, in the perpendicular direction, is concurrent to the laser facet. Therefore, the distance between the laser facet and the position of the minimum beam width, in Fig. 6.12, corresponds to the astigmatism value. Note that the horizontal scales are different

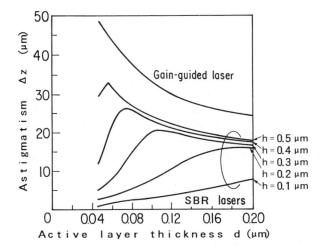

FIGURE 6.13. Calculated astigmatism Δz as a function of d and h.

between (a) and (b). It is thus clear that an SBR laser has a smaller astigmatism than a gain-guided laser.

The astigmatism value is affected by the dimensions of the device configuration, for example, the active layer thickness d and the distance h between the active layer and the n-GaAs absorbing layer in the region outside the stripe. Figure 6.13 shows the calculated astigmatism as a function of d and h.[41,48] In an SBR laser, the astigmatism increases when the values of d and h are large. This is because the effective refractive-index difference ΔN_{eff} between the stripe and the outside region decreases. A decrease in ΔN_{eff} for a large d is due to an increase in the optical confinement factor of the active layer. When the optical mode is strongly confined within the active layer, the influence of the n-GaAs light-absorbing layer decreases and thus the subsequent loss-guiding effect becomes small.

The relation between astigmatism and the active layer thickness for a gain-guided laser is opposite to that for an SBR laser, as shown in Fig. 6.13. This difference originates in the relation between the gain profile and the active layer thickness. In gain-guided lasers, the current spreading width depends on the threshold current density, which is determined by the active layer thickness. Consequently, the astigmatism determined by the gain distribution depends on the active layer thickness d. The astigmatism of the SBR laser with a large h value shows a similar dependence on d, as shown in Fig. 6.13. This is because the effective refractive-index difference ΔN_{eff} for an SBR structure with large h is almost zero and the optical mode is formed essentially by the gain-guiding mechanism.

6.3.2. Heterobarrier Blocking InGaAlP Lasers

The large valence-band discontinuity ΔE_V between InGaAlP and GaAs, described in Sec. 6.2.4, can be utilized to form a current confinement structure. Figure 6.11(c) shows the transverse-mode stabilized structure of an InGaAlP laser utilizing the heterobarrier effect for current blocking purposes.[40] This laser is thus designated a heterobarrier-blocking (HBB) laser. As shown in Fig. 6.6, the current does not flow in the layer structure shown in Fig. 6.6(a), which corresponds to the region outside the stripe of the HBB laser. Conversely, a current easily flows in the stripe region which has a layer structure corresponding to that shown in Fig. 6.6(b).

The optical characteristics of an HBB laser are almost identical to those of an SBR laser, because their complex refractive-index configurations and the optical confinement mechanisms are essentially the same. With regard to the fabrication process, HBB lasers are simpler to grow than SBR lasers, because an HBB laser structure requires only a two-step MOCVD growth process. High-power operation[49] and short-wavelength oscillation[50] have been reported for lasers with an HBB structure as well as with an SBR structure.

6.3.3. Real-Index-Guided InGaAlP Lasers

The SBR and HBB lasers shown in Figs. 6.11(b) and 6.11(c) utilize the loss effect of the GaAs layer outside the stripe, for transverse mode stabilization. Thus, a stable fundamental-transverse mode oscillation can be realized for structures with relatively large ($\sim 5 \mu$m) stripe widths, because the higher order mode is cut off by the loss effect. A structure with a smaller stripe width is preferable for reducing the astigmatism value. However, it is difficult to reduce the stripe width of loss-guided lasers such as an SBR and HBB, because the threshold current increases with a decreasing stripe width due to the GaAs loss effect on the fundamental mode.

The structure shown in Fig. 6.11(d) is a real-index-guided InGaAlP laser with a slab-coupled waveguide configuration.[51] In this laser, the lateral mode confinement is accomplished by a change in the real part of the refractive index distribution. An optical guide layer which has a higher refractive index than that of the cladding layers is introduced sufficiently close to the active layer inside the stripe so that an effective refractive index change is created along the junction plane. The layers near the active layer have no loss in both the stripe and outside the stripe regions. Therefore, the stripe width can be reduced to a small value without increasing the threshold current. This also leads to a reduction in astigmatism. A small astigmatism value of less than 2 μm has been reported.[51]

6.3.4. Other Structures for InGaAlP Lasers

More complex laser structures used in the conventional semiconductor lasers have been examined to apply to InGaAlP lasers. New structures peculiar to the InGaAlP material systems have also been investigated.

Optically pumped operation of a distributed feedback (DFB) laser[52] and current-injected operation of a distributed Bragg reflector (DBR) laser[53] have been reported for short-wavelength InGaAlP lasers. Vertical cavity surface emitting lasers with GaAlAs/AlAs DBRs have also been reported.[54,55] Real-index-guided InGaAlP lasers have been realized by using a planar native-oxide buried-mesa structure.[56] Lateral leaky waveguide structures for fundamental transverse-mode stabilization have been demonstrated.[57] One-step MOCVD-grown index-guided InGaAlP lasers have been fabricated by utilizing simultaneous doping of Zn and Se on a stepped substrate.[58] The realization of these various structures, which has resulted from the improvements in the MOCVD growth technology, indicates the potential flexibility of device design for high-performance InGaAlP lasers.

6.4. HIGH-POWER OPERATION

High-power operation is required for various applications, especially in optical disk systems. The reduction of the active layer thickness is a simple and effective method for obtaining high-power operation. However, it causes a carrier overflow from the active layer to the p-cladding layer due to an increase in the injected carrier concentration required for laser oscillation. Such carrier overflow is enhanced at high temperatures. For practical application of high-power laser diodes, the temperature characteristics of the device are very important. High-power operation is required not only at room temperature but also at higher temperatures. Therefore, a device design for preventing carrier overflow is needed to realize high power and high-temperature operation.

6.4.1. Optical Power Density

Catastrophic optical damage (COD) of the laser facet is a dominant factor limiting the high-power operation of InGaAlP laser diodes. According to experimental data for InGaAlP laser diodes, the optical power density inside the laser facet, which induces COD, has been found to be in the range 1–5 MW/cm^2. The device structure for realizing a high-power operation should be designed so as to reduce the optical power density below the above figure.

The optical power density for the double-heterostructure is proportional to the following P_{d0} value:[35]

$$P_{d0} = \frac{\Gamma}{d}\left(\frac{1+R}{1-R}\right), \tag{6.22}$$

where d is the active layer thickness, Γ is the optical confinement factor, and R is the facet reflectivity. In the above equation, Γ/d gives the optical power density in the active layer for unit output power. The term $(1+R)/(1-R)$ represents the ratio of the optical power inside the laser facet to that outside. Therefore, P_{d0} represents the optical power density inside the cavity at the laser facet normalized by the output power emitted from the facet. Figure 6.14 shows the dependence of Γ, R, and P_{d0} on the active layer thickness d for a double heterostructure with an InGaP active layer and In$_{0.5}$(Ga$_{0.3}$Al$_{0.7}$)$_{0.5}$P cladding layers.[35] As shown in Fig. 6.14(c), P_{d0} is nearly proportional to d in the small d region. Thus, it is clear that the employment of a thin active layer has a large effect on the reduction in optical power

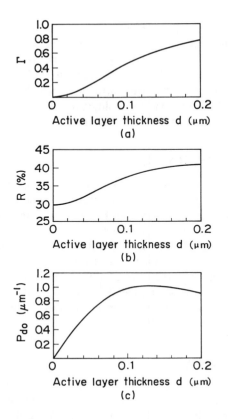

FIGURE 6.14. Dependence of Γ, R, and P_{d0} on active layer thickness d for a double hetero-structure with an InGaP active layer and $In_{0.5}(Ga_{0.3}Al_{0.7})_{0.5}P$ cladding layers.

density and thus on the prevention of COD at the laser facet.

6.4.2. Carrier Overflow Reduction

In a laser with a thin active layer, the optical confinement factor Γ is reduced as is shown in Fig. 6.14(a). Therefore, carrier concentration n_{th} required for laser oscillation increases, as understood from Eq. (6.2) and the relation between the carrier concentration n and the current density J, which is represented by the following equation:

$$J = \frac{qd}{\tau_n} n, \tag{6.23}$$

where τ_n is the carrier lifetime. An increase in the n_{th} value leads to an increase in $\phi_n + \phi_p$ [see Eqs. (6.11) and (6.12)] in the active layer, resulting in a reduction in

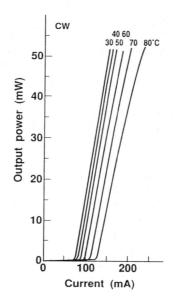

FIGURE 6.15. Example of the light-output vs current characteristics of a transverse-mode stabilized InGaAlP laser with a strained active layer.

the heterobarrier height δ_P, as represented in Eq. (6.9). This causes a carrier overflow and a deterioration of the temperature characteristics, as described in Sec. 6.2.3. Therefore, it is necessary to increase ΔE_g, or decrease Φ_P, in order to retain a sufficient barrier height δ_P.

A highly doped p-cladding layer has a large effect on decreasing Φ_P and thus on increasing the heterobarrier height, as shown in Fig. 6.3. An increase in the ΔE_g value can be realized by increasing the cladding layer bandgap and/or decreasing the active layer band gap. The band-gap energy values for the $In_{0.5}(Ga_{1-x}Al_x)_{0.5}P$ material system, lattice-matched to GaAs, are smallest for $x = 0$ and largest for $x = 0.7$ within the extent of the direct-transition region. Therefore, it has been difficult to increase the band-gap difference to more than that in the case for an $In_{0.5}Ga_{0.5}P$ active layer and $In_{0.5}(Ga_{0.3}Al_{0.7})_{0.5}P$ cladding layers. A compressively strained $In_{0.5+\delta}Ga_{0.5-\delta}P$ active layer[59-64] has been introduced to obtain a larger band-gap difference. Such a strained active layer can be employed for a thin active layer with a thickness less than the critical thickness value, which fortunately is also preferable for high-power operation. The employment of a strained active layer also reduces the threshold current density, due to a band structure change caused by the strained layer effect.[65,66]

Figure 6.15 shows an example of the light-output versus current characteristics of a transverse-mode stabilized InGaAlP laser fabricated by MOCVD employing a thin (15 nm) strained active layer.[46] The indium composition of the active layer was 0.62, which corresponds to a lattice mismatch value of 1%. The cavity length was

800 μm. The employment of such a long cavity has the effect of reducing $\phi_n + \phi_p$ and thus of reducing the carrier overflow.[35,60,64] It also has the effect of reducing the thermal resistance. The oscillation wavelength of this laser was 698 nm. The maximum cw output power exceeded 100 mW at room temperature and high-power operation at 50 mW was maintained even at a high temperature of 80 °C, as shown in Fig. 6.15.

Carrier overflow reduction and resulting improvements in the temperature characteristics for high-power InGaAlP lasers have also been achieved by using a multiquantum well (MQW) structure[67] and/or a multiquantum barrier (MQB) structure.[68] These improvements lead to high reliability of the device. In general, the device lifetime depends on the ambient temperature, output power, and operating current.[44] According to experimental data, these dependencies are consistent with a strong correlation between the device lifetime and operation current density.[60,64] The lifetime increases with decreasing operation current density. Therefore, carrier overflow reduction by the aforementioned methods is very effective to produce highly reliable high-temperature operation.

6.4.3. Window Structure

The employment of a window structure is an effective method for preventing COD and thus for obtaining high-power operation. In InGaAlP lasers, Zn diffusion is utilized to form a transparent window region at the output facet. Zn diffusion enhances the disordering of the InGaAlP natural superlattice and has the effect of increasing the band-gap energy in the Zn-diffused region. High-power operations using window structures have been reported.[68–70]

Fabrication processes for window structure lasers are somewhat complicated as compared to usual device structures because of the additional process to form the window region. A relatively simple process using a self-selective Zn diffusion[71] has been reported for realizing a window-structure in InGaAlP lasers.[70]

6.5. SHORT-WAVELENGTH OPERATION

The shortening of the oscillation wavelength requires a larger band-gap active layer. As the band-gap energy of the InGaAlP cladding layer has a limit determined by the material system lattice matched to the GaAs substrate, the band-gap difference between the active and cladding layers becomes small with shortening wavelength. This causes an enhancement of carrier overflow, resulting in a significant deterioration of the temperature characteristics. Therefore, a reduction in the carrier overflow is required, as in the case of high-power operation, to obtain a short-wavelength InGaAlP laser operating at and above room temperature.

6.5.1. Methods for Short-Wavelength Operation

Various methods for improving the temperature characteristics of short-wavelength InGaAlP lasers have been developed. A highly doped p-type cladding layer has a marked effect on the prevention of carrier overflow and thus on the improvement of the temperature characteristics, as described in Secs. 6.2.3 and 6.4.2. Such a high doping effect, in the p-cladding layer, has been confirmed from experimental results. The maximum cw operation temperature has been improved by increasing the acceptor concentration of the p-cladding layer. InGaAlP lasers oscillating in the 630 nm band at the relatively high temperature of 50 °C have been realized by this method.[34,50,72]

An off-angle substrate and a multiquantum barrier (MQB) structure are also effective for reducing the carrier overflow. Threshold current reduction by employing a multiquantum well (MQW) and/or a strained active layer is also a requirement for short-wavelength operation.

An increase in the Al composition of the active layer, the simplest method for shortening the oscillation wavelength, causes a substantial increase in the threshold current density due to an increasingly large nonradiative recombination. The employment of an MQW structure, an off-angle substrate, and a tensile strained active layer are effective for obtaining a large band-gap active layer with a small Al composition.

The MQW structure has been used in InGaAlP lasers from the early stages of their development.[73-75] Short-wavelength InGaAlP lasers with a low threshold current have been realized by employing an MQW. The shortest wavelength cw oscillation so far achieved, of 555 nm at 77 K, was realized by using an InGaP/InGaAlP MQW structure.[76] The combination of an MQW structure with other methods such as a strained active layer, an off-angle substrate, and an MQB structure has enabled further improvements in the performance of InGaAlP lasers. Well-designed structures and high-quality crystal growth for the MQW layers obtained by the MOCVD method have led to the realization of highly reliable InGaAlP MQW lasers.[77-80]

6.5.2. Effects of an Off-Angle Substrate

The employment of an off-angle substrate has various advantages for improving the laser characteristics. One is an increase in the band-gap energy which results from the suppression of the ordering in InGaAlP crystals grown on a misoriented substrate.[81-86] Figure 6.16 shows the photoluminescence (PL) peak energy of InGaP as a function of the substrate tilt angle from (100) orientation.[84] As the substrate is tilted towards the [011] direction, the energy increases with increasing tilt angle. Therefore, a large band-gap active layer can be obtained by using a smaller Al composition material. The band-gap of the cladding layers can also be increased. Thus, electron overflow can be effectively suppressed due to the larger band-gap difference for a set wavelength.

A high acceptor concentration in the p-type cladding layer is also obtained by using an off-angle substrate.[84,85,87] According to experimental results, Zn incorporation and the net acceptor concentration strongly depend on the tilt angle of the substrate. This dependence is similar to that for the PL peak energy. Both the Zn

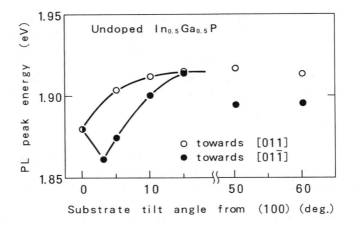

FIGURE 6.16. PL peak energy of InGaP as a function of substrate tilt angle from the (100) orientation.

concentration and the net acceptor concentration increase with increasing the tilt angle from (100) towards the [011] direction.[84]

Experimental results indicating an improvement in the heterointerface properties of quantum wells grown on off-angle substrates have been reported.[85,88] Full width at half-maximum (FWHM) value of the PL spectra for InGaP/InGaAlP single quantum wells has shown strong dependence on the substrate misorientation. The FWHM value decreased with increasing misorientation from (100) towards the [011] direction.[88] This result is considered to indicate that the interface smoothness and abruptness are improved by employing off-angle substrates.

Another effect of the off-angle substrate is the reduction of deep levels, which are related to the residual oxygen concentration. Experimental data have been reported showing a reduction in the oxygen concentration by using an off-angle substrate.[89]

A remarkable improvement in the temperature characteristics of InGaAlP lasers has been achieved by employing an off-angle substrate technique. Short-wavelength and high-temperature operations have been reported for InGaAlP lasers grown on misoriented substrates.[78,85,86,88,90,91]

6.5.3. Multiquantum Barrier Structure

The concept of a multiquantum barrier (MQB) was first proposed by Iga[92] and has been applied to short-wavelength InGaAlP lasers by several groups.[93–101] The MQB is a semiconductor multilayer structure which acts as a barrier for electrons. The overflow electrons from the active layer to the p-cladding layer are reflected by the effect of electron interference in the multiquantum barrier region. Theoretical calcu-

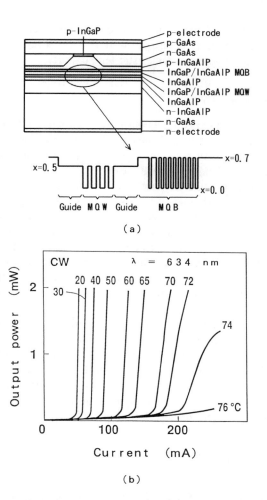

FIGURE 6.17. (a) Transverse-mode stabilized InGaAlP laser with an MQB structure. (b) Light-output vs current characteristics of the InGaAlP MQB laser with the structure shown in (a) and grown on a 15°-off substrate.

lations of the electron reflection spectra have shown that the MQB has a high reflectivity even for electrons with an energy higher than the barrier height of the p-cladding layer.[92,93,99,101]

Figure 6.17(a) shows an example of a transverse-mode stabilized InGaAlP laser with an MQB structure. In this example, ten wells and barriers were introduced into the p-type cladding layer. The active layer consisted of an MQW structure with four well layers. Figure 6.17(b) shows the light-output versus current characteristics of an InGaAlP MQB laser with the structure shown in Fig. 6.17(a) grown on a 15°-off substrate.[101] The oscillation wavelength was 634 nm and a maximum operation temperature as high as 74 °C was obtained. Experimental results comparing lasers with and without an MQB structure showed that the MQB laser had a lower threshold current and a higher characteristic temperature.

6.5.4. Tensile Strain Effect

The employment of a strained active layer has a remarkable effect on the reduction of the threshold current density and thus on the improvement of the temperature characteristics. The benefits of both compressive and tensile strain effects have been confirmed for short-wavelength InGaAlP lasers.[76–78,97,98,100,102–111] For the application to short-wavelength InGaAlP lasers, a tensile strained active layer has an advantage wherein the shortening of the oscillation wavelength is easily realized. The tensile strain effect brings about an increase in the quantum energy level without reducing the well width or increasing the Al fraction of the wells. Therefore, a relatively large optical confinement, and thus a high gain, is achieved for short-wavelength InGaAlP lasers. A reduction in the Al fraction of the wells is also preferable for short-wavelength operation because the problem of the increase in the threshold current, due to nonradiative centers related to the presence of Al, is reduced.

Several experimental results for short-wavelength InGaAlP lasers with tensile strained quantum wells are reported.[98,100,104–111] Remarkable effects of threshold current reduction and improvements in the temperature characteristics have been confirmed for InGaAlP lasers oscillating in the 630 nm band. The strained QW technique and other techniques described previously are very effective for realizing reliable 630 nm band InGaAlP lasers, which can take the place of conventional He-Ne gas lasers in a variety of applications.

ACKNOWLEDGMENTS

The author would like to thank M. Okajima, K. Nitta, M. Ishikawa, K. Itaya, H. Sugawara, Y. Nishikawa, M. Suzuki, J. Rennie, and M. Watanabe for providing a part of the figures for this chapter and for their useful discussions. He would also like to thank M. Azuma and Y. Uematsu for their support.

REFERENCES

1. S. Yamamoto, H. Hayashi, T. Hayakawa, N. Miyauchi, S. Yano, and T. Hijikata, Appl. Phys. Lett. **41**, 796 (1982).
2. A. Usui, T. Matsumoto, M. Inai, I. Mito, K. Kobayashi, and H. Watanabe, Jpn. J. Appl. Phys. **24**, L163 (1985).
3. T. Chong and K. Kishino, IEEE Photon. Technol. Lett. **2**, 91 (1990).
4. K. Kobayashi, S. Kawata, A. Gomyo, I. Hino, and T. Suzuki, Electron. Lett. **21**, 931 (1985).
5. M. Ikeda, Y. Mori, H. Sato, K. Kaneko, and N. Watanabe, Appl. Phys. Lett. **47**, 1027 (1985).
6. M. Ishikawa, Y. Ohba, H. Sugawara, M. Yamamoto, and T. Nakanisi, Appl. Phys. Lett. **48**, 207 (1986).
7. M. Kazumura, I. Ohta, and I. Teramoto, Jpn. J. Appl. Phys. **22**, 654 (1983).

8. H. Asahi, Y. Kawamura, and H. Nagai, J. Appl. Phys. **53**, 4928 (1982).
9. H. Tanaka, Y. Kawamura, S. Nojima, K. Wakita, and H. Asahi, J. Appl. Phys. **61**, 1713 (1987).
10. T. Hayakawa, K. Takahashi, M. Hosoda, S. Yamamoto, and T. Hijikata, Jpn. J. Appl. Phys. **27**, L1553 (1988).
11. Y. Ohba, M. Ishikawa, H. Sugawara, M. Yamamoto, and T. Nakanisi, J. Cryst. Growth **77**, 374 (1986).
12. K. Kobayashi, I. Hino, A. Gomyo, S. Kawata, and T. Suzuki, IEEE J. Quantum Electron. **QE-23**, 704 (1987).
13. M. Ikeda, K. Nakano, Y. Mori, K. Kaneko, and N. Watanabe, J. Cryst. Growth **77**, 380 (1986).
14. A. Gomyo, K. Kobayashi, S. Kawata, I. Hino, T. Suzuki, and T. Yuasa, J. Cryst. Growth **77**, 367 (1986).
15. C. Nozaki, Y. Ohba, H. Sugawara, S. Yasuami, and T. Nakanisi, J. Cryst. Growth **93**, 406 (1988).
16. T. Suzuki, A. Gomyo, S. Iijima, K. Kobayashi, S. Kawata, I. Hino, and T. Yuasa, Jpn. J. Appl. Phys. **27**, 2098 (1988).
17. T. Suzuki, A. Gomyo, and S. Iijima, J. Cryst. Growth **93**, 396 (1988).
18. E. Morita, M. Ikeda, O. Kumagai, and K. Kaneko, Appl. Phys. Lett. **53**, 2164 (1988).
19. O. Ueda, M. Takechi, and J. Komeno, Appl. Phys. Lett. **54**, 2312 (1989).
20. M. Kondow, H. Kakibayashi, and S. Minagawa, J. Cryst. Growth **93**, 412 (1988).
21. M. O. Watanabe and Y. Ohba, Appl. Phys. Lett. **50**, 906 (1987).
22. M. A. Rao, E. J. Caine, H. Croemer, S. I. Long, and D. I. Babic, J. Appl. Phys. **61**, 643 (1987).
23. T. Kobayashi, K. Taira, F. Nakamura, and H. Kawai, J. Appl. Phys. **65**, 4898 (1989).
24. D. Biswas, N. Debbar, P. Bhattacharya, M. Razeghi, M. Defour, and F. Omnes, Appl. Phys. Lett. **56**, 833 (1990).
25. J. Chen, J. R. Sites, I. L. Spain, M. J. Hafich, and Y. Robinson, Appl. Phys. Lett. **58**, 744 (1991).
26. C. T. H. F. Liedenbaum, A. Valster, A. L. G. J. Severence, and G. W.'t Hooft, Appl. Phys. Lett. **57**, 2698 (1990).
27. J. S. Roberts, G. B. Scott, and J. P. Gowers, J. Appl. Phys. **52**, 4018 (1981).
28. H. Tanaka, Y. Kawamura, and H. Asahi, J. Appl. Phys. **59**, 985 (1986).
29. F. Stern, IEEE J. Quantum Electron. **QE-9**, 290 (1973).
30. M. Ishikawa, H. Shiozawa, K. Itaya, G. Hatakoshi, and Y. Uematsu, IEEE J. Quantum Electron. **27**, 23 (1991).
31. P. Rees, H. D. Summers, and P. Blood, Appl. Phys. Lett. **59**, 3521 (1991).
32. K. Kishino, A. Kikuchi, Y. Kaneko, and I. Nomura, Appl. Phys. Lett. **58**, 1822 (1991).
33. S. H. Hagen, A. Valster, M. J. B. Boermans, and J. van der Heyden, Appl. Phys. Lett. **57**, 2291 (1990).
34. G. Hatakoshi, K. Itaya, M. Ishikawa, M. Okajima, and Y. Uematsu, IEEE J. Quantum Electron. **27**, 1476 (1991).
35. G. Hatakoshi, K. Nitta, K. Itaya, Y. Nishikawa, M. Ishikawa, and M. Okajima, Jpn. J. Appl. Phys. **31**, 501 (1992).
36. D. P. Bour, D. W. Treat, R. L. Thornton, R. S. Geels, and D. F. Welch, IEEE J. Quantum Electron. **29**, 1337 (1993).
37. G. Hatakoshi, M. Kurata, E. Iwasawa, and N. Motegi, Trans. IEICE Jpn. **E71**, 923 (1988).
38. K. Itaya, M. Ishikawa, and G. Hatakoshi, Jpn. J. Appl. Phys. **32**, 1919 (1993).
39. M. Honda, M. Ikeda, Y. Mori, K. Kaneko, and N. Watanabe, Jpn. J. Appl. Phys. **24**, L187 (1985).
40. K. Itaya, M. Ishikawa, Y. Watanabe, K. Nitta, G. Hatakoshi, and Y. Uematsu, Jpn. J. Appl. Phys. **27**, L2414 (1989).
41. G. Hatakoshi and Y. Uematsu, Int. J. Optoelectronics **7**, 359 (1992).
42. G. Hatakoshi, M. Suzuki, N. Motegi, M. Ishikawa, and Y. Uematsu, Trans. IEICE Jpn. **E71**, 315 (1988).
43. H. Okuda, M. Ishikawa, H. Shiozawa, Y. Watanabe, K. Itaya, K. Nitta, G. Hatakoshi, Y. Kokubun, and Y. Uematsu, IEEE J. Quantum Electron. **25**, 1477 (1989).
44. M. Ishikawa, H. Okuda, K. Itaya, H. Shiozawa, and Y. Uematsu, Jpn. J. Appl. Phys. **28**, 1615 (1989).
45. O. J. F. Martin, G. L. Bona, and P. Wolf, IEEE J. Quantum Electron. **28**, 2582 (1992).
46. G. Hatakoshi, K. Nitta, Y. Nishikawa, K. Itaya, and M. Okajima, Proc. SPIE **1850**, 388 (1993).
47. M. Ishikawa, Y. Ohba, Y. Watanabe, H. Nagasaka, H. Sugawara, M. Yamamoto, and G. Hatakoshi, in *Extended Abstracts of the 18th Conference on Solid State Devices and Materials* (Japan Society of Applied Physics, Tokyo, 1986), pp. 153–156.

48. K. Nitta, K. Itaya, M. Ishikawa, Y. Watanabe, G. Hatakoshi, and Y. Uematsu, Jpn. J. Appl. Phys. **28**, L2089 (1989).
49. K. Itaya, Y. Watanabe, M. Ishikawa, G. Hatakoshi, and Y. Uematsu, Appl. Phys. Lett. **56**, 1718 (1990).
50. K. Itaya, M. Ishikawa, and Y. Uematsu, Electron. Lett. **26**, 839 (1990).
51. M. Okajima, Y. Watanabe, Y. Nishikawa, K. Itaya, G. Hatakoshi, and Y. Uematsu, IEEE J. Quantum Electron. **27**, 1491 (1991).
52. M. Korn, T. Koerfer, A. Forchel, and P. Roentgen, Electron. Lett. **26**, 614 (1990).
53. D. H. Jang, Y. Kaneko, and K. Kishino, Electron. Lett. **28**, 428 (1992).
54. J. A. Lott and R. P. Schneider, Jr., Electron. Lett. **29**, 830 (1993).
55. K. Tai, K. F. Huang, C. C. Wu, and J. D. Wynn, Electron. Lett. **29**, 1314 (1993).
56. F. A. Kish, S. J. Caracci, S. A. Maranowski, N. Holonyak, Jr., K. C. Hsieh, C. P. Cuo, R. M. Fletcher, T. D. Osentowski, and M. G. Craford, J. Appl. Phys. **71**, 2521 (1992).
57. I. Kidoguchi, S. Kamiyama, M. Mannoh, Y. Ban, and K. Ohnaka, Appl. Phys. Lett. **62**, 2602 (1993).
58. C. Anayama, H. Sekiguchi, M. Kondo, H. Sudo, T. Fukushima, A. Furuya, and T. Tanahashi, Appl. Phys. Lett. **63**, 1736 (1993).
59. K. Nitta, K. Itaya, Y. Nishikawa, M. Ishikawa, M. Okajima, and G. Hatakoshi, Appl. Phys. Lett. **59**, 149 (1991).
60. K. Nitta, K. Itaya, Y. Nishikawa, M. Ishikawa, M. Okajima, and G. Hatakoshi, Jpn. J. Appl. Phys. **30**, 3862 (1991).
61. H. B. Serreze, Y. C. Chen, and R. G. Waters, Appl. Phys. Lett. **58**, 2464 (1991).
62. T. Katsuyama, I. Yoshida, I. Shinkai, J. Hashimoto, and H. Hayashi, Appl. Phys. Lett. **59**, 3351 (1991).
63. D. F. Welch and D. R. Scifres, Electron. Lett. **27**, 1915 (1991).
64. K. Nitta, M. Okajima, Y. Nishikawa, K. Itaya, and G. Hatakoshi, Electron. Lett. **28**, 1069 (1992).
65. J. Hashimoto, T. Katsuyama, J. Shinkai, I. Yoshida, and H. Hayashi, Appl. Phys. Lett. **58**, 879 (1991).
66. A. Valater, C. J. van der Poel, M. N. Finke, and M. J. B. Boermans, Conf. Dig. of 13th IEEE Int. Semiconductor Laser Conf., (IEEE, New York, 1992), paper G-1.
67. Y. Ueno, H. Fujii, H. Sawano, K. Kobayashi, K. Hara, A. Gomyo, and K. Endo, IEEE J. Quantum Electron. **29**, 1851 (1993).
68. S. Arimoto, M. Yasuda, A. Shima, K. Kadoiwa, T. Kamizato, H. Watanabe, E. Omura, M. Aiga, K. Ikeda, and S. Mitsui, IEEE J. Quantum Electron. **29**, 1874 (1993).
69. Y. Ueno, H. Fujii, K. Kobayashi, K. Endo, A. Gomyo, K. Hara, S. Kawata, T. Yuasa, and T. Suzuki, Jpn. J. Appl. Phys. **29**, L1666 (1990).
70. K. Itaya, M. Ishikawa, G. Hatakoshi, and Y. Uematsu, IEEE J. Quantum Electron. **27**, 1496 (1991).
71. M. Ishikawa, M. Suzuki, Y. Nishikawa, K. Itaya, G. Hatakoshi, Y. Kokubun, and Y. Uematsu, Inst. Phys. Conf. Ser. **106**, 575 (1989).
72. M. Ishikawa, H. Shiozawa, Y. Tsuburai, and Y. Uematsu, Electron. Lett. **26**, 211 (1990).
73. M. Ikeda, A. Toda, K. Nakano, Y. Mori, and N. Watanabe, Appl. Phys. Lett. **50**, 1033 (1987).
74. J. M. Dallasasse, D. W. Nam, D. G. Deppe, N. Holonyak, Jr., R. M. Fletcher, C. P. Kuo, T. D. Osentowski, and M. G. Craford, Appl. Phys. Lett. **53**, 1826 (1988).
75. S. Kawata, K. Kobayashi, H. Fujii, I. Hino, A. Gomyo, H. Hotta, and T. Suzuki, Electron. Lett. **24**, 1489 (1988).
76. A. Valater, M. N. Finke, M. J. B. Boermans, J. M. M. van der Heijden, C. J. G. R. Spreuwenberg, and C. T. H. F. Liedenbaum, Conf. Digest of 12th IEEE Int. Semiconductor Laser Conf., (IEEE, New York, 1990) paper PD-12.
77. A. Valster, C. J. van der Poel, M. N. Finke, and M. J. B. Boermans, Electron. Lett. **28**, 144 (1992).
78. S. Honda, H. Hamada, M. Shono, R. Hiroyama, K. Yodoshi, and T. Yamaguchi, Electron. Lett. **28**, 1365 (1992).
79. M. Watanabe, M. Okajima, K. Itaya, K. Nitta, Y. Nishikawa, and G. Hatakoshi, Jpn. J. Appl. Phys. **31**, L1399 (1992).
80. M. Watanabe, J. Rennie, M. Okajima, and G. Hatakoshi, Appl. Phys. Lett. **63**, 1486 (1993).
81. M. Ikeda, E. Morita, A. Toda, T. Yamamoto, and K. Kaneko, Electron. Lett. **24**, 1094 (1988).
82. S. Minagawa and M. Kondow, Electron. Lett. **25**, 758 (1989).
83. K. Kobayashi, Y. Ueno, H. Hotta, A. Gomyo, K. Tada, K. Hara, and T. Yuasa, Jpn. J. Appl. Phys. **29**, L1669 (1990).
84. M. Suzuki, Y. Nishikawa, M. Ishikawa, and Y. Kokubun, J. Cryst. Growth **113**, 127 (1991).

85. H. Hamada, M. Shono, S. Honda, R. Hiroyama, K. Yodoshi, and T. Yamaguchi, IEEE J. Quantum Electron. **27**, 1483 (1991).
86. A. Kikuchi and K. Kishino, Appl. Phys. Lett. **60**, 1046 (1992).
87. S. Minagawa and M. Kondow, Electron. Lett. **25**, 413 (1989).
88. M. Watanabe, J. Rennie, M. Okajima, and G. Hatakoshi, Electron. Lett. **29**, 250 (1993).
89. M. Suzuki, K. Itaya, Y. Nishikawa, H. Sugawara, and G. Hatakoshi, Inst. Phys. Conf. Ser. **129** 465 (1992).
90. M. Shono, H. Hamada, S. Honda, R. Hiroyama, K. Yodoshi, and T. Yamaguchi, Electron. Lett. **28**, 905 (1992).
91. T. Tanaka, H. Yanagisawa, S. Yano, and S. Minagawa, Electron. Lett. **29**, 24 (1993).
92. K. Iga, H. Uenohara, and F. Koyama, Electron. Lett. **22**, 1008 (1986).
93. T. Takagi, F. Koyama, and K. Iga, IEEE J. Quantum Electron. **27**, 1511 (1991).
94. T. Takagi, F. Koyama and K. Iga, Electron. Lett. **27**, 1081 (1991).
95. K. Kishino, A. Kikuchi, Y. Kaneko, and I. Nomura, Appl. Phys. Lett. **58**, 1822 (1991).
96. J. Rennie, M. Watanabe, M. Okajima, and G. Hatakoshi, Electron. Lett. **28**, 150 (1992).
97. H. Hamada, K. Tomonaga, M. Shono, S. Honda, K. Yodoshi, and T. Yamaguchi, Electron. Lett. **28**, 1834 (1992).
98. J. Rennie, M. Okajima, M. Watanabe, and G. Hatakoshi, Electron. Lett. **28**, 1950 (1992).
99. A. Furuya and H. Tanaka, IEEE J. Quantum Electron. **28**, 1977 (1992).
100. M. Shono, S. Honda, T. Ikegami, Y. Bessyo, R. Hiroyama, H. Kase, K. Yodoshi, T. Yamaguchi, and T. Niina, Electron. Lett. **29**, 1010 (1993).
101. J. Rennie, M. Okajima, M. Watanabe, and G. Hatakoshi, IEEE J. Quantum Electron. **29**, 1857 (1993).
102. C. J. Chang-Hasnain, R. Bhat, and M. A. Koza, Electron. Lett. **27**, 1553 (1991).
103. D. P. Bour, D. W. Treat, R. L. Thornton, R. D. Bringans, T. L. Paoli, R. S. Geels, and D. F. Welch, IEEE Photon. Technol. Lett. **4**, 1081 (1992).
104. D. F. Welch, T. Wang, and D. R. Scifres, Electron. Lett. **27**, 693 (1991).
105. D. P. Bour, D. W. Treat, R. L. Thornton, T. L. Paoli, R. D. Bringans, B. S. Krusor, R. S. Geels, D. F. Welch, and T. Y. Wang, Appl. Phys. Lett. **60**, 1927 (1992).
106. R. S. Geels, D. P. Bour, D. W. Treat, R. D. Bringans, D. F. Welch, and D. R. Scifres, Electron. Lett. **28**, 1043 (1992).
107. R. S. Geels, D. F. Welch, D. R. Scifres, D. P. Bour, D. W. Treat, and R. D. Bringans, Electron. Lett. **28**, 1810 (1992).
108. T. Tanaka, H. Yanagisawa, S. Yano, and S. Minagawa, Electron. Lett. **29**, 606 (1993).
109. T. Tanaka, H. Yanagisawa, M. Takimoto, S. Yano, and S. Minagawa, Electron. Lett. **29**, 722 (1993).
110. H. D. Summers and P. Blood, Electron. Lett. **29**, 1007 (1993).
111. M. Watanabe, J. Rennie, M. Okajima, and G. Hatakoshi, Appl. Phys. Lett. **63**, 1486 (1993).

CHAPTER 7

Semiconductor Lasers With Wide-Gap II–VI Materials

A. V. Nurmikko* and R. L. Gunshor[†]

*Division of Engineering and Department of Physics, Brown University,
Providence, Rhode Island
[†]School of Electrical Engineering, Purdue University,
West Lafayette, Indiana

7.1. INTRODUCTION

The entry of the wide band-gap II–VI semiconductors into the laser arena is a very recent and exciting one, with research results increasingly indicative of technological promise now emerging from laboratories. On the other hand, the field of wide band-gap II–VI semiconductors itself has a long history, extending at least to the early part of this century with the use of ZnO in a point contact diode. (ZnO, incidentally, is still used today as a binding compound in automobile tires, while transition metal-doped ZnS has long been used as a phosphor in cathode-ray tubes exploiting a high cathodoluminescence efficiency.) In the 1960s, interband optical spectroscopy of ZnSe, ZnTe, ZnS, CdSe, CdTe, and related mixed crystal compounds near their band-gap photon energies was instrumental in the study and discovery of the new exciton and polariton phenomena (that is, the solid state physicist's hydrogen atom, and its propagation in the crystalline lattice, respectively). On the other hand, while the device prospects for compact, directly electrically excited light emitters across the visible spectrum were recognized early, they remained frustrated for three decades due to the poor electrical and electronic properties of the bulk materials. While a large body of important basic work was accumulated during this period, the lack of applications caused many (industrial) research laboratories to drop work in the field, which to some extent retreated into relative academic obscurity. In contrast, research with narrow-gap materials, notably $Hg_{1-x}Cd_xTe$, succeeded in establishing this II–VI semiconductor as the prime material in infrared detector applications, notably in the important 10-μm wavelength region.

In the 1980s, a few research groups ventured into the wide-gap II–VIs through the new nonequilibrium epitaxial growth techniques, notably by molecular beam epitaxy (MBE), ushering in a "second coming" of the field.[1] Two main consequences of the nonequilibrium epitaxial approach to II–VI semiconductor synthesis were: (i) improvement of impurity and defect control, which had a profound impact in isolating and identifying reasons for the difficulty in achieving useful doping, and (ii) the realization of heterostructures such as quantum wells and superlattices. The study of optical physics in the quantum wells (QW) was soon underway, and a number of wide-gap II–VI QW and superlattice systems were detailed by optical spectroscopy within half a dozen years, with ZnSe, ZnTe, and CdTe as the primary starting compounds. It was soon realized that some of the QW choices would offer significant advantages for radiative recombination, partly due to the enhanced electron-hole overlap that influenced the interband transition rate, and partly due to the efficient carrier collection which reduced the detrimental effects of nonradiative traps in the bulklike portions of the heterostructures. These benefits were demonstrated, e.g., by Jeon et $al.$ in the room-temperature optical pumping of a (Zn,Cd)Se/ZnSe MQW laser.[2]

A major advance in the control of electrical properties occurred in 1990 when Park et $al.$[3] and Ohkawa et $al.$[4] independently showed how p-type doping of ZnSe could be accomplished by using nitrogen as the dopant element within the MBE growth, thereby paving the road for useful pn-junction heterostructures (for additional comments on doping, see Sec. 7.2). This development overcame a principal hurdle that work on bulk materials had simply been unable to surmount. For example, while ZnSe could be doped n type to provide a reasonable electron density (say, up to 10^{18} cm^{-3} and beyond), p-type doping was apparently "forbidden." Curiously, in the case of ZnTe, it had been the n-type doping that was practically impossible while p-type doping was "allowed." In seminal events, combining the pn junction and the new quantum well heterostructures, Haase et $al.$[5] demonstrated blue-green diode laser operation from a (Zn,Cd)Se/ZnSe/Zn(S,Se) device in 1991, which was also achieved independently by the authors' group.[6,7] To date, about a dozen research groups worldwide have succeeded in showing some form of blue-green diode laser operation under pulsed electrical injection, based on related and similar heterostructures which are discussed below. Threshold current densities below 1 kA/cm^2 at room temperatures have been realized, with threshold currents reaching down to the mA range for short-ridge waveguide structures. The reader should be aware of the fact that both the device physics and engineering of these new laser structures are subjects of intense current research efforts; hence the examples of device performance summarized below in this chapter will most likely require substantial revision on a time scale of one year. At this point, cw operation at room temperature has been achieved at bias conditions on the order of 5.5–6 V and 10 mA as discussed below; however, the rapid device degradation seen under these conditions now poses a major immediate technical challenge for the II–VI community from the standpoint of commercial viability. At this writing, microscopic origins of the device degradation have been elucidated, at least in part, and active research is producing potential solutions for the control of the underlying defects. On the other hand, the new short-wavelength lasers offer a rich opportunity to research device physics in a class of semiconductors which have their own fascinating idiosyncrasies.

In this chapter we employ illustrative examples to highlight the characteristics of these short-wavelength II–VI emitters. Since the other chapters in this book are devoted to III–V semiconductor lasers, we will try to draw the contrast with the II–VI compounds rather sharply, to emphasize their special properties which lead to practical difficulties but also to opportunities. From a fundamental point of view, ZnSe and related II–VI semiconductors differ from, e.g., GaAs in that their covalent bonding contains a distinctly ionic component. Consequently, a wide range of physical properties must be examined in their possible impact on diode laser design and technological prospects. For example, the strong coupling of electronic excitation to the lattice has an important influence on the scattering processes which enter into determining gain bandwidth as well as carrier mobilities. Likewise, the mechanical properties of the wide-gap II–VI semiconductors are such that the propagation of defects such as native vacancies and dislocations is expected to occur more readily than in a more purely covalent material. The issue of dislocations is very relevant inasmuch as lattice mismatch strain and its control is a major challenge for epitaxial growth in the multilayer II–VI diode lasers.

The chapter is organized as follows. In Sec. 7.2, we introduce the range of compounds from which current light-emitting structures are being fabricated, by considering the basic quantum well and heterojunction physics of ZnSe-based heterostructures. We focus chiefly on optical properties, but also address key aspects of electron and hole "vertical transport." Section 7.3 shows examples of current device characteristics and performance, both for LEDs and diode lasers. In Sec. 7.4 we consider the physical manifestation of gain in II–VI lasers in the context of spontaneous and stimulated emission by quasi-two-dimensional electron-hole pairs. Concluding remarks are made in Sec. 7.5. on the opportunities for future research and development in bringing the new blue-green emitters into the realm of practical devices.

The examples used in this chapter to illustrate the present status of the field provide, of necessity, only a limited sampling of an active field. In attempting to give some continuity to the subject matter within this limited space, we have sometimes had to omit references to numerous important contributions by colleagues. We apologize for such omissions.

7.2. QUANTUM WELL AND HETEROJUNCTION PHYSICS

7.2.1. Electronic Confinement and Interaction with Phonons

Figure 7.1 shows the standard design map for heteroepitaxy of II–VI compounds in relation to some cubic III–V and IV–VI semiconductors, by plotting the band gap (at room temperature) against the lattice constant. Both lattice matching and

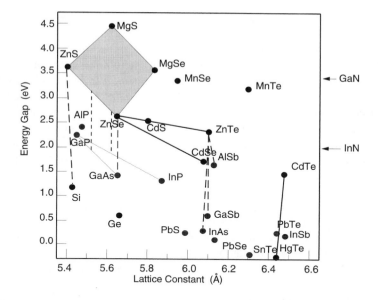

FIGURE 7.1. Energy band gap vs lattice constant for common II–VI, III–V, IV, and IV–VI semiconductors in the cubic phase. Solid lines indicate some of the ternary combinations and dashed lines suggest near lattice matching onto specific substrates. Note that the values for neither MgSe nor MgS have not been been experimentally determined in the cubic phase and they appear as extrapolations in the graph. Band gaps for the wide gap III–V (noncubic) GaN and InN are also shown.

quantum well prospects can be roughly appreciated from this map, although little insight into band offsets for the QWs (i.e., partitioning of the band-gap difference between the conduction and valence bands) is gained in the figure. For reference, band gaps for (noncubic) GaN and InN are also included, as these semiconductors are being actively researched as an alternative route to short-wavelength visible light emitters. At present, as there are no practically useful ZnSe-based substrates available (although several research and development efforts are underway), the epitaxy is performed predominantly on GaAs substrates, either directly or through the use of a GaAs buffer layer, including the choice of InGaAs for lattice matching to ZnSe.[8] However, the growth of ZnSe on InGaP buffer layers by gas source epitaxy has also been demonstrated quite recently,[9] thereby widening the potential scope for lattice-matched choices for the II–VI "superstructures." First efforts are also underway to improve and perfect ZnSe single-crystal substrates for epitaxy. Generally, we can expect much more exploration into the use of both III–V and II–VI (and perhaps other) semiconductor mixed crystal substrates in facilitating the exploration of an increasingly wide range of short visible and near UV light emitters.

To date, at least a dozen wide band-gap II–VI heterostructures have been investigated in terms of their electronic confinement. The primary objective is the determination of the conduction and valence band offsets, almost entirely by optical spectroscopy. Thus, for example, $CdTe/Cd_{1-x}Mn_xTe$ and $ZnSe/Zn_{1-x}Mn_xSe$ were determined early to be weakly of type I (i.e., with electrons strongly confined within the QW but holes only weakly so), whereas the ZnTe/ZnSe band alignment is of type II (electrons and holes confined in ZnSe and ZnTe, respectively).[1] As a striking example of the brute force of quantum confinement effects in a rather strong type-I QW, spectroscopy in the CdTe/MnTe system demonstrated how the interband transition energy, measured by means of low-temperature photoluminescence, could be increased from the bulk value of CdTe in the near infrared (for wide QWs) throughout the visible, into the blue, for well thicknesses as narrow as three monolayers (≈ 1 nm).[1]

In terms of the blue-green pn-junction light emitters which are discussed in this chapter, the most common choice for the quantum well (QW) has been $Zn_{1-x}Cd_xSe$, typically with Cd concentration of $x \approx 0.15-0.25$. This results in a lattice mismatch with ZnSe barrier layers on the order of 1%, and somewhat larger for ZnS_ySe_{1-y} with $y \leqslant 0.07$, allowing for an empirical critical thickness of about 10–20 nm. As discussed elsewhere in this volume (see Chap. 1), the critical thickness gives a measure of the maximum layer thickness that can be anticipated to accommodate the lattice mismatch strain by elastic distortion of the lattice, before the onset of strain relaxing dislocations. The common cation/anion argument (or "myth" in some quarters) in defining the band offsets is found not to be unreasonable in the wide-gap II–VIs, so that for the ZnCdSe/ZnSe junction, most of the band offset occurs in the conduction band as the cation Cd brings down the conduction band edge to define the QW layer. The useful type I nature of this QW system was first verified in magneto-optical experiments in which the diluted magnetic semiconductor (Zn,Mn)Se was used as the barrier layer material.[12] The biaxial lattice mismatch strain leads to a uniaxial component which enhances this confinement by splitting the heavy-light hole degeneracy so that the effective mass in the growth direction ($m_{h\perp}$) is heavy-hole-like, while the in-plane mass ($m_{h\parallel}$) and its dispersion resembles that of the light hole near $k \approx 0$ (see inset of Fig. 7.2). Consequently, the in-plane density of states is reduced from the bulk value for ZnCdSe, and aids in reducing the threshold inversion density required for gain in the conventional degenerate electron-hole plasma model. In that sense the ZnCdSe/ZnSe is quite analogous to the InGaAs/GaAs QW system. One distinction is that precise effective mass values for the p-type materials in our case are not accurately known. Indirectly, however, magneto-optical spectroscopy of the so-called 1S and 2S exciton states in the ZnCdSe/ZnSe QW has indicated the following in-plane effective mass values: $m_e = 0.17m_0$ and $m_{h\parallel} \approx 0.2m_0$.[10]

Optical spectroscopy of the (Zn,Cd)Se/ZnSe QWs has also yielded approximate values for the band offsets from the comparison of observed interband transition energies (either from absorption or luminescence). For the Cd concentration $x \approx 0.24$, for example, one has the following values: $\Delta E_c \approx 180$ meV and $\Delta E_v \approx 60$ meV, where the valence band well is mainly induced by the strain component.[10] Comparable values have also been obtained by other authors.[11]

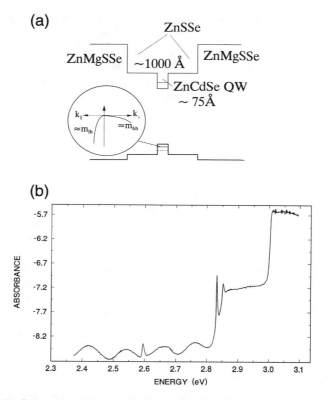

FIGURE 7.2. (a) Schematic of the conduction and valence bands in the SCH diode laser heterostructure composed of a ZnCdSe QW, ZnSSe optical/electrical confinement layer, and the ZnMgSSe outer cladding. (b) Absorbance of a typical laser laser structure (in direction of growth) at $T = 10$ K, showing the relationship between the (excitonic) energy band gaps of the constituent materials. The low-energy oscillations are due to interference effects.

Replacing the ZnSe barriers by ZnS_ySe_{1-y} deepens the valence band further, increasing the QW confinement. In addition, for $y \approx 0.07$, a lattice match with a GaAs substrate is obtained at the growth temperature, with a pseudomorphic structure remaining to cryogenic temperatures. While precise values for the valence band offset of the ZnCdSe/ZnSSe QW depend on the details of the strain distribution, the valence band probably deepens to about 100 meV.[13] In this case, one is in the regime where the energy separation between the lowest heavy-hole subbands in the quantum well E_{h1-h2} is on the order of kT at room temperature (the $n = 1$ heavy-light hole separation is split by the strain by at least 50 meV). This implies a sizable population of the second-hole subband in a laser structure; not an optimal circumstance. The situation has an impact both on the gain and leakage currents expected in a laser diode and calls for further "bandgap engineering" for optimizing the carrier confinement.

Quite recently, the extension of the MBE growth into the II–VI quaternaries has made it possible to realize pseudomorphic heterostructures which are particularly useful for diode lasers. In this case one uses an element from the column IIa, Mg, to increase the band gap further in the compound $Zn_{1-x}Mg_xS_ySe_{1-y}$ (the elements Zn and Cd are column IIb cations). The group at Sony Laboratories first showed how the ternary $Zn_{1-x}Mg_xSe$ could be used as a QW barrier material in a ZnSe QW diode laser.[14] The choice of a quaternary compound allows much flexibility in the joint control of the band gap and the lattice constant. With the inclusion of $Zn_{1-x}Mg_xS_ySe_{1-y}$, a *pseudomorphic* separate confinement heterostructure (SCH) for a blue-green diode laser becomes possible, as indicated schematically in the energy band diagram of Fig. 7.2(a) and by the shaded area in Fig. 7.1. The group at Philips Laboratories (Briarcliff Manor, NY) obtained room-temperature threshold current densities with short electrical pulse injection ($\tau_p \approx 50$ ns) below 500 A/cm^2 from an SCH device,[15] which was the lowest of any reported in the II–VI lasers to date. This basic structure, with some modifications including a particular electrical contacting scheme, was also employed independently in the authors' laboratories to demonstrate room-temperature high duty cycle operation and for recent characterization studies;[16,17] very similar structures were also realized by the groups at 3M (St. Paul, MN) and at Sony Laboratories.[18] Further discussion of the design and performance of these lasers is deferred to Sec. 7.3, including the achievement of room-temperature cw operation. Figure 7.2(b) displays the relationship of the band-gap energies between the ZnCdSe QW, ZnSSe optical waveguide layer, and the ZnMgSSe outer barrier layer, respectively, as measured in optical absorption of a *pn*-junction device structure at $T = 10$ K.[18] At this temperature both the thin QW layer and the ZnSSe layer show a well-defined excitonic absorption peak (at energy $h\nu = E_g - E_x$, where E_x is the exciton binding energy), which attests to the crystal quality. The excitonic band edge of the quaternary is also sharp, but the thick layers ($\sim 2\ \mu$m total) attenuate the optical probe so heavily that the peak itself cannot be seen in the figure. While the concept of a SCH-QW diode laser is, of course, not new in itself, it should be emphasized that its pseudomorphic implementation in the wide-gap II–VIs necessarily requires a quaternary (see Fig. 7.1). This makes the epitaxy more challenging since (unlike, e.g., the III–V materials) none of the II–VI elements have unity sticking coefficient in the MBE growth, so that compositional control is inherently more difficult with an increasing number of constituent atoms. Undoubtedly, other group IIa alkaline earth elements will also be investigated as possible future ingredients in wide-gap heterostructures for alternative material possibilities, especially with the need to shorten the emission wavelengths further.

An illustration of the present quaternary material quality within a SCH-SQW pseudomorphic diode laser structure from the authors laboratories is shown in Fig. 7.3,[16] where the x-ray rocking curves highlight the $Zn_{0.91}Mg_{0.09}S_{0.12}Se_{0.88}$ cladding layers in (a). The full width at half maximum of the (400) diffraction peaks from the quaternary layers of the laser structure as narrow as 44 arcsec were obtained, a very satisfactory result. [The weak features on the high angle side of the GaAs buffer layer peak are attributed to the Zn(S,Se) layers.] Figure 7.3(b) shows the layered details of the entire multilayer structure in a cross-sectional bright field transmission

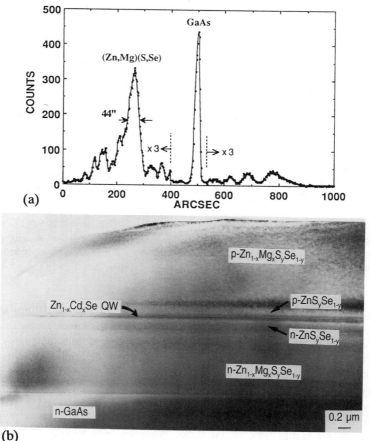

FIGURE 7.3. (a) X-ray rocking curves from a SCH laser diode structure of Fig. 2(b), showing the quality of the (thick) quaternary layers. (b) The cross-sectional bright-field TEM image indicating the absence of significant threading/misfit dislocations and stacking faults (after Ref. 15).

electron microscope image, including the ZnCdSe QW and the ZnSSe optical waveguide layers. On this spatial scale, the absence of threading/misfit dislocations and stacking faults is also qualitatively consistent with the pseudomorphic condition.

One consequence of the polar nature of ZnSe-based II–VI semiconductors is their very strong coupling with the longitudinal optical (LO) phonons via the Frohlich interaction. The LO-phonon interaction presents major consequences to a room-temperature light emitter: (a) broadening of optical linewidths, and (b) decrease of the carrier mobility, especially for holes. The former is illustrated in Fig. 7.4, which shows the experimental linewidth of the exciton resonance in a ZnCdSe/

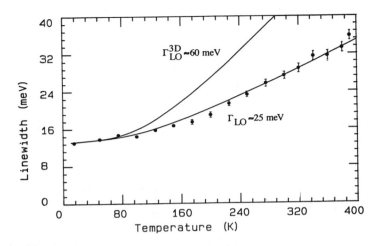

FIGURE 7.4. Linewidth of the $n = 1$ HH exciton resonance in a ZnCdSe/ZnSe QW as a function of temperature. The temperature independent portion is due to an inhomogeneously broadened resonance [after Ref. 10].

ZnSe QW (data points), as a function of temperature.[10] The solid lines are fits to simple theory of LO-phonon interaction in the 3D and 2D cases, respectively, with the effective strength of the coupling (Γ_{LO}) expressed in units of energy. At low temperatures the linewidth is of inhomogeneous origin, determined by alloy potential fluctuations within ZnCdSe (an intrinsic property) and QW thickness variations (extrinsic). At elevated temperatures, the Frohlich interaction (scattering of the electron-hole pair by the phonons) manifests itself by a pronounced increase in the exciton linewidth, to approximately $\Gamma \sim 30$ meV at room temperature. It should be noted that to reach this "low" value (when compared with bulk ZnSe), one already takes advantage of the exciton QW confinement whereupon the exciton binding energy increases to a value exceeding the LO-phonon energy. Elsewhere, in p-ZnSe, mobility measurements have recently yielded values on the order of $\mu \approx 35$–39 cm^2/Vsec at $T = 300$ K,[19] and lower.[20] Such low values (even if containing extrinsic scattering contributions) underscore the severity of the polar mode (LO-phonon) scattering for holes when viewed through a transport experiment. (In unstrained bulk material, of course, the dominant hole state is the heavy mass one corresponding to the $J = 3/2$ state at $k = 0$.)

A pedagogical illustration of the impact of the interaction between LO phonons and e-h pairs on a QW laser is shown in Fig. 7.5, in which the calculated gain spectra in a simple electron-hole plasma (EHP) model is shown for a (Zn,Cd)Se/ Zn(S,Se) case at room temperature. In this calculation, where material parameters listed above have been used, the gain spectrum is obtained with a convolution of the population difference (Fermi factors) with a Lorentzian lineshape broadening: $g(\omega) \sim \int (f_c - f_v)(\Gamma/[(\omega - \omega_0)^2 - \Delta^2]) d\omega_0$. All the scattering is included in the linewidth broadening parameter Δ which is used as an empirical variable. The

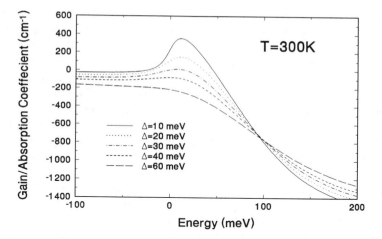

FIGURE 7.5. Gain spectrum calculated for a (Zn,Cd)Se/Zn(S,Se) QW at $T = 300$ K, demonstrating the influence of line width broadening (2Δ) due to electron-hole pair scattering. An injection level of 2×10^{12} cm^{-2} has been assumed.

assumed $e\text{-}h$ pair density in a $L_w = 10$ nm well has been taken as $n_s = 2\times10^{12}$ cm^{-2}, a value which is in the range of experimental conditions. In this example, note how the sharp 2D density of states edge is quickly eliminated and how for this injection density, no gain is available when Δ exceeds a value of about 20 meV. In fact, the simple convolution expression loses physical validity as Δ becomes comparable or exceeds the Fermi energies in the problem, so that Fig. 7.5 represents only a schematic illustration (a more rigorous examination of the linewidth broadening must also call into examination the Lorentzian lineshape itself[21]). The exciton linewidth information from Fig. 7.4 gives a direct optical measurement about the value for Δ, which for simplicity has been here assumed as being energy independent. (Screening of the LO-phonon interaction is not yet significant at this injection level.) Clearly, then, the influence of LO-phonon scattering on diode laser gain must be considered in the wide-gap II–VI semiconductors; for comparison, the scattering rate (broadening) in GaAs or InGaAs is a factor of 6–8 less for comparable conditions. We return to this subject in Sec. 7.4.

7.2.2. Electrical Contacts and Vertical Transport

When comparing with the III–V, IV–VI, or group IV semiconductors, the historical difficulties of flexible doping of II–VI materials resulted in a very limited data base about transport properties. While much research is presently underway about the nature of (especially acceptor) impurity states in ZnSe and its alloys, there are still many incompletely understood issues related, for example, to vertical transport in the new blue-green diode laser heterostructures. The questions include the fabrication of ohmic contacts to p-ZnSe, transport within the pn-junction QW heterojunctions (such as ZnMgSSe/ZnSSe/ZnCdSe), and across the n-GaAs/n-ZnSe (or equivalent III–V/II–VI) heterovalent heterojunction. Complicating experimental analysis, problems of the quality of epitaxial material and interfaces, including aspects of defect formation and device degradation, add an extrinsic element to an equation which for the time being has many unknowns. Yet, the overall problem and present challenge might be stated in a simplified way as follows: the total vertical impedance in the present diode lasers is still higher than acceptable from a practical technological viewpoint. Until recently, voltages on the order of 10 V were required to reach laser thresholds, implying a device resistance on the order of tens of ohms. Significant progress has, however, occurred at this writing as detailed in the next section on recent progress.

In this section we focus on the problem of electrical (ohmic) contacts, since much recent work has been expended in this area, while leaving other aspects of vertical transport uncovered. Undoubtedly, much more information about the vertical transport will be forthcoming from ongoing research efforts. One common denominator which underlies the present difficulties in general is the still rather low level of p-type doping in the ZnSe-based materials. Useful hole concentrations on the order of 3×10^{17} cm^{-3} can now be routinely produced in ZnSe with nitrogen as the dopant of choice during MBE growth. This is a key factor in having made it possible to demonstrate the diode laser, but it is presently unclear whether such (moderate) concentrations can be maintained for the wider gap ZnSSe and ZnMgSSe compounds, since the acceptor binding energy for these is larger. In preliminary work, it has been found that the binding energy increases from approximately 92 meV in ZnSe to approximately 177 meV in ZnMgSSe (with a corresponding increase in band gap from about 2.8 to 3.0 eV).[22a] Fundamentally, the probability of acceptor atoms forming deep levels (with possible local distortion in the bonding configuration by interaction of the electronic and lattice states) is expected to increase with the polarity (band gap). For example, there is already evidence that such deeper, lattice-relaxed levels do occur in ZnMgSSe. In this article we will not review the large amount of ongoing work in this very important and active area, experimental and theoretical, where the aim is to improve the microscopic understanding of the acceptor states at an atomic level. The reader is referred, e.g., to the proceedings of the 1991 and 1993 Conferences of the II–VI Compound Semiconductors.[22b]

Obtaining a low-resistance contact between a metal and a wide band-gap semiconductor presents a fundamental dilemma, that of electrically bridging two materials that are vastly dissimilar in terms of their electronic band structures. The problem scales, again, in its degree of difficulty with the band gap of the semiconductor. In the case of ZnSe and its alloys, the relative ease of achieving heavy n-type doping (up to 10^{19} cm^{-3} and beyond), together with the position of the surface Fermi level, presents a relatively low barrier for electron injection when many common metals (such as indium) are used. Furthermore, in typical diode laser structures, electron injection occurs from n-GaAs into the n-type ZnSe-based material so that the problem is transformed to a question about the heterovalent heterointerface. Some potential energy discontinuity is present, perhaps up to 300 meV for GaAs/ZnSe (and a few hundred meV for GaAs/ZnMgSSe), that, together with interface traps, contributes to a finite heterojunction impedance (hence a voltage drop). While this is not the major hurdle at present, issues about the influence of the microstructure at the heterointerface, i.e., interface states, remain open and are the subject of ongoing research. The main challenges exist at the p-ZnSe/metal contact, where three contemporary approaches are being followed. These are as follows: (i) the growth by low-temperature epitaxy (< 200 °C) of a highly p-type ZnSe conducting contact layer by nitrogen doping,[23] (ii) the implementation of a graded p-ZnTe/ZnSe heterostructure to facilitate an ohmic contact,[24] and (iii) the deposition of a narrow band-gap semimetallic layer of p-HgSe.[25] In this chapter we highlight second approach used in the authors' laboratories, while noting that at this writing it is undetermined whether any of these contact schemes are sufficiently robust and offer low enough contact resistance to withstand the requirements of a *practical* (cw) blue-green diode laser device.

We have recently reported that the same approach which has been found effective for the p-doping of ZnSe and its alloys by nitrogen, can be used to obtain a very high degree of p-doping within the MBE growth of ZnTe.[26] Free-hole concentrations approaching 10^{19} cm^{-3} have been readily achieved. Since, e.g., palladium, platinum, and gold (and their multilayers) form a low-resistance contact to such highly doped p-ZnTe epilayers, one is naturally led to consider the use of ZnTe as an intermediate "electrical buffer" for contacting to p-ZnSe. However, the large valence band offset ($\Delta E_v \approx 1$ eV) between ZnTe and ZnSe layers forms a barrier to the hole injection at a p-ZnTe/p-ZnSe interface (ZnTe/ZnSe is a type-II QW). The consequence of this is illustrated clearly in the current voltage characteristics in Fig. 7.6(c) below. A possible solution for removing the expected energy barrier in the valence band (in reality a potential energy "spike" due to the electrostatic charge redistribution) is to introduce a graded bandgap p-ZnSe$_{1-y}$Te$_y$ layer in which the Te concentration varies from $y = 0$ to 1 across such a contact layer. Due to the practical difficulty of controlling the Te concentration in an MBE-grown Zn(Se,Te) alloy (selenium and tellurium compete for surface incorporation), a "pseudograded" p-type, strained ZnTe/ZnSe ultrathin layer structure (2 nm per cell) was designed and grown, with the ZnTe and ZnSe layer thicknesses in each cell varying to approximate a graded band-gap material.[24] Although the overall "superlattice" thickness (≈ 34 nm) exceeds substantially the critical thickness for the 7% lattice mismatch between ZnSe and ZnTe, electron microscopy shows that the dislocations in the contact layer do not propagate into the adjacent quaternary in a SCH-SQW diode laser structure. Results of conductivity measurements of p-ZnSe epilayers for three different contacting schemes consisting of (i) Au/pZnTe/p-Zn(Se,Te)/p-ZnSe, housing the graded pseudoalloy; (ii) direct Au/p-ZnSe; and (iii) an Au/p-ZnTe/p-ZnSe heterostructure are compared in Fig. 7.6. We note that a similar structure, also composed of a short period p-ZnSe/ZnTe superlattice has been implemented by Hiei *et al.*;[24a] however, these authors have interpreted their results in terms of resonant tunneling processes. As seen in Fig. 7.6(a), the graded

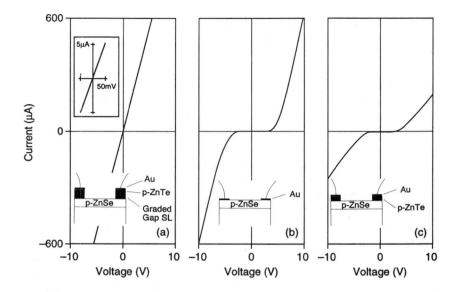

FIGURE 7.6. Current-voltage characteristics at room temperature of contact arrangements whose schematics are shown in the inset: (a) Au/Zn(Te,Se) graded gap layer/p-ZnSe, (b) Au/p-ZnSe, and (c) Au/p-ZnTe/p-ZnSe (after Ref. 24).

contact appears to be quite ohmic, showing a straight line through the origin. The inset shows that the I–V characteristics maintain the same slope even at a few mV from the origin. Figure 7.6(b) shows the characteristics of the contact formed by gold deposited onto an as-grown ZnSe:N epilayer. The I–V characteristic corresponds to two back-to-back Schottky diodes; the observed turn on voltage is the reverse bias breakdown voltage. Increasing the doping level is expected to reduce the "turn-on" voltage of the contact. As already mentioned above, Fig. 7.6(c) shows the I-V characteristic when p-ZnTe is used to inject holes into the ZnSe epilayer in the absence of the graded region (so that the ZnSe/ZnTe valence band offset forms the junction). The specific contact resistance for a structure such as (i) but with Au replaced by Pd was determined to be as low as 3×10^{-4} Ω cm^2, which can be considered acceptable with LEDs but is somewhat high for laser diode devices. One might well expect an improvement in performance of this contact scheme as the growth is modified to more closely approach a continuous alloy grading and as p-doping levels in the selenide layers are increased.

7.3. BLUE-GREEN LEDS AND DIODE LASERS

In this section we examine the device features and contemporary performance of the blue-green light emitters. It is very important for the reader to realize that at the present stage (early 1994), both the LEDs and the diode lasers are still being studied at research laboratories in terms of basic device design, and that neither yet represents a viable technology. While there is considerable optimism about the further advances that can be expected with both types of devices, the crucial issue of reliability is only beginning to be addressed. Continued improvement in underlying material quality and ability to control defects, electronic and mechanical, are examples of key issues in this context.

7.3.1. Light-Emitting Diodes

The conventional approach to LEDs is through the use of bulk or thick heterojunctions. Until recently, the only commercially available blue LEDs have been based on SiC, with electrical-to-optical efficiencies on the order of 10^{-4}. The devices are fabricated from bulk SiC which, as an indirect gap material, is unsuitable for a diode laser. Due to the indirect nature and the presence of a variety of crystal defects and polytypes, the emission from the SiC LEDs is spectrally quite broad (tens of nm). Within the past couple of years, epitaxially grown GaN has shown promise as an LED, and bright although spectrally broad (up to mW) pn-junction emitters at efficiencies of about 10^{-2} are now being produced in Japan.[27] The dominant radiative transition in the blue (≈ 450 nm) involves the Mg acceptor state in the p side of the junction, making it unlikely that a (bulk) epilayer configuration would lead to a diode laser. However, good progress in heterostructures is now also being made with AlN, GaN, and InN, so that the prospects of a diode laser demonstration are rapidly improving. In many ways, the wide-gap nitrides represent a selection of semiconductors complementary to the II–VI compounds for light emitters at short visible wavelengths.

In contrast to the current wide heterostructure designs in nitride-based LEDs, the authors believe that a QW configuration is fundamentally important for an LED in the wide-gap II–VI compounds. This is based in part on the strong electron-hole pair Coulomb pairing effects which can enhance the radiative transitions in ZnSe-based quantum wells up to room temperature. In an LED, the basic requirement is for radiative recombination to exceed the competing nonradiative processes. Since in a narrow QW ($L_W \approx 5$ nm), the quasi-2D exciton binding energy (40 meV) can exceed the LO-phonon energy (≈ 31 meV), the dissociation of excitons into free e-h pairs is suppressed or much reduced by inelastic ionizing collisions with the phonons. Instead, the scattering leads to a population of exciton-excited states, which can effectively supply the Coulomb correlated e-h pairs for radiative recombination at the exciton wave vector $K = 0$ if nonradiative processes due to defects, etc., can be inhibited. Figure 7.7(a) shows the emission spectra of a ZnCdSe/ZnSe QW LED at $T = 77$ and 300 K, and compares these with the absorption spectrum of the lowest exciton resonance ($n = 1$ HH exciton).[28] The room-temperature emission peaks at $= 494$ nm. The spectral coincidence alone strongly suggests that the recombination involves excitons. Figure 7.7(b) shows a photograph of a seven-segment test display device in which the optical emission from the ZnCdSe QWs is retrieved through a transparent top electrode material.

An example of the temperature-dependent electron-hole pair radiative lifetime is

(a)

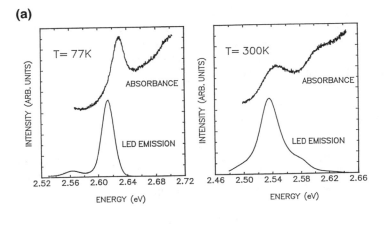

(b)

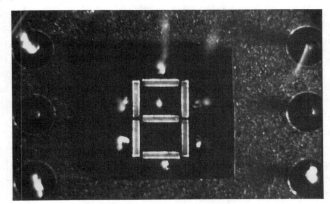

FIGURE 7.7. (a) Comparison of absorption and cw spectra near the $n = 1$ HH exciton resonance of a (Zn,Cd)Se/Zn(S,Se) QW LED at $T = 77$ K and $T = 300$ K (after Ref. 28). (b) Photograph of a seven segment numeric display, fabricated from the same material.

shown in Fig. 7.8, for the structure whose optical absorption was shown in Fig. 7.2(b). These time-resolved photoluminescence measurements indicate that the lifetime increases from approximately 180 ps at $T = 77$ K to nearly 900 ps at room temperature.[17] Estimates for the free e-h pair recombination process (in the effective mass approximation, and without Coulomb interaction) suggest an underlying radiative decay (oscillator strength) which is as much as one order of magnitude weaker than observed. In these experiments one must ensure sufficiently high levels of photoexcitation so that nonradiative recombination centers related, e.g., to impurities and other point defects are saturated. In fact, from a practical point of view,

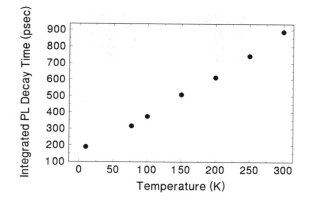

FIGURE 7.8. Photoluminescence decay time of the SCH-QW structure of Fig. 2(a), obtained from time-resolved spectroscopy at the spectrally integrated $n = 1$ QW transition (after Ref. 16).

the present situation is such that under the typically low level of injection that characterizes an LED, the radiative efficiency of the ZnSe-based QWs, while strikingly high at cryogenic temperatures (approaching 100% internal conversion), typically drops by one to two orders of magnitude by room temperature. This drop is presumably due to the presence of the miscellaneous point and other structural defects in the material which capture electron-hole pairs prior to their reaching the QW ground electronic states. Further material improvement is now being witnessed that has produced blue-green ZnCdSe/ZnSe QW LEDs at room temperature at quantum efficiencies on the order of 10^{-2}, for example at Sony Laboratories. At the same time, the application of, e.g., the graded band-gap contacting technique gives the LEDs a relatively low "turn-on" voltage of 2.5–3 V, so that once the room-temperature radiative efficiency improves further, the blue LEDs based on II–VI QWs should have good prospects for reaching the commercial marketplace. One contemporary development is the use of ZnSe substrates and homoepitaxy for higher brightness II–VI green LEDs, including efforts by Schetzina and co-workers at North Carolina State University (in which ZnSeTe forms the active region, with benefits as discussed next).

The strong coupling between electronic and lattice excitations in the wide-gap II–VI semiconductors can be exploited also to extend the wavelength range of the QW emitters in LED applications. The idea is based on the role of Te as an isoelectronic center in ZnSe, where trapping of excitons has been known to strongly red shift the emission wavelengths.[29] In our interpretation, small Te clusters initially provide a locally (atomic scale) attractive potential, chiefly for the holes. Upon capture, the interaction with the lattice leads to configurational coordinate changes so that the hole's self energy is effectively renormalized as much as 300 meV, depending on the details of the Te cluster. The practical implementation of the idea for QW light emitters is conveniently realized by imbedding monolayers of ZnTe within the ZnSe-based QWs, as depicted schematically in Fig. 7.9.[30] Recently, such

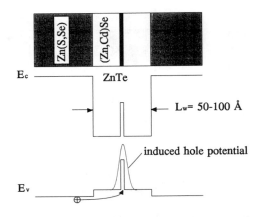

FIGURE 7.9. Schematic of planar isoelectronic doping of a (Zn,Cd)Se/ZnSe QW by ultrathin ZnTe "sheets." The self-trapping of the hole is depicted by deepening of the valence band local potential well.

pn-junction heterostructures have provided green and yellow-green emission,[31] thereby spanning the range for an LED from about 490 to 560 nm in structures containing no Te, and one and two monolayers of ZnTe within the QW, respectively. When coupled with the inclusion of Mn as a luminescent center, whose d-electron states can also effectively capture free e-h pairs and whose emission energy is influenced by the crystal field arrangement, both yellow and red emissions[32] can, in principle, be obtained from such pn-junction heterostructures. A real possibility therefore exists for multicolor LEDs over the entire visible range from the ZnSe-based heterostructures, which are compatible with multilayer growth within a single epitaxial chamber. Furthermore, due to the high density of Te isoelectronic centers which can be embedded in the active QW region, it may be possible to achieve diode laser operation from such structures as well.

Another potentially important approach to green LEDs has been recently proposed and reported in a design where a n-CdSe/p-ZnTe heterojunction, separated by a graded $Cd_{1-x}Mg_xSe$ injecting region allows low-voltage injection of electrons into the heavily doped p-ZnTe for radiative recombination.[33] This scheme, which circumvents the difficulty of realizing a ZnTe pn junction, is an example of the versatility that is offered by the "bandstructure engineering" techniques which are very likely to be increasingly widely implemented in the wide band-gap II–VI semiconductors. Such approaches may be especially relevant to LEDs which lack many of the additional constraints and structural limitations imposed in diode lasers.

7.3.2. Diode Lasers

A historical schematic which broadly summarizes the blue-green diode laser structures realized to date is sketched in Fig. 7.10, showing the evolution towards the pseudomorphic SCH-QW configuration [from (a) to (d)]. The advantages of such

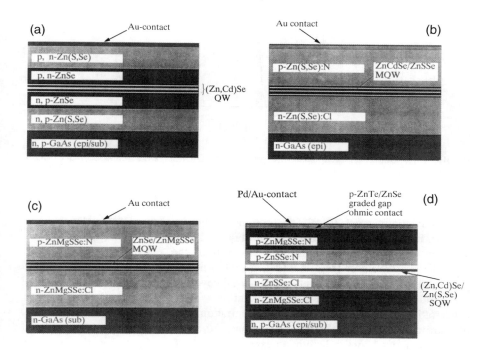

FIGURE 7.10. Schematic of blue-green diode laser structures in rough chronological order in their development since 1991, from (a) to (d).

joint electronic/optical confinement are, of course, very well known, but limitations of lattice matching constraints prevented its realization in the ZnSe-based emitters until the introduction of the quaternary components. Other pseudomorphic configurations such as the graded index design, which have been hampered by strain-induced defects in II–VI lasers,[34] are only now becoming possible. A high dislocation density ($\approx 10^7$ per cm^2 or more) was present in the structures that were first demonstrated as diode lasers in 1991 [Fig. 7.10(a)]. On the other hand, configurations such as the pseudomorphic structure in Fig. 7.10(b) had poor optical confinement factors (the optical intensity overlap with the confined electronic states of the QW, $\Gamma \sim 10^{-2}$ and less). In either case, it was difficult to reach room-temperature operation, although with reflective end facet coatings, pulsed threshold current densities of about 1 kA/cm^2 were obtained.[23,35] Even at low duty cycles ($\sim 10^{-3}$), however, such devices operated only a few minutes. On the other hand, cw operation at $T = 77$ K was relatively easily accomplished, producing output powers on the order of a mW.

The structure of Fig. 7.10(c) was introduced by the group at Sony Laboratories in 1992 to show wavelength shortening effects by the choice of the wide band-gap ternary ZnMgSe as a barrier material for a ZnSe QW,[18] but the operation was still limited to cryogenic temperatures, probably due to both inadequate optical and

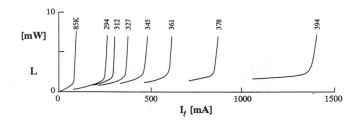

FIGURE 7.11. Output characteristics of a gain guided ZnCdSe/ZnSSe/ZnMgSSe SCH-SQW diode laser, operated under low duty cycle pulsed injection (5×10^{-4}; $\tau \approx 50$ nsec) (after Ref. 15).

electronic confinement (recall the common anion argument for valence band offsets). Indicating marked improvement, Fig. 7.11 shows the output characteristics obtained by the group at Phillips in 1993 from the ZnCdSe/ZnSSe/ZnMgSSe SCH-SQW diode laser [Fig. 7.10(d)], operated under low duty cycle, very short pulsed injection ($\tau \approx 50$ ns) but reaching well beyond room temperature.[15] In these gain-guided lasers, with the Cd concentration typically $x \approx 0.15$ to 0.20, sulfur concetration $y \approx 0.07$, and the Mg concentration $x' \approx 0.1$ to 0.2, the room-temperature current injection density was measured as approximately 500 A/cm^2 for devices of about 1 mm in length, an excellent figure of merit. Differential quantum efficiency of about 17% was also obtained. While reliable values for the index of refraction for ZnMgSSe are being measured, a plausible value for the electronic/optical confinement factor is about $\Gamma \approx 0.03$ for typical quantum-well thicknesses of $L_W \approx 7.5$ with thickness of ~ 200 nm for the ZnSSe optical guiding layer. Very similar diode-laser structures were fabricated in the authors' laboratories, but now incorporating the graded gap ZnSeTe contact layer,[16] as well as by the groups at 3M and Sony. Both single and multiple quantum-well structures have been fabricated; the relative insensitivity of the threshold current versus the number of QWs may be an indicator of the role of leakage currents in these heterostructures. At present, given the relative uncertainty about the values for band offsets, band bending at the heterointerfaces, and the details of overall vertical transport, the issue of leakage clearly calls for more systematic experimental and theoretical studies. The commonly chosen composition of the SCH-QW structure, and that of the ZnCdSe QW within, places the room-temperature laser wavelength into the green range of 510–530 nm; however, in a ZnCdSe/Zn$_{1-x'}$Mg$_{x'}$S$_{y'}$Se$_{1-y'}$/Zn$_{1-x''}$Mg$_{x''}$S$_{y''}$Se$_{1-y''}$ heterostructure, emission wavelengths as short as 480 nm have been obtained under pulsed excitation at Sony Laboratories. By and large, these devices are presently test structures designed for feasibility studies, whose evaluation will pave the way for shortening the emission wavelengths deeper into the blue by appropriate adjustment of the compositional ratios and the choice of substrate.

The benchmark development with the same SCH/SQW lasers, namely the observation of operation under *continuous* electrical injection, with the device lasting about one second, came from Sony Laboratories.[36] A relatively high voltage (~ 14

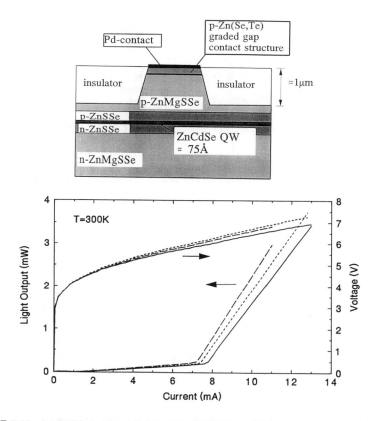

FIGURE 7.12. (top): Schematic of the ZnCdSe/ZnSSe/ZnMgSSe SCH-SQW ridge waveguide laser structure. (bottom): Optical output power vs current of a 500-μm-long device at room temperature under *continuous-wave* injection and the corresponding current vs voltage traces (after Ref. 37).

V), however, was required in these gain-guided devices at $\sim$ 520 nm wavelength. Further improvement of the SCH-QW device performance has been realized in index-guided ridge waveguide (mesa) structures, shown schematically in the top part of Fig. 7.12. The use of a self-aligned process is the most logical fabrication approach, while the mesa itself is presently produced by dry etching techniques in various laboratories. Generally, the processing of the II–VI devices presents many challenges, since the low-temperature conditions of molecular beam epitaxy ($T \approx 300$ °C) require low-temperature processing techniques. Furthermore, much care needs to be taken to minimize the introduction of mechanical defects during the handling of the devices. The bottom portion of Fig. 7.12 shows an example of the performance characteristics for three nominally identical laser devices from the authors' group, under *room-temperature continuous-wave conditions*, configured as ridge waveguide devices ($\sim$ 4.2 μm wide) into which the graded band-gap ohmic

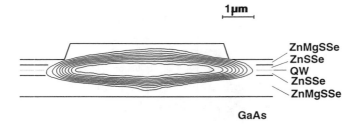

FIGURE 7.13. Contour plots of the near field emission pattern of the ridge waveguide laser. The contours correspond to 10% increments in intensity. The outline of the mesa and constituent layers are included (from Ref. 36a).

contacts were also incorporated.[37] With reflective facet coatings in these $\sim$ 500-μm-long devices, laser threshold currents in the range of 5–7 mA and voltages of about 5.8 V have been obtained. Such devices have been operated up to several tens of seconds; a time span clearly well short of any real application requirement, yet indicative of a significant rate of progress in the field within the time span of a couple of years. Single transverse-mode characteristics have been obtained in this geometry for both single and multiple quantum-well devices with average output powers up to 10 mW. Differential quantum efficiencies of about 50% have also been reached. . The benefits of index guiding by the mesa structure are evident in the near field emission pattern of Fig. 7.13 which was recorded by combining a CCD camera with a microscope. The contour plots of the optical intensity distribution, incremental in steps of 10% in this figure, show the predominance of the lowest transverse mode. Recently, the group at 3M has reported excellent device characteristics on the SCH-SQW diode lasers in very narrow ridge waveguide structures (< 2 μm wide) housing the p^+-ZnSe contacts,[38] including very well-defined fundamental transverse-mode and threshold currents as low as $I_{th} = 20$ mA. With high-reflectivity facet coatings, values as low as $I_{th} = 2.5$ mA have been reached in the 3M devices of about 300 μm in length, the lowest reported in a II–VI laser to date, with duty cycles at room temperature of several tens of percent; however, the devices were still very short lived under these conditions. It is very likely that the performance figures of merit of the SCH-QW diode lasers will continue to develop rapidly, especially in terms of the threshold voltage and the device lifetime under continuous-wave conditions.

A major engineering obstacle in the further development of the blue-green II–VI light emitters, beyond optimizing the heterostructure optical/electronic confinement and improving the p-type doping, etc., concerns the question of device degradation and lifetime. As mentioned above, the electrical dissipation in the present LED and laser devices is still unacceptably high, so that early failure and obstacles for cw demonstration (of the laser) are not surprising. However, beyond that, the mechanical properties of the relatively ionic II–VI materials suggest, for one

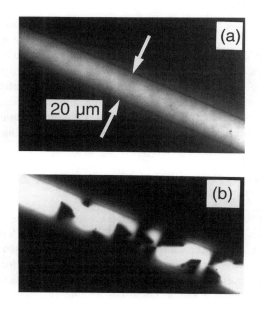

FIGURE 7.14. Photograph of optical emission through a transparent stripe top contact of a SCH-QW diode laser under cw injection below laser threshold (J 100 A/cm^2); (a) at time $t = 0$, and (b) after 30 min. The white areas correspond to the pattern of blue-green spontaneous emission originating from the QW layer in the structure.

example, a more vigorous dislocation dynamics than in III–V materials. This fundamental fact, coupled with the low MBE growth temperature, implies that the the control of defect-induced device degradation poses an inherently larger challenge. Since the availability of room-temperature devices is very recent, the study of device failure modes and their identification has only began in the ZnSe-based diode lasers. At this preliminary stage, there is some evidence that morphological defects which occur very early during epitaxy act as nucleation centers for dislocations which, in conjunction with point defects acting as nonradiative recombination centers in the active QW layer, in turn accumulate at a finite rate even under relatively modest current densities (such as in a LED). The coalescing of defected microareas into macroscopic regions leads to an accelerated reduction in the optical emission from the device, impairment of the vertical transport, and to a subsequent failure. Figure 7.14 is an illustration from the authors' laboratories of such defect formation in a gain-guided SCH-QW diode laser structure, observed in an optical microscope from within the QW layer.[39] The device was equipped with a transparent top electrode and the spontaneous emission emanating in the vertical direction from the QW was photographed. In this example, the initially uniform emission in the fresh device [Fig. 7.14(a)] has significantly deteriorated during approximately $\frac{1}{2}$ h at a level of 100 A/cm^2 of electrical injection, *below* that of laser threshold [Fig.

7.14(b)]. The large dark areas within the stripe, with geometrical features along principal crystallographic planes, show nearly total quenching of optical (and significant deterioration of electrical) activity. We have also observed that random pulsations in the temporal output of a laser are typical of a degraded device. Recently, the group at 3M has conducted an initial but systematic study, including transmission electron microscope analysis, of the microstructure associated with this type of device failure to conclude that the development of macroscopic regions of high-density dislocations is a serious impediment to cw lasing.[40]

7.4. PHYSICS OF STIMULATED EMISSION IN ZnSe-BASED QWS: COULOMB INTERACTION

The physical idea where gain and stimulated emission is retrieved from a degenerate free-electron hole plasma (EHP) predates even the actual demonstration of the first semiconductor lasers three decades ago. Usually in III–V semiconductor lasers we can make allowances for the many-body, electron/hole Coulomb interactions to first order, by simply renormalizing (reducing) the band gap according to a carrier density dependence which is also a function of the dimensionality of the system. In practical terms, even the gap normalizations are rarely explicitly discussed in the semiconductor device literature. For the ZnSe-based QWs, gain calculations for an EHP can be readily performed in the usual effective mass approximation, including the effects of the lattice mismatch strain on $k = 0$ valence band states. In analogy to the InGaAs/GaAs strained layer system, the uniaxial strain component yields a light hole-like, in-plane effective mass and hence a low density of states (versus the bulk) which aids in reducing the e-h pair density required to reach the transparency condition. Here we do not review such routine calculations, although we mention again the important complication presented by the large linewidth broadening which was briefly addressed in Sec. 7.2. Instead, we raise the basic question about the possible importance of the electron-hole Coulomb interactions in the formation of gain and stimulated emission in the ZnSe-based QWs. As pointed out already, the unusual condition in the ZnSe-based QWs is that the exciton binding energy can be made large enough so that the LO-phonon-induced dissociation process into the free e-h pair continuum is much weakened at noncryogenic temperatures. This circumstance, which is found neither in bulk ZnSe nor, e.g., in GaAs QWs, allows the excitons to be present in principle not only in absorption but also *in emission* at device temperatures, as argued in Sec. 7.3 for LEDs.

In this section we consider available experimental evidence about the nature of stimulated emission in the ZnCdSe QW structures. We emphasize that the conventional EHP model for gain is not excluded, and is indeed likely to dominate under high-injection conditions at room temperature and above. The question we raise is about the possible component in the stimulated emission process by the pair-wise electron-hole Coulomb correlations, and under what conditions such effects can occur in a real laser device. Device goals aside, the issues are of fundamental interest from the standpoint of many-body physics in lower dimensions in (weakly) disordered systems. However, since the subject is still not fully developed experimentally or theoretically, the discussion below does not represent a definitive proof and represents a particular viewpoint.

In bulk II–VI semiconductors studied at low lattice temperatures and under high optical excitation, work more than a decade ago showed that stimulated emission (viewed as amplified spontaneous emission) was due to excitons recombining in a strong scattering regime.[41] Electronic and LO-phonon scattering processes were evoked to explain spectral anomalies in the emission wavelengths and to make the connection to bulk excitons. The circumstance is fundamentally different in the new ZnSe-based QWs due to the added robustness of the 2D excitons.

7.4.1. Optical Pumping Studies

That excitons are important in stimulated emission in the (Zn,Cd)Se QWs at cryogenic temperatures has been demonstrated through optical pumping studies.[42,43] For example, a form of photoexcitation spectroscopy of the stimulated emission in edge-emitting structures was applied by the authors (in collaboration with the group at the University of Notre Dame) to show that, with photon energy of excitation near the $n = 1$ HH exciton resonance, laser emission through cleaved end facets commences readily, once the excitation source is tuned directly into resonance. Especially at a low temperature ($T \sim 10$ K), such an experiment shows clearly that a gas of cold 2D excitons has formed the gain for the stimulated emission process. These observations have been carried out to temperatures of 220 K.[44]

Direct further evidence for the presence of "excitonic gain" in the (Zn,Cd)Se MQW structures has been obtained recently by using time-resolved spectroscopic methods.[45] In these pump-probe experiments a form of photomodulation spectroscopy is performed through the QW section so that the probe pulses measure the optical constants near the $n = 1$ HH exciton, following intense short-pulse excitation at time $t = 0$. Figure 7.15 shows the results of one such experiment at $T = 77$ K for a structure of 30 Å QW thickness, where the absorption spectrum measured by the probe is displayed in the absence of excitation (establishing the exciton resonance profile), and for a 10-ps time delay under conditions of resonance excitation by the pump. Note that while most of the exciton resonance is maintained (as an absorptive entity), large reduction of absorption as well as *actual presence of gain* is realized on the low-energy tail of the resonance. In a sample prepared with cleaved end facets, laser emission (in a perpendicular direction) ensues at a photon energy which is consistent with the position of measured maximum gain (see inset). The experimental accuracy is such that the actual measurement of the gain coefficient ($g \approx 1200$ cm^{-1} for the QW material at $h\nu = 2.588$ eV) is possible. The details of observed dynamical behavior can be understood on the basis of broadening of the exciton absorption line by exciton-exciton collisions, followed by energy relaxation to the low-energy tail, where gain develops. Both from the standpoint of the typical electron-hole pair densities at laser threshold ($\sim 5 \times 10^{11}$ cm^{-2} at 77 K) as well as the experimental observation of the remaining presence of (a

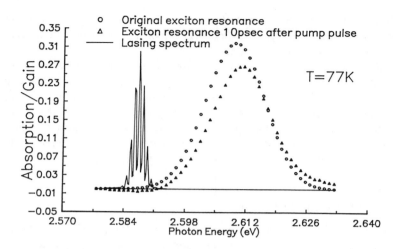

FIGURE 7.15. Example of time-resolved pump-probe spectroscopy in a (Zn,Cd)Se/ZnSe QW at $T = 77$ K. Note the appearance of gain on the low-energy edge of the $n = 1$ HH exciton resonance within 10 ps following excitation by the pump pulse (after Ref. 45).

partially saturated) exciton resonance in pump-probe experiments, we deduce that a free electron-hole plasma is clearly *not* present at cryogenic temperatures (e.g., at $T = 77$ K).

We now outline a very simple atomiclike phenomenological model for excitonic gain in the (Zn,Cd)Se/ZnSe QWs, which assumes an inhomogeneous, isolated $n = 1$ HH exciton resonance. The inhomogeneous broadening is due to a combination of alloy potential (compositional) fluctuations and QW interface roughness, which allows a spatially random exciton distribution. If a given localization site or a volume defined by the mean free path can only be occupied by one exciton, that is, that the exchange interaction between the electrons and holes of different excitons is considered to be infinitely large, the population inversion condition for such excitons is $f - (1 - f) = 2f - 1 > 0$, where f is the probability of the state being occupied. (The validity of this assumption will be examined below for a high-temperature situation where homogeneous broadening dominates.) The argument is a variation of the phase-space filling (PSF) idea,[46] but now applied to an *inhomogeneously* broadened system.

In terms of a highly simplified three-energy-level scheme, illustrated in Fig. 7.16, electron-hole pair or exciton energy relaxation occurs from initially excited states $|X\rangle$ to $n = 1$ HH exciton states $|X'\rangle$ where gain is possible in terms of induced emission to the ground state $|0\rangle$. The relaxation of the e-h pair or the exciton population to such states is quite rapid in the (Zn,Cd)Se/ZnSe system (on a ps timescale). The gain/absorption coefficient then can be written as the following, if steady state is assumed as

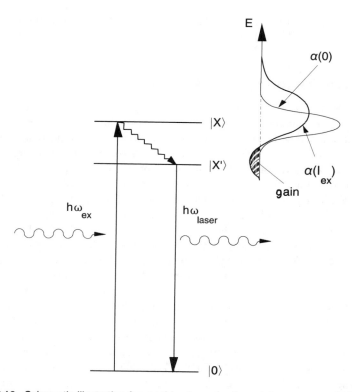

FIGURE 7.16. Schematic illustration for reaching inversion in an inhomogeneously broadened exciton resonance for the (Zn,Cd)Se/ZnSe QW.

$$g(E) = \int_{-\infty}^{\infty} D_i(E')D_h(E-E')\left(\frac{2}{\exp[(E'-\mu)/kT]+1}-1\right)dE', \quad (1)$$

where $D_i(E)$ is the inhomogeneous, and $D_h(E)$ the homogeneous lineshape function. The normalized total density of excitons n is given by

$$n = \int_{-\infty}^{\infty} D_i(E') \frac{1}{\exp[(E'-\mu)/kT]+1} dE'. \quad (2)$$

The energies in both equations are defined with respect to the center of the inhomogeneous exciton resonance. The quantity $g(E)$ is normalized to the peak value of the absorption coefficient in unexcited material, while the exciton density n is normalized to the maximum density of excitons $N_{\max}$, a good estimate of which is given by the phase-space filling criterion for the entire inhomogeneous line, $N_{\max} = n_c \approx (a^2{}_B)^{-1} \approx 6 \times 10^{12} \, \text{cm}^{-2}$ in our system. For more detailed discussion about a model calculation of gain in this model, the reader is referred to Ref. 45. A more complete theory would especially need to consider the many-body aspects of the exciton-exciton interaction, including the issue of effect of temperature (on the homogeneous broadening).

7.4.2. Gain Spectroscopy of Diode Lasers

Initial work presently in progress is also focusing on means to directly measure and understand gain spectra in the new diode lasers, including the issue of the mechanism of stimulated emission. Here again, the availability of room-temperature diode lasers is so recent that much more gain spectroscopy and characterization will undoubtedly be carried out in the near term. Nonetheless, we wish to illustrate to the reader some of the gain characteristics which have been identified in early studies on the ZnCdSe/ZnSSe/ZnMgSSe SCH-QW lasers. For example, there is the important practical question of the influence of scattering at room temperature, both by carrier-carrier and carrier-LO-phonon interactions, which substantially modifies the ideal 2D density of states "step." Recent work, where the gain has been calculated in the ZnCdSe QW in the usual degenerate e-h plasma model, has ignored the broadening,[47,48] an assumption which can be often justified in III–V semiconductor lasers but not in the present case. In the light of the identification of the important electron-hole pair Coulomb interaction discussed in the previous section, a fundamental question must also be asked in the II–VI diode lasers about the relevance and role of such pair correlations and other many-body effects (at room temperature). It is important to point out that the relevant energy reference scale, i.e., the 2D exciton Rydberg, is much larger in the ZnSe-based semiconductors than in the GaAs-based materials. Typically, $E_x \approx 40$ meV $>$ kT, $h\nu_{LO}$ for the former, while $E_x \approx 10$ meV $<$ kT, $h\nu_{LO}$ for the latter.

Experimental work in the authors' group has focused on the gain spectroscopy in the ZnCdSe/ZnSSe/ZnMgSSe separate confinement diode laser structures. The two most common methods for the measurement of gain versus injection current relationships in semiconductor lasers involve (a) the measurement of the threshold current for a series of devices by varying either the cavity length or resonator mirror reflectivity (so that no spectral information is obtained), and (b) the so-called Hakki–Paoli method (from which spectral lineshapes are obtained).[49] Both of these methods have similar disadvantages in that the absolute value of gain requires an accurate knowledge of the optical losses, optical confinement factor, and the leakage current. The particular approach for the ZnSe-based QW lasers that we have found useful makes use of the correlation between (top) spontaneous emission and edge-stimulated emission spectra, which was first employed by Henry et al. to obtain the actual gain spectra for a GaAs double heterostructure laser.[50] It has been subsequently used to construct detailed gain spectra for GaAs QW diode lasers as well.[51] In the present instance, gain-guided devices, with 20-μm-wide stripes were fabricated with a top-transparent electrode of indium-tin oxide.[28] Spectroscopy was performed by recording both the emission through the top electrode and the stimulated (or spontaneous) emission through the cleaved end facets, as shown schematically in the top portion of Fig. 7.17, which also shows the layered heterostructure.

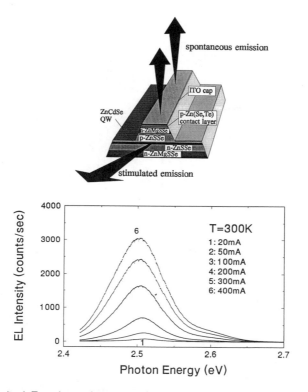

FIGURE 7.17. (top) Experimental geometry for the gain spectroscopy in the ZnCdSe/ZnSSe/ZnMgSSe SCH QW diode laser structures. (bottom) The spontaneous emission spectra at $T = 300$ K, obtained through the transparent top electrode at different levels of current injection.

The bottom part shows an example of the top emission spectra, as a function of the injection current at room temperature of a device where the active region was composed of three quantum wells. The well thickness was $L_w = 7.5$ nm and that of the separating ZnSSe barriers was $L_b = 10$ nm. The room-temperature pulsed threshold current for this device was $I = 280$ mA, corresponding to a density $J = 1.4$ kA/cm^2, i.e., approximately 470 A/cm^2 per QW.

The formulation of gain in the approach by Henry *et al.* is independent of the details of the gain mechanism or the nature of the electronic states that participate in the radiative process, and draws from the fundamental connection between gain and absorption by detailed balance arguments. This leads to the following explicit relationship between the gain and experimental spontaneous emission spectra, $S(E)$, where the separation between the quasi-Fermi levels, ΔE_F, is also experimentally obtained from the spontaneous emission spectrum at threshold conditions from the known position of laser emission:

$$g(E, \Delta E_F) = C^* \frac{S(E)}{E^2} \left[1 - \exp\left(\frac{E - \Delta E_F}{kT} \right) \right].$$

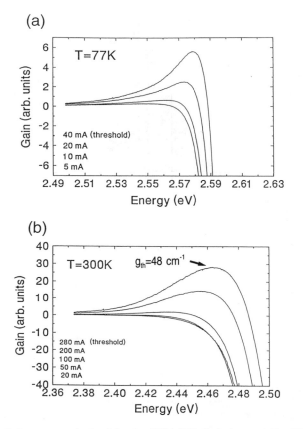

FIGURE 7.18. Gain spectra obtained for the SCH QW diode laser at T = 77 and 300 K for different levels of current injection (after Ref. 17).

In this expression C includes fundamental constants and experimental amplitude calibration factors. In the limit where the homogeneous broadening is negligible, the expression is exact. However, when substantial broadening is present, the quantitative use of this formulation for extracting the value for the gain coefficient can lead to errors. Spectral lineshapes are much more immune to the effects of broadening, however.

With input provided directly from our experiments, Fig. 7.18 shows the results for the gain spectrum of the typical SCH QW diode laser at T = 77 K and at room temperature, as a function of the injection current. We have found it more reliable to obtain the calibration of the gain coefficient calibration from the cavity length dependence of the threshold current. On the low-energy side, the spectra deviate strongly from the step function for an ideal 2D density of states and indicate a substantial broadening, as well as the presence of Coulomb interaction (enhanced oscillator strength). A larger spectral window measurement is shown in Fig. 7.19, at

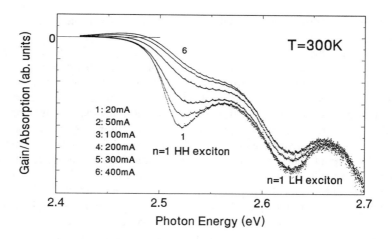

FIGURE 7.19. A wider spectral view of the gain/absorption spectrum of the same SCH device, showing the effects of injection on the HH and LH exciton states from below to above the laser threshold (after Ref. 17).

room temperature, where the unambiguous presence of both the $n = 1$ HH and LH exciton resonances is quite evident at photon energies of $h\nu = 2.52$ and 2.625 eV, respectively. The identification of the LH state is made clear by analyzing the polarization of the edge emission, which shows the HH state emitting in the TE polarization, and the LH in the TM polarization. The uniaxial strain in the ZnCdSe QW (in our pseudomorphic structures) splits the HH and LH bands such that the in-plane mass for the HH state is relatively light ($m_{\parallel} \approx 0.33$), while that for the LH state is heavier ($m_{\parallel} \approx 0.5$). Precise mass values require taking into account the mixing between the HH and LH states which occurs away from $k = 0$.

The experimentally derived gain and absorption spectra in Figs. 7.18 and 7.19 provide a heretofore largely unavailable glimpse into the operation of the SCH diode laser. The gain spectrum of Fig. 7.18 at room temperature shows a striking, nearly exponential tail, which has an overall extension approaching 100 meV. (As one consequence, in a low-loss case, this leads to the onset of lasing at longer wavelengths.) There are at least three contributions to the low energy tail: (i) alloy potential fluctuations, which add an inhomogeneous contribution of approximately 6–7 meV for the Cd composition in our structures ($x \sim 0.15$); (ii) the Frohlich interaction which is a major contributor to homogeneous linewidth broadening of optical absorption resonances in the ZnSe-based QWs at room temperature. This effect can add on the order of 30–50 meV to the linewidth (depending on the details of the QW configuration, e.g., in terms of the particular hole states that are involved); and (iii) the Coulomb interaction (carrier scattering) in the many-particle electron-hole gas, either direct or intermediated by phonons, which makes a contribution at higher carrier densities. (Under typical II–VI laser diode conditions, *e-h* pair densities are not yet high enough to significantly screen the Frohlich interaction.)

While leaving the more detailed analysis of the gain spectral lineshape elsewhere, we note that the result in Fig. 7.18 contains dominant contributions from processes (ii) and (iii). The contribution from the LO-phonon interaction has been already referred to, including its prior experimental quantification.[6] The importance of carrier-carrier interaction can be directly seen in the linewidth of the spontaneous emission spectrum in the bottom panel of Fig. 7.17. The role of the carrier-carrier Coulomb interaction can be addressed either in terms of a pair-wise contribution (excitonlike bound states) and that in a many-body Fermi liquid (free EHP). The former was already shown to be important in stimulated emission at cryogenic temperatures in the optical pumping experiments on ZnCdSe QWs in Fig. 7.16. At room-temperature conditions the distinctions between the pair-wise and many-body interactions are much more difficult to detail experimentally, and may in fact be to some extent unrecognizable in practice. However, that electron-hole Coulomb pair-wise effects remain important in our diode lasers is strongly suggested by the data in Fig. 7.19 where at low levels of injection, the HH and LH exciton resonances are initially clearly visible. Qualitatively at least, increase in the injection level appears to nearly saturate the exciton resonance, before gain appears on the low-energy portion of the spectrum. Even at saturation, where the exciton resonance merges into the EHP continuum, the pair-wise correlations remain strong, in a many-body analog to the continuum states of a single exciton. We note that recent re-examination of radiative recombination in GaAs and InGaAs QWs has pointed out the relevance of the pair-wise correlations at relatively high temperatures and carrier densities, where these effects have been previously ignored.[52] The ZnSe-based QWs present, of course, much more excitonic circumstances by comparison. The issues of many-body interactions present, therefore, a unique opportunity for experimental and theoretical study of quasi-2D electron-hole systems in the blue-green diode lasers, and a future extension to quantum wire configurations.

7.5. SUMMARY REMARKS

In this review we have addressed a range of topics which are relevant to present research on the new blue-green II–VI emitters. The field is moving rapidly, so that predictions about future pace are difficult to make. The milestone of a low-voltage (by previous standards), room-temperature cw operation has been demonstrated in 1993, but the lasers are still very short lived. Clearly, much progress is required, e.g., in improving the vertical transport, the p-type doping, and QW designs for larger gain. Device fabrication techniques need much improvement, taking into account the rather low temperature processing requirements (typically $T < 250\,°C$) that also need to be mechanically gentle. Better control of crystalline defects, including choices of substrate and buffer layers, and their identification in the device degradation process (including the influence of processing) is a subject that has now become very relevant as technological prospects of LEDs and diode lasers

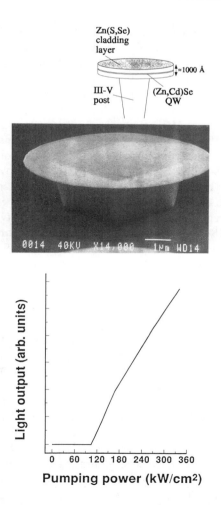

FIGURE 7.20. Scanning electron microscope images of the (Zn,Cd)Se/Zn(S,Se) microresonators. Top: An individual 7-μm-diam device with its structure shown in the inset. Bottom: Output characteristics under optical pumping at room temperature (after Ref. 54).

need be evaluated realistically. At the same time, opportunities for much innovation and new research is awaiting researchers in this expanding field. For example, lateral optical confinement in the stripe geometry lasers by impurity-induced disordering has been recently demonstrated in optically pumped structures.[53] Extension of emission deeper into the blue and near UV is being pursued, as well as studies of microresonator laser configurations, such as the whispering gallery mode structures.[54] Figure 7.20 shows the electron microscope image of such an early structure, housing a ZnCdSe QW, where lasing has been recently observed under pulsed optical pumping conditions at room temperature. Work on surface-emitting lasers is also beginning and may offer specific advantages for the wide-gap II–VI diode lasers. Finally, from a technological viewpoint, we note that potentially important developments are taking place with the MOCVD growth of the II–VI heterostructures, where initial reports of LED and possible diode laser operation are emerging.[55] Encouraging progress has also been reported with metalo-organic molecular beam epitaxy (MOMBE) of ZnSe,[56] a technique which has potential advantages over solid-source MBE methods.

The authors are grateful to members of their groups who made major contributions to those portions of the work carried out in their laboratories, as described in this article. They are J. Ding, H. Jeon, M. Hagerott, M. Hovinen, P. Kelkar, V. Kozlov, and A. Salokatve at Brown; D. Grillo, Y. Fan, J. Han, Li He, and M. Ringle at Purdue. The research at Brown University and Purdue University was supported by the Defense Advanced Projects Agency, under the University Research Initiative program (Grant No. 218-25015XX). The work was also supported by the National Science Foundation (at Brown DMR-9112329 and DMR-9121747; at Purdue DMR-9202957) and by the Air Force Office of Scientific Research (F49620-92-J-0440).

7.6. NOTES ADDED IN PROOF

Since this manuscript was completed, a number of advances relevant to the subject matter of this chapter have occurred. In the area of p-type doping of the quaternary, observations of persistent photoconductivity have firmed the assumption that this wide gap alloy (E_g–3.0 eV) possesses both a hydrogenic shallow and a deeper lattice relaxed state.[57] In terms of the diode laser performance, the group at Sony Laboratories has extended the cw operation of the ZnCdSe/ZnSSe/ZnMgSSe SCH device at $\lambda = 520$ nm up to an hour, apparently as a result of reduction of the stacking fault-like defects that occur near the GaAs/ZnSe heterovalent interface during epitaxial growth.[58] These defects, in turn, have been the subject of intense study in terms of their role in the light emitting device degradation.[59,60] Presently, it appears that the partial (threading) dislocations that are associated with these defects and which penetrate into the active QW layer, nucleate and induce the generation of subsequent dislocation networks in the strained ZnCdSe QW under a high current density.

In other areas of progress, the implementation of SCH diode laser structures which contain ZnSe as the active layer, laser emission at $\lambda = 463$ nm has been obtained in the authors' group[61] and at $\lambda = 471$ nm at Sony Laboratories.[62] In a new direction, a vertical cavity blue-green surface emitting laser has also been recently demonstrated by optical pumping techniques.[63]

REFERENCES

1. For a review see, e.g., R. Gunshor, L. Kolozdiejski, and A. Nurmikko, *Semiconductors and Semimetals*, edited by T. Pearsall (Academic, San Diego, 1990), Vol. 33, pp. 337–409.

2. H. Jeon, J. Ding, A. V. Nurmikko, H. Luo, N. Samarth, J. Furdyna, W. Bonner, and R. Nahory, Appl. Phys. Lett. **57**, 2413 (1990).
3. R. M. Park, M. B. Troffer, C. M. Rouleau, J. M. De Puydt, and M. A. Haase, Appl. Phys. Lett. **57**, 2127 (1990).
4. K. Ohkawa, T. Karasawa, and O. Yamazaki, Jpn J. Appl. Phys. **30**, L152 (1991).
5. M. Haase, J. Qiu, J. DePuydt, and H. Cheng, Appl. Phys. Lett. **59**, 1272 (1991).
6. H. Jeon, J. Ding, W. Patterson, A. V. Nurmikko, W. Xie, D. C. Grillo, M. Kobayashi, and R. L. Gunshor, Appl. Phys. Lett. **59**, 3619 (1991).
7. H. Jeon, J. Ding, A. V. Nurmikko, W. Xie, D. C. Grillo, M. Kobayashi, and R. L. Gunshor, Appl. Phys. Lett. **60**, 2045 (1992).
8. D. C. Grillo, W. Xie, M. Kobayashi, R. L. Gunshor, G. C. Hua, N. Otsuka, H. Jeon, J. Ding, and A. V. Nurmikko, J. Electron. Mater. **22**, 441 (1993).
9. L. Kolodziejski, K. Lu, E. Ho, C. A. Coronado, P. A. Fisher, and G. S. Petrich, Proc. Int. Conf. II–VI Compounds, Newport, RI, September 1993.
10. J. Ding, N. Pelekanos, A. V. Nurmikko, H. Luo, N. Samarth, and J. Furdyna, Appl. Phys. Lett. **57**, 2885 (1990); N. T. Pelekanos, J. Ding, M. Hagerott, A. V. Nurmikko, H. Luo, N. Samarth, and J. Furdyna, Phys. Rev. B **45**, 6037 (1992).
11. H. J. Lozykowski and V. K. Shastri, J. Appl. Phys. **69**, 3235 (1991).
12. W. Walecki, A. Nurmikko, N. Samarth, H. Luo, and J. Furdyna, Appl. Phys. Lett. **57**, 466 (1990).
13. Y.-h. Wu, K. Ichino, Y. Kawakami, S. Fuijita, and S. Fujita, Jpn. J. Appl. Phys. **31**, 1737 (1992).
14. H. Okuyama, T. Miyajima, Y. Morinaga, F. Hiei, M. Ozawa, and K. Akimoto, Electron. Lett. **28**, 1798 (1992).
15. J. M. Gaines, R. R. Drenten, K. W. Haberern, T. Marshall, P. Mensz, and J. Petruzzello, Appl. Phys. Lett. **62**, 2462 (1993).
16. D. C. Grillo, Y. Fan. J. Han, H. Li, R. L. Gunshor, M. Hagerott, H. Jeon, A. Salokatve, G. Hua, and N. Otsuka, Device Research Conference, June 1993, Santa Barbara, CA and Appl. Phys. Lett. **63**, 2723 (1993).
17. J. Ding, M. Hagerott, A. Salokatve, H. Jeon, A. Nurmikko, D. C. Grillo, J. Han, H. Li, and R. L. Gunshor, J. Crystal Growth **138**, 719 (1994); Phys. Rev. B **50**, 5787 (1994).
18. S. Itoh, H. Okuyama, S. Matsumoto, N. Nakayama, T. Ohata, T. Miyajima, A. Ishibashi, and K. Akimoto, Electron. Lett. **29**, 766 (1993).
19. Y. Fan, J. Han, J. Saraie, R. L. Gunshor, M. M. Hagerott, and A. V. Nurmikko, Appl. Phys. Lett. **63**, 1812 (1993); P. M. Mensz, S. Herko, K. W. Habarern, J. Gaines, and C. Ponzoni, Appl. Phys. Lett. **63**, 2800 (1993).
20. Z. Yang, K. A. Bowers, J. Ren, Y. Lansari, J. W. Cook, Jr., and J. F. Schetzina, Appl. Phys. Lett. **61**, 2671 (1993).
21. M. Asada, in *Quantum Well Lasers*, edited by P. Zory (Academic, San Diego, 1993) pp. 97–130.
22. (a) J. Han, Y. Fan, M. Ringle, L. He, D. Grillo, R. Gunshor, G. Hua, and N. Otsuka, J. Cryst. Growth **138**, 464 (1994). (b) 6th International Conference on II–VI Compounds, J. Crystal Growth **117**, 1 (1992); J. Crystal Growth **138**, 1 (1994).
23. C. T. Walker, J. M. DePuydt, M. A. Haase, J. Qiu, and H. Cheng, Physica B **185**, 27 (1993).
24. Y. Fan, J. Han, L. He, J. Saraie, R. L. Gunshor, M. Hagerott, H. Jeon, and A. V. Nurmikko, Appl. Phys. Lett. **61**, 3160 (1992); F Hiei, M. Ikeda, M. Ozawa, T. Miyajima, A. Ishibashi, and K. Akimoto, Electron. Lett. **29**, 878 (1993).
24a. F. Hiei, M. Ikeda, M. Ozawa, T. Miyajima, A. Iskibashi, and K. Akimoto, Electron. Lett. **29**, 878 (1993).
25. J. Ren, Y. Lanzari, J. W. Cook, Jr., and J. F. Schetzina, J. Electron. Mater. **22**, 973 (1993).
26. J. Han, T. Stavrinides, M. Kobayashi, M. Hagerott, and A. V. Nurmikko, Appl. Phys. Lett. **62**, 840 (1993).
27. S. Nakamura, T. Mukai, and M. Senoh, Appl. Phys. Lett. **64**, 1687 (1994).
28. M. Hagerott, H. Jeon, J. Ding, A. V. Nurmikko, W. Xie, D. C. Grillo, M. Kobayashi, and R. L. Gunshor, Appl. Phys. Lett. **60**, 2825 (1992).
29. D. Lee, A. Mysyrowicz, A. V. Nurmikko, and B. F. Fitzpatrick, Phys. Rev. Lett. **58**, 1475 (1987); G. W. Iseler and A. J. Strauss, J. Lumin. **3**, 1 (1970); O. Goede and D. Hennig, Phys. Status Solidi b **119**, 261 (1983); S. Permogorov, A. Reznitsky, S. Verbin, and V. Lysenko, Solid State Comm. **47**, 5 (1983).
30. L. A. Kolodziejski, R. L. Gunshor, Q. Fu, D. Lee, A. V. Nurmikko, and N. Otsuka, Appl. Phys. Lett. **52**, 1080 (1988); Q. Fu, D. Lee, A. V. Nurmikko, R. L. Gunshor, and L. A. Kolodziejski, Phys. Rev B **39**, 3173 (1989).

31. M. Hagerott, J. Ding, H. Jeon, A. V. Nurmikko, Y. Fan, L. He, J. Han, He, J. Saraie, R. L. Gunshor, G. C. Hua, and N. Otsuka, Appl. Phys. Lett. **62**, 2108 (1993).
32. A. V. Nurmikko, Q. Fu, D. Lee, R. L. Gunshor, and L. A. Kolodziejski, Proc. 19th Int. Conf. Physics of Semiconductors, Warsaw (1988), p. 1523.
33. M. C. Phillips, M. W. Wang, J. F. Swenberg, J. O. McCaldin, and T. C. McGill, Appl. Phys. Lett. **61**, 1962 (1992).
34. e.g., R. Legras, Le Si Dang, C. Bodin, J. Cibert, F. Marcenat, G. Feuillet, J. L. Pautrat, D. Herve, and E. Molva, J. Electron. Mater. **23**, 313 (1994).
35. H. Jeon, M. Hagerott, J. Ding, A. V. Nurmikko, D. C. Grillo, W. Xie, M. Kobayashi, and R. L. Gunshor, Opt. Lett. **18**, 125 (1993).
36. N. Nakayama, S. Itoh, K. Nakano, H. Okuyama, M. Ozawa, A. Ishibashi, M. Ikeda, and Y. Mori, Electron. Letters **29**, 1488 (1993); *ibid.* **29**, 2194 (1993).
37. A. Salokatve, H. Jeon, J. Ding, M. Hovinen, A. Nurmikko, D. C. Grillo, J. Han, H. Li, R. L. Gunshor, C. Hua, and N. Orsuka, Electron. Lett. **29**, 2192 (1993); note that the correct threshold voltage in this paper should have been 5.8 V.
38. M. A. Haase, P. F. Baude, M. S. Hagedorn, J. Qiu, J. DePuydt, H. Cheng, S. Guha, G. E. Hofler, and B. J. Wu, Appl. Phys. Lett. **63**, 2315 (1993).
39. Minna Hovinen (unpublished data).
40. S. Guha, J. M. DePuydt, M. A. Haase, J. Qiu, and H. Cheng, Appl. Phys. Lett. **63**, 3607 (1993).
41. C. Klingshirn and H. Haug, Phys. Rep. **70**, 315 (1981).
42. J. Ding, H. Jeon, T. Ishihara, H. Luo, N. Samarth, and J. Furdyna, Conference on Quantum Optoelectronics, Salt Lake City, March 1991, Opt. Soc. Am. Tech. Dig. **7**, 134 (1991); J. Ding, H. Jeon, T. Ishihara, H. Luo, N. Samarth, and J. Furdyna, Surf. Sci. **259**, 616 (1992).
43. K. Ando, A. Ohki, and S. Zembutsu, Jpn J. Appl. Phys. **31**, L1362 (1992); Y. Kawakami, B. C. Cavenett, K. Ichino, Sz. Fujita, and Sg. Fujita, Jpn. J. Appl. Phys. **32**, L730 (1993).
44. J. Ding, H. Jeon, T. Ishihara, A. V. Nurmikko, H. Luo, N. Samarth, and J. Furdyna, Phys. Rev. Lett. **60**, 1707 (1992).
45. J. Ding, T. Ishihara, M. Hagerott, H. Jeon, and A. V. Nurmikko, Phys. Rev. B **47**, 10528 (1993).
46. R. Zimmerman, K. Kilimann, W. D. Kraeft, D. Kremp, and G. Ropke, Phys. Status Solidi B **90**, 175 (1982); S. Schmitt-Rink, D. S. Chemla, and D. A. B. Miller, Adv. Phys. **38**, 89 (1989).
47. D. Ahn, T. K. Yoo, and S. L. Chuang, Jpn. J. Appl. Phys. **31**, L556 (1992).
48. R. L. Aggarwal, J. J. Zayhowski, and B. Lax, Appl. Phys. Lett. **62**, 2899 (1993).
49. B. W. Hakki and T. L. Paoli, J. Appl. Phys. **46**, 1299 (1975).
50. C. H. Henry, R. A. Logan, and F. R. Merritt, J. Appl. Phys. **51**, 3042 (1980).
51. M. P. Kesler and C. Harder, Appl. Phys. Lett. **57**, 123 (1990); P. Blood, A. I. Kucharska, J. P. Jacobs, and K. Griffiths, J. Appl. Phys. **70**, 1144 (1991).
52. P. Michler, A. Hangleiter, A. Moritz, V. Harle, and F. Scholz, Phys. Rev. B **47**, 1671 (1993); A. Hangleiter, Phys. Rev. B **48**, 9146 (1993).
53. T. Yokogawa, P. D. Floyd, and J. L. Merz, J. Crystal Growth **138**, 564 (1994).
54. M. Hovinen, J. Ding, A. V. Nurmikko, D. C. Grillo, Y. Fan. J. Han, H. Li, and R. L. Gunshor, Appl. Phys. Lett. **63**, 3128 (1993).
55. Shigeo Fujita, J. Crystal Growth **138**, 737 (1994).
56. C. A. Coronado, E. Ho, and L. A. Kolodziejski, J. Cryst. Growth **127**, 323 (1993).
57. J. Han, M. Ringle, Y. Fan, R. L. Gunshor, and A. V. Nurmikko, Appl. Phys. Lett. **65**, 3230 (1994).
58. A. Ishibashi, IEEE LEOS Conference, Boston, MA (November 1994).
59. S. Guha, H. Cheng, M. A. Haase, J. M. DePuydt, J. Qiu, B. J. Wu, and G. E. Hofler, Appl. Phys. Lett. **65**, 801 (1994).
60. G. C. Hua, N. Otsuka, D. C. Grillo, Y. Fan, J. Han, M. D. Ringle, R. L. Gunshor, M. Hovinen, and A. V. Nurmikko, Appl. Phys. Lett. **65**, 1331 (1994); M. Hovinen, J. Ding, A. Salokatve, A. V. Nurmikko, G. C. Hua, D. C. Grillo, Li He, J. Han, M. Ringle, and R. L. Gunshor, J. Appl. Phys., **77**, 4150 (1995).
61. D. C. Grillo, J. Han, M. Ringle, G. Hua, R. L. Gunshor, P. Kelkar, V. Kozlov, H. Jeon, and A. V. Nurmikko, Electr. Lett. **25**, 2131 (1994).
62. H. Okuyama, E. Kato, S. Itoh, N. Nakayama, T. Ohata, and A. Ishibashi, Appl. Phys. Lett., **66**, 656 (1995).
63. H. Jeon, V. Kozlov, P. Kelkar, A. V. Nurmikko, D. C. Grillo, J. Han, M. Ringle, and R. L. Gunshor, Electr. Lett., **31**, 106 (1995).

CHAPTER 8

Semiconductor Laser Amplifiers

Govind P. Agrawal

The Institute of Optics, University of Rochester, Rochester, New York

8.1. INTRODUCTION

An important application of semiconductor lasers is as an optical source in fiber-optic communication systems (see Chap. 10). Indeed, many kinds of semiconductor lasers (e.g., InGaAsP lasers operating in the wavelength range 1.3–1.6 μm) were developed with this application in mind. However, the transmission distance of any fiber-optic communication system is limited by fiber loss and dispersion.[1] For long-haul lightwave systems this limitation is overcome by periodic regeneration of the optical signal at repeaters, where the optical signal is converted into the electric domain by using a receiver and then regenerated by using a transmitter. Such regenerators become quite complex and expensive for multichannel lightwave systems. Although regeneration of the optical signal is necessary for dispersion-limited systems, loss-limited systems could benefit considerably if electronic repeaters were replaced by much simpler, and potentially less expensive, optical amplifiers which amplify the optical signal directly.

Several kinds of optical amplifiers were studied and developed during the 1980s.[1–5] The technology has matured enough that the use of optical amplifiers in fiber-optic communication systems is likely to become widespread during the 1990s. Two classes of optical amplifiers that have reached an advanced stage of development are semiconductor laser amplifiers (SLAs) and erbium-doped fiber amplifiers (EDFAs). Even though EDFAs at present are almost certain to be used as in-line amplifiers in undersea lightwave systems, the small size and the electrical-pumping properties of SLAs make them attractive for many applications involving optoelectronic integration. Examples of such applications include photonic switching, optical interconnects, and photonic integrated circuits for optical transmitters and receivers.[4,5] Since SLAs are likely to play an important role in future applications of semiconductor lasers, this chapter is devoted to them.

All lasers act as amplifiers close to but before reaching threshold. Indeed, laser amplifiers have been extensively studied in the context of solid-state and dye lasers, and their properties are invariably discussed in any textbook on lasers.[6,7] It comes therefore as no surprise that the research on SLAs started[8-11] soon after the invention of semiconductor lasers in 1962. However, it was only during the 1980s that SLAs were developed for practical applications, mainly associated with lightwave telecommunication systems.[12-25] This chapter reviews the current status of SLAs and is organized as follows. To provide the background material, Sec. 8.2 discusses general concepts such as gain bandwidth, gain saturation, and spontaneous-emission noise that are common to all optical amplifiers. Design aspects of SLAs are considered in Sec. 8.3 where the difference between the Fabry–Perot and the traveling-wave SLAs is outlined. Section 8.4 focuses on the amplification characteristics of SLAs including amplifier bandwidth, saturation power, noise figure, and polarization sensitivity. Amplification of ultrashort optical pulses in SLAs is discussed in Sec. 8.5 with emphasis on spectral changes introduced by self-phase modulation occurring as a result of gain saturation. Section 8.6 considers interchannel cross-talk resulting from cross saturation and four-wave mixing during multichannel amplification. The potential applications of SLAs are discussed in Sec. 8.7. The last section concludes the chapter with a discussion of future prospects.

8.2. GENERAL CONCEPTS

Optical amplifiers amplify incident light through stimulated emission, the same mechanism used by lasers. Indeed, an optical amplifier is nothing but a laser without feedback. Its main ingredient is the optical gain realized when the amplifier is pumped (optically or electrically) to achieve population inversion. The optical gain in general depends not only on the frequency (or wavelength) of the incident signal, but also on the local beam intensity at any point inside the amplifier. Details of the frequency and intensity dependence of the optical gain depend on the amplifier medium. To illustrate the general concepts, let us consider the case in which the gain medium is modeled as a homogeneously broadened two-level system.[6,7] The gain coefficient of such a medium can be written as

$$g(\omega) = \frac{g_0}{1 + (\omega - \omega_0)^2 T_2^2 + P/P_s}, \tag{8.1}$$

where g_0 is the peak value of the gain determined by the pumping level of the amplifier, ω is the optical frequency of the incident signal, ω_0 is the atomic transition frequency, and P is the optical power of the signal being amplified. The saturation power P_s depends on the gain-medium parameters such as the fluorescence time T_1 and the transition cross section; its expression for SLAs is given in Sec. 8.4. The parameter T_2 in Eq. (8.1) is known as the dipole relaxation time[6,7] and is typically quite small (~ 0.1 ps for SLAs). The fluorescence time T_1 is also called the population relaxation time and is related to the carrier lifetime (~ 0.1–1 ns) in the case of SLAs. Equation (8.1) can be used in a discussion of important characteristics of optical amplifiers, such as the gain bandwidth, amplification factor, and the output saturation power. We begin by considering the case in which $P/P_s \ll 1$ throughout the amplifier. This is referred to as the unsaturated region, since the gain remains unsaturated during amplification.

8.2.1. Gain Spectrum and Bandwidth

By neglecting the term P/P_s in Eq. (8.1), the gain coefficient becomes

$$g(\omega) = \frac{g_0}{1 + (\omega - \omega_0)^2 T_2^2}. \tag{8.2}$$

This equation shows that the gain is maximum when the incident frequency ω coincides with the atomic transition frequency ω_0. The gain reduction for $\omega \neq \omega_0$ is governed by a Lorentzian profile that is a characteristic of homogeneously broadened two-level systems.[6,7] As discussed later, the gain spectrum of SLAs deviates considerably from the Lorentzian profile. The *gain bandwidth* is defined as the full width at half-maximum (FWHM) of the gain spectrum $g(\omega)$. For the Lorentzian spectrum, the gain bandwidth is given by $\Delta\omega_g = 2/T_2$ or by

$$\Delta\nu_g = \frac{\Delta\omega_g}{2\pi} = \frac{1}{\pi T_2}. \tag{8.3}$$

As an example, $\Delta\nu_g \sim 3$ THz for SLAs if we use Eq. (8.3) with $T_2 \sim 0.1$ ps. Amplifiers with a relatively large bandwidth are preferred for optical communication systems, since the gain is then nearly constant over the entire bandwidth of even a multichannel signal. Large-bandwidth amplifiers are, of course, necessary for amplification of ultrashort optical pulses.

The concept of amplifier bandwidth is commonly used in place of the gain bandwidth. The difference between the two becomes clear when one considers the amplifier gain G, also known as the amplification factor and defined as

$$G = P_{\text{out}}/P_{\text{in}}, \tag{8.4}$$

where P_{in} and P_{out} are the input and output powers of the continuous-wave (cw) signal being amplified. We can obtain an expression for G by using

$$\frac{dP}{dz} = gP(z), \tag{8.5}$$

where $P(z)$ is the optical power at a distance z from the input end. A straightforward integration with the initial condition $P(0) = P_{\text{in}}$ shows that the signal power grows exponentially as

$$P(z) = P_{\text{in}} \exp(gz). \tag{8.6}$$

By noting that $P(L) = P_{\text{out}}$ and using Eq. (8.4), the amplification factor for an amplifier of length L is given by

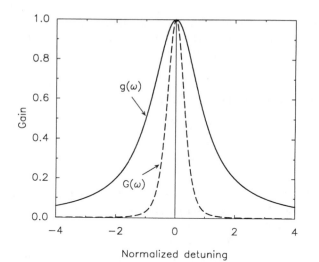

FIGURE 8.1. Lorentzian gain profile $g(\omega)$ and the corresponding amplifier-gain spectrum $G(\omega)$ for an amplifier modeled as a two-level atomic system. The gain bandwidth and the amplifier bandwidth correspond to the full width at half-maximum (FWHM) of the two spectra.

$$G(\omega) = \exp[g(\omega)L],\qquad(8.7)$$

where the frequency dependence is shown explicitly. Both the amplifier gain $G(\omega)$ and the gain coefficient $g(\omega)$ are maximum when $\omega = \omega_0$ and decrease with the signal detuning $\omega - \omega_0$. However, $G(\omega)$ decreases much faster than $g(\omega)$ because of the exponential dependence of G on g. The amplifier bandwidth $\Delta \nu_A$ is defined as the full width at half maximum of $G(\omega)$ and is related to the gain bandwidth $\Delta \nu_g$ by

$$\Delta \nu_A = \Delta \nu_g \left(\frac{\ln 2}{\ln G_0 - \ln 2} \right)^{1/2},\qquad(8.8)$$

where $G_0 = \exp(g_0 L)$ was assumed to be larger than 2. As expected, the amplifier bandwidth is smaller than the gain bandwidth, and the difference depends on the amplifier gain itself. Figure 8.1 shows the gain profile $g(\omega)$ and the amplification factor $G(\omega)$ by plotting g/g_0 and G/G_0 as a function of $(\omega - \omega_0)T_2$ for $G_0 = 10$. Even for such a low value of G_0, $\Delta \nu_A$ is smaller than $\Delta \nu_g$ by a factor of 2.3. It becomes smaller by almost a factor of 10 for an amplifier with 30 dB peak gain.

8.2.2. Gain Saturation

The origin of gain saturation lies in the power dependence of the gain coefficient in Eq. (8.1). When $P \ll P_s$, $g(\omega)$ reduces to Eq. (8.2) and is referred to as the small-signal gain since the incident signal power should be small and remain small during amplification. Since g is reduced when P becomes comparable to P_s, the

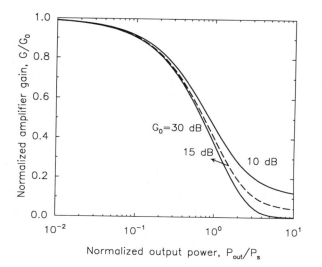

Normalized output power, P_{out}/P_s

FIGURE 8.2. Saturated amplifier gain G as a function of the output power (normalized to the saturation power) for several values of the unsaturated amplifier gain G_0.

amplification factor G is also expected to decrease. To simplify the discussion, let us consider the case in which incident signal frequency is exactly tuned to the atomic transition frequency ω_0 to maximize the small-signal gain. The detuning effects can be incorporated in a straightforward manner. By substituting g from Eq. (8.1) in Eq. (8.5), we obtain

$$\frac{dP}{dz} = \frac{g_0 P}{1 + P/P_s}.$$ (8.9)

This equation can be easily integrated over the amplifier length.[6] By using the initial condition $P(0) = P_{in}$ together with $P(L) = P_{out} = GP_{in}$, we obtain the following implicit relation for the large-signal amplifier gain:

$$G = G_0 \exp\left(-\frac{G-1}{G}\frac{P_{out}}{P_s}\right),$$ (8.10)

where $G_0 = \exp(g_0 L)$ is the unsaturated value of the amplification factor $(P_{out} \ll P_s)$.

Equation (8.10) shows that the amplification factor G decreases from its unsaturated value G_0 when P_{out} becomes comparable to P_s. Figure 8.2 shows the saturation characteristics by plotting G as a function of P_{out}/P_s for several values of G_0. A quantity of practical interest is the output saturation power P_{out}^s, defined as the output power for which the amplifier gain G is reduced by a factor of 2 (or by 3 dB) from its unsaturated value G_0. Setting G equal to $G_0/2$ in Eq. (8.10) yields

$$P_{\text{out}}^{S} = \frac{G_0 \ln 2}{G_0 - 2} P_s. \tag{8.11}$$

P_{out}^{s} is smaller than P_s by about 30%. Indeed, noting that $G_0 \gg 2$ in practice ($G_0 = 1000$ for 30 dB amplifier gain), $P_{\text{out}}^{s} \cong (\ln 2)P_s \approx 0.69P_s$. As seen in Fig. 8.2, P_{out}^{s} becomes nearly independent of G_0 for $G_0 > 20$ dB.

8.2.3. Amplifier Noise

All laser amplifiers degrade the signal-to-noise ratio (SNR) of the amplified signal because of spontaneous emission that adds to the signal during its amplification. The SNR degradation is quantified through a parameter F_n, called the amplifier noise figure in analogy with the electronic amplifiers and defined as[1]

$$F_n = \frac{(\text{SNR})_{\text{in}}}{(\text{SNR})_{\text{out}}}, \tag{8.12}$$

where SNR refers to the electrical power generated when the signal is converted to electrical current by using a photodetector. In general, F_n would depend on several detector parameters which govern the shot noise and thermal noise associated with the detector.[1] One can obtain a simple expression for F_n by considering an ideal detector whose performance is limited by the shot noise only.

Consider an amplifier with amplification factor G so that the output power is related to the input power by $P_{\text{out}} = GP_{\text{in}}$. The SNR of the input signal is given by

$$(\text{SNR})_{\text{in}} = \frac{\langle I \rangle^2}{\sigma_s^2} = \frac{(RP_{\text{in}})^2}{2q(RP_{\text{in}})\Delta f} = \frac{P_{\text{in}}}{2h\nu\Delta f}, \tag{8.13}$$

where q is the electron charge, $h\nu$ is the photon energy, $\langle I \rangle = RP_{\text{in}}$ is the average photocurrent, $R = q/h\nu$ is the responsivity of an ideal photodetector with unit quantum efficiency, and

$$\sigma_s^2 = 2q(RP_{\text{in}})\Delta f \tag{8.14}$$

is the shot noise[1] obtained by neglecting the contribution of the dark current for a detector of bandwidth Δf. To evaluate the SNR of the amplified signal, one should add the contribution of spontaneous emission to the receiver noise.

The spectral density of spontaneous emission is constant over the entire gain bandwidth of the amplifier and can be written as[1]

$$S_{\text{sp}}(\nu) = (G-1)n_{\text{sp}}h\nu, \tag{8.15}$$

where $S_{\text{sp}}(\nu)$ represents the two-sided spectral density. The parameter n_{sp} is called the spontaneous-emission factor or population-inversion factor. Its value is 1 for amplifiers with complete population inversion (all atoms in the upper state), but becomes > 1 when the population inversion is incomplete. For a two-level system

$$n_{sp} = N_2/(N_2-N_1), \tag{8.16}$$

where N_1 and N_2 are the atomic populations in the lower and upper states respectively. The effect of spontaneous emission is to add fluctuations to the amplified power which are converted to current fluctuations during the photodetection process.

The dominant contribution to the receiver noise comes from beating of spontaneous emission with the signal. This beating phenomenon is similar to heterodyne detection used in the context of coherent communication systems: Spontaneously emitted radiation mixes coherently with the amplified signal at the photodetector and produces a heterodyne component of the photocurrent. The variance of the photocurrent can then be written as[1]

$$\sigma^2 = 2q(RGP_{in})\Delta f + 4(RGP_{in})(RS_{sp})\Delta f, \tag{8.17}$$

where the first term is due to shot noise and the second term results from signal-spontaneous emission beating. All other contributions to the receiver noise have been neglected for simplicity. The SNR of the amplified signal is thus given by

$$(SNR)_{out} = \frac{\langle I \rangle^2}{\sigma^2} = \frac{(RGP_{in})^2}{\sigma^2} \approx \frac{GP_{in}}{4S_{sp}\Delta f}, \tag{8.18}$$

where the last relation was obtained by neglecting the first term in Eq. (8.17) and is valid for $G \gg 1$.

The amplifier noise figure can now be obtained by substituting Eqs. (8.13) and (8.18) in Eq. (8.12). If we also use Eq. (8.15) for S_{sp}, we obtain

$$F_n = 2n_{sp}(G-1)/G \approx 2n_{sp}. \tag{8.19}$$

This equation shows that the SNR of the amplified signal is degraded by a factor of 2 (or 3 dB) even for an ideal amplifier for which $n_{sp} = 1$. For most practical amplifiers, F_n exceeds 3 dB and is typically in the range 4–8 dB. For its application in optical communication systems, an optical amplifier should have F_n as low as possible.

8.3. AMPLIFIER DESIGN

8.3.1. Fabry–Perot Amplifiers

The amplifier characteristics discussed in Sec. 8.2 assumed an ideal optical amplifier without any feedback from its input and output ports. Such amplifiers are called traveling-wave (TW) amplifiers to emphasize that the amplified radiation travels in the forward direction only. Semiconductor lasers experience a relatively large feedback because of reflections occurring at the cleaved facets (~ 32% reflectivity). They can be used as amplifiers when biased below threshold, but the effects of the Fabry–Perot (FP) cavity formed through reflections at the facets must be included in their analysis. Amplifiers of this kind are called FP amplifiers. The amplification factor is obtained by using the standard theory of FP interferometers and is given by[13]

$$G_{FP}(\nu) = \frac{(1-R_1)(1-R_2)G(\nu)}{(1-G\sqrt{R_1 R_2})^2 + 4G\sqrt{R_1 R_2}\,\sin^2[\pi(\nu-\nu_m)/\Delta\nu_L]}, \quad (8.20)$$

where R_1 and R_2 are the facet reflectivities; $\nu_m = mc/2nL$ with m an integer representing the longitudinal-mode frequencies for an SLA of length L and the mode index n; and $\Delta\nu_L = c/2n_g L$, with n_g as the group index, is the longitudinal-mode spacing, also known as the free-spectral range of the FP cavity. The single-pass amplification factor G corresponds to that of a TW amplifier and is given by Eq. (8.7) when gain saturation is negligible. Indeed, G_{FP} reduces to G when $R_1 = R_2 = 0$.

As evident from Eq. (8.20), $G_{FP}(\nu)$ peaks whenever ν coincides with one of the cavity-resonance frequencies ν_m and drops sharply in between them. The amplifier bandwidth is thus determined by the sharpness of the cavity resonance. One can calculate the amplifier bandwidth by calculating the detuning $\nu-\nu_m$ for which G_{FP} drops by 3 dB from its peak value. The result is

$$\Delta\nu_A = \frac{2\Delta\nu_L}{\pi}\sin^{-1}\left[\frac{1-G\sqrt{R_1 R_2}}{(4G\sqrt{R_1 R_2})^{1/2}}\right]. \quad (8.21)$$

To achieve a large amplification factor, $G\sqrt{R_1 R_2}$ should be quite close to 1. As seen from Eq. (8.21), the amplifier bandwidth is then a small fraction of the free spectral range of the FP cavity (typically, $\Delta\nu_L \sim 100$ GHz and $\Delta\nu_A < 10$ GHz). Such a small bandwidth makes FP amplifiers less suitable than TW amplifiers for lightwave system applications, and they are primarily used for signal-processing applications.

8.3.2. Traveling-Wave Amplifiers

Traveling-wave-type SLAs can be made if the reflection feedback from the end facets is suppressed. A simple way to reduce the reflectivity is to coat the facets with an antireflection coating. However, the facet reflectivity must be extremely small ($< 10^{-3}$) for the SLA to act as a TW amplifier. Furthermore, the minimum reflectivity depends on the amplifier gain itself. One can estimate the tolerable value of the facet reflectivity by considering the maximum and minimum values of G_{FP} from Eq. (8.20) near a cavity resonance. It is easy to verify that their ratio is given by

$$\Delta G = \frac{G_{FP}^{max}}{G_{FP}^{min}} = \left(\frac{1+G\sqrt{R_1 R_2}}{1-G\sqrt{R_1 R_2}}\right)^2. \quad (8.22)$$

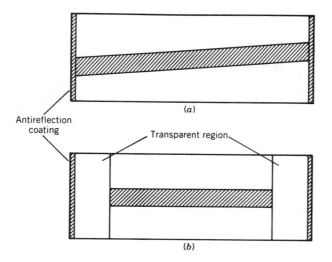

FIGURE 8.3. (a) Tilted-stripe and (b) and buried-facet structures for nearly traveling-wave (TW) semiconductor laser amplifiers.

If ΔG exceeds 3 dB, the amplifier bandwidth is set by the cavity resonances rather than by the gain spectrum. To keep $\Delta G < 2$, the facet reflectivities should satisfy the condition

$$G\sqrt{R_1 R_2} < 0.17. \qquad (8.23)$$

It is customary to characterize the SLA as a TW amplifier when Eq. (8.23) is satisfied. An SLA designed to provide a 30 dB amplification factor $(G = 1000)$ should have facet reflectivities such that $\sqrt{R_1 R_2} < 1.7 \times 10^{-4}$.

Considerable effort is required to produce antireflection coatings with reflectivities less than 0.1%.[26-29] Even then, it is difficult to obtain low facet reflectivities in a predictable and regular manner. For this reason, alternative techniques have been developed to reduce the reflection feedback in SLAs. In one method, the active-region stripe is tilted from the facet normal, as shown in Fig. 8.3(a). Such a structure is referred to as the angled-facet or the tilted-stripe structure.[30-34] The reflected beam at the facet is physically separated from the forward beam because of the angled facet. Some feedback can still occur but the combination of an antireflection coating (reflectivity $\sim 1\%$) and the tilted stripe can produce reflectivities below 10^{-3} (as small as 10^{-4} with design optimization). In an alternative scheme[35,36] [Fig. 8.3(b)] a transparent window region is inserted between the active-layer ends and the facets. The optical beam spreads in this window region before arriving at the semiconductor-air interface. The reflected beam spreads even further and does not couple much light into the thin active layer. Such a structure is called the buried-facet or the window-facet structure and has provided reflectivities as small as 10^{-4} when used in combination with antireflection coatings.[35,36]

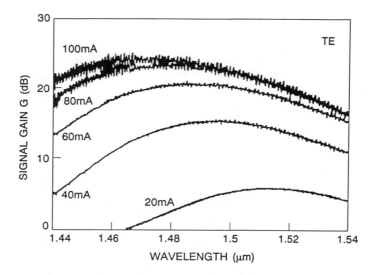

FIGURE 8.4. Amplifier gain vs signal wavelength at several currents for a semiconductor laser amplifier whose facets are antireflection coated to reduce reflectivity to about 0.04% (after Ref. 18 © 1988 IEEE).

8.4. AMPLIFIER CHARACTERISTICS

8.4.1. Amplifier Bandwidth

The amplification factor of SLAs is given by Eq. (8.20). Its frequency dependence results mainly from the frequency dependence of $G(\nu)$ when the condition (8.23) is satisfied. The measured amplifier gain exhibits ripples reflecting the effects of residual facet reflectivities. Figure 8.4 shows the wavelength dependence of the amplifier gain measured through the amplified spontaneous emission.[18] The facet reflectivities of the SLA were about 4×10^{-4}. Condition (8.23) is well satisfied since $G\sqrt{G_1 R_2} \approx 0.04$ for this amplifier. Gain ripples are negligibly small because the SLA in this case operates in a nearly TW mode. The 3 dB amplifier bandwidth is about 70 nm (9 THz). This bandwidth reflects the relatively broad gain spectrum, $g(\omega)$, of SLAs. The gain spectrum of semiconductor lasers is far from being Lorentzian, and its calculation requires a knowledge of the band-structure details such as the joint density of states associated with the semiconductor material of the active region. The gain bandwidth can exceed 100 nm for quantum-well SLAs. As discussed in Sec. 8.2.1, the amplifier bandwidth is smaller than the gain bandwidth. In the case of SLAs the amplifier bandwidth may depend on the facet reflectivities. Indeed, one should use Eq. (8.21) if condition (8.23) is not satisfied. For a nearly TW-type SLA, the amplifier bandwidth can be estimated by using Eq. (8.8).

8.4.2. Saturation Power

To discuss gain saturation, we consider the case of a cw input beam whose wavelength nearly coincides with the gain peak. In most cases of practical interest the peak gain can be assumed to increase linearly with the carrier population N as[25]

$$g = (\Gamma \sigma_g / V)(N - N_0), \qquad (8.24)$$

where Γ is the confinement factor, σ_g is the differential gain coefficient, V is the active volume, and N_0 is the value of N at which the active region of the SLA becomes transparent (the rate of stimulated emission becomes large enough to compensate for the material loss). The gain has been reduced by Γ to account for spreading of the waveguide mode outside the gain region of SLAs. The carrier population N changes with the injection current I and the signal power P according to the well-known carrier rate equation.[25] By expressing the photon number in terms of the optical power, this equation can be written as

$$\frac{dN}{dt} = \frac{I}{q} - \frac{N}{\tau_c} - \frac{\sigma_g(N - N_0)}{\sigma_m h \nu} P, \qquad (8.25)$$

where τ_c is the carrier lifetime and σ_m is the cross-section area of the waveguide mode. In the case of a cw beam, or pulses of duration much longer than τ_c, the steady-state value of N can be obtained by setting $dN/dt = 0$ in Eq. (8.25). When the solution is substituted in Eq. (8.24), the optical gain is found to saturate in accord with

$$g = \frac{g_0}{1 + P/P_s}, \qquad (8.26)$$

where the small-signal gain g_0 is given by

$$g_0 = (\Gamma \sigma_g / V)(I \tau_c / q - N_0) \qquad (8.27)$$

and the saturation power P_s is defined as

$$P_s = h \nu \sigma_m / (\sigma_g \tau_c). \qquad (8.28)$$

A comparison of Eqs. (8.1) and (8.26) shows that the SLA gain saturates in the same way as that of a two-level system. We can therefore use the analysis of Sec. 8.2.2. In particular, the output saturation power P_{out}^s is obtained by using P_s from Eq. (8.28) in Eq. (8.11) and is given

$$P_{\text{out}}^s = \frac{G_0 \ln 2}{G_0 - 2} \frac{h \nu \sigma_m}{\sigma_g \tau_c}. \qquad (8.29)$$

Figure 8.5 shows experimentally measured gain-saturation characteristics of both TW- and FP-type SLAs.[24] In general, the output saturation power for FP-type SLAs (~ 0.1 mW) is much smaller than that of TW amplifiers (~ 5–10 mW) because of

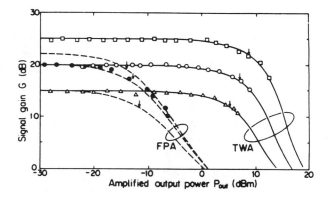

FIGURE 8.5. Theoretical and experimental signal gain as a function of the amplified output power for a TW (solid lines) and a FP (dashed lines) amplifier (after Ref. 18).

the resonant nature of amplification. Many applications require SLAs with large values of the output saturation power and therefore use TW amplifiers.

8.4.3. Noise Figure

The noise figure F_n of SLAs is larger than the minimum value of 3 dB for several reasons. The dominant contribution comes from the population-inversion factor n_{sp}, which represents the ratio of spontaneous to net stimulated emission rate.[17] For SLAs, n_{sp} is obtained from Eq. (8.16) by replacing N_2 and N_1 by N and N_0, respectively. An additional contribution results from nonresonant internal losses α_{int} (such as free-carrier absorption or scattering loss) which reduce the available gain from g to $g - \alpha_{int}$. By using Eq. (8.19) and including this additional contribution, the noise figure can be written as[24]

$$F_n = 2\left(\frac{N}{N-N_0}\right)\left(\frac{g}{g-\alpha_{int}}\right). \tag{8.30}$$

Residual facet reflectivities increase F_n by an additional factor that can be approximated by $1+R_1 G$, where R_1 is the reflectivity of the input facet.[17] In most TW amplifiers $R_1 G \ll 1$, and this contribution can be neglected. Typical values of F_n for SLAs are in the 5–7 dB range.[21]

Equation (8.30) for the amplifier noise figure is based on the analysis of Sec. 8.2.3 and neglects several sources of noise. For a more accurate analysis of amplifier noise, one should consider how the photocurrent generated at a detector is effected by the process of amplification. The amplified signal received by the photodetector can be written as

$$P_{amp} = GP_s+P_{sp}, \tag{8.31}$$

where G is the amplifier gain, P_s is the input optical signal, and P_{sp} is the spontaneous-emission noise power added to the signal such that

$$P_{sp} = S_{sp}\Delta\nu_{sp}. \tag{8.32}$$

The spectral density S_{sp} is given by Eq. (8.15) and $\Delta\nu_{sp}$ is the effective bandwidth of spontaneous emission; $\Delta\nu_{sp}$ can be approximated by the amplifier bandwidth in practice.

The photocurrent generated at the detector can be written as

$$I(t) = I_p + i(t), \tag{8.33}$$

where $I_p = RP_{amp}$ is the average current and $i(t)$ represents current fluctuations which include the effects of shot noise, thermal noise, and spontaneous-emission noise. The variance ($\sigma^2 = \langle i^2 \rangle$) of current fluctuations can be written as[17,23]

$$\sigma^2 = \sigma_T^2 + \sigma_s^2 + \sigma_{sp\text{-}sp}^2 + \sigma_{sig\text{-}sp}^2 + \sigma_{s\text{-}sp}^2, \tag{8.34}$$

where σ_T^2 is the thermal noise and the remaining four terms are given by[1]

$$\sigma_s^2 = 2q[R(GP_s + P_{sp}) + I_d]\Delta f, \tag{8.35}$$

$$\sigma_{sp\text{-}sp}^2 = 4R^2 S_{sp}^2 \Delta\nu_{opt}\Delta f, \tag{8.36}$$

$$\sigma_{sig\text{-}sp}^2 = 4R^2 GP_s S_{sp}\Delta f, \tag{8.37}$$

$$\sigma_{s\text{-}sp}^2 = 4qRS_{sp}\Delta\nu_{opt}\Delta f, \tag{8.38}$$

where $R = \eta q/h\nu$ is the photodetector responsivity, η is the quantum efficiency, and Δf is the receiver bandwidth. The shot-noise term σ_s^2 is the same as in Eq. (8.14) except that P_{sp} has been added to GP_s to account for the shot noise generated by spontaneous emission. The three contributions $\sigma_{sp\text{-}sp}^2$, $\sigma_{sig\text{-}sp}^2$, and $\sigma_{s\text{-}sp}^2$ originate from beating of spontaneous emission against itself, signal, and shot noise, respectively. Their origin has been discussed extensively in both classical and quantum-mechanical contexts.[17,23] Since the signal and spontaneous emission noise are not at the same optical frequency, the two can beat against each other to produce fluctuations in the detector current. Spontaneous emission can also beat against itself because it spans a wide frequency range governed by $\Delta\nu_{sp}$. All such beating terms contribute to the receiver noise and must be included. In Eqs. (8.12)–(8.38) Δf is the receiver (electrical) bandwidth, whereas $\Delta\nu_{opt}$ is the optical bandwidth of the spontaneous-emission noise. $\Delta\nu_{opt}$ can be different from $\Delta\nu_{sp}$ if an optical filter is placed before the photodetector to reduce the amount of spontaneous emission reaching the photodetector. In the absence of an optical filter $\Delta\nu_{opt} = \Delta\nu_{sp}$. When an optical filter is used, $\Delta\nu_{opt}$ is related to the filter bandwidth. For a filter with a Lorentzian response, $\Delta\nu_{opt} = (\pi/4)\Delta\nu_f$, where $\Delta\nu_f$ is the filter bandwidth (FWHM).

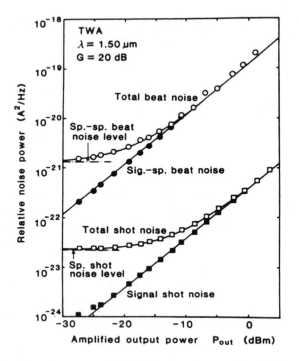

FIGURE 8.6. Relative noise power of different noise sources as a function of the amplified output power for a 1.5 μm TW amplifier operating at 20 dB gain. An ètalon filter with a free-spectral range of 1.5 THz and a finesse of 11 was used to suppress the contribution of $\sigma^2_{\text{sp-sp}}$ and $\sigma^2_{s\text{-sp}}$ (after Ref. 18 © 1987 IEEE).

Figure 8.6 shows the measured relative noise powers (per unit bandwidth) as a function of the amplified output power for a 1.5-μm InGaAsP amplifier when the SLA is operated with 20 dB gain.[18] An FP filter with 1.5 THz bandwidth and a finesse of 11 was placed in front of the avalanche photodiode in order to reduce the spontaneous-emission power reaching the detector. The experimental results can be understood by comparing the magnitude of various terms in Eq. (8.34). For this purpose, it is useful to substitute S_{sp} from Eq. (8.15), use $R = \eta q / h\nu$ and Eq. (8.19), and write Eqs. (8.35)–(8.38) in terms of the noise figure F_n as

$$\sigma^2_s = 2q^2 \eta G P_s \Delta f / h\nu, \tag{8.39}$$

$$\sigma^2_{\text{sp-sp}} = (q\eta G F_n)^2 \Delta \nu_{\text{opt}} \Delta f, \tag{8.40}$$

$$\sigma^2_{\text{sig-sp}} = 2(q\eta G)^2 F_n P_s \Delta f / h\nu, \tag{8.41}$$

$$\sigma^2_{s\text{-sp}} = 2q^2 \eta G F_n \Delta \nu_{\text{opt}} \Delta f, \tag{8.42}$$

where RP_{sp} and I_d were neglected in Eq. (8.35) as they contribute negligibly to the shot noise. A comparison of Eqs. (8.39) and (8.41) shows that σ_s^2 can be neglected in comparison with σ_{sig-sp}^2 as it is smaller by a large factor $\eta G F_n$. Experimental data of Fig. 8.6 confirm this conclusion since the measured shot noise is smaller by about 20 dB compared with the signal-spontaneous beat noise. Similarly, a comparison of Eqs. (8.40) and (8.42) shows that σ_{s-sp}^2 can be neglected in comparison with σ_{sp-sp}^2. The dominant contribution to the photocurrent noise comes from σ_{sig-sp}^2 and σ_{sp-sp}^2, both of which increase with the noise figure. At high output powers the total beat noise is dominated by the signal-spontaneous beat noise as also observed experimentally in Fig. 8.6.

8.4.4. Polarization Sensitivity

An undesirable characteristic of SLAs is the sensitivity of their gain to polarization. The amplifier gain G differs for the transverse-electric (TE) and transverse-magnetic (TM) polarizations by as much as 5–8 dB since the confinement factor Γ and the effective mode index are different for the TE and TM polarizations because of different physical dimensions of the width and the thickness of the active layer. This feature makes the amplifier gain dependent on the polarization state of the input beam, a property undesirable for lightwave system applications where the polarization state changes with propagation along the fiber (unless polarization-preserving fibers are used). Several schemes have been devised to reduce the polarization sensitivity.[36–41] Clearly an SLA will become nearly polarization insensitive if the width and the thickness of the active layer are made the same. A gain difference of less than 1.3 dB between TE and TM polarizations has been realized by making the active layer 0.26 μm thick and 0.4 μm wide.[36] Another scheme makes use of a large-optical-cavity structure so that the confinement factors are nearly the same for TE and TM modes; a gain difference of less than 1 dB has been obtained with such a structure.[37]

Several other schemes reduce the polarization sensitivity by using two amplifiers[38] or two passes through the same amplifier.[39] Figure 8.7 shows three appropriate configurations. In Fig. 8.7(a) the TE-polarized signal in one amplifier becomes TM polarized in the second amplifier, and vice versa. If both amplifiers have identical gain characteristics, the twin-amplifier configuration provides signal gain that is independent of the signal polarization. A drawback of the series configuration is that residual facet reflectivities lead to mutual coupling between the two amplifiers. In the parallel configuration shown in Fig. 8.7(b) the incident signal is split into a TE- and a TM-polarized signal, each of which is amplified by separate amplifiers. The amplified TE and TM signals are then combined to produce the amplified signal with the same polarization as that of the input beam.[38] The double-pass configuration of Fig. 8.7(c) passes the signal through the same amplifier twice, but the polarization is rotated by 90° between the two passes.[39] Since the amplified signal propagates in the backward direction, a 3-dB fiber coupler is needed to separate it from the incident signal. In spite of a 6 dB loss occurring at the fiber

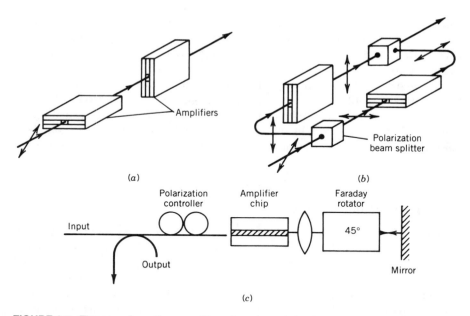

FIGURE 8.7. Three configurations used to reduce the polarization sensitivity of semiconductor laser amplifiers: (a) twin amplifiers in series, (b) twin amplifiers in parallel, and (c) double pass through a single amplifier.

coupler (3 dB for the input signal and 3 dB for the amplified signal) this configuration provides high gain from a single amplifier, as the same amplifier supplies gain on the two passes.

8.4.5. Multiquantum-Well Amplifiers

The amplification characteristics of SLAs can be improved depending on the application requirements by suitable design changes. For example, the output saturation power can be increased by increasing the mode cross section in Eq. (8.28) through an increase in the active-region width. However, when the width increases beyond a certain value ($> 2\ \mu$m), the SLA begins to support multiple lateral modes, a feature that degrades the beam quality. A solution is provided by the use of a tapered waveguide in which the width of the active region increases continuously from input to the output end. Such a tapered waveguide can increase the output saturation power to as high as 100 mW without excitation of higher-order lateral modes.[42] Tapered SLAs are discussed further in Sec. 8.7.3 in the context of high-power lasers.

Recently considerable attention has focused on SLAs whose active region consists of multiple quantum wells (MQWs).[43–51] Such amplifiers have a number of attractive properties resulting from the quantum-confinement effects discussed in Chapter 1. A 1.5-μm MQW amplifier with four InGaAs 8-nm quantum wells separated by 10-nm InGaAsP barrier layers exhibited a noise figure of 4.4 dB while operating with 19 dB gain.[47] In another 1.55-μm MQW amplifier, the measured noise figure was only 3.9 dB for a single-pass gain of 22 dB.[48] Theoretical calculations show that the noise figure can be reduced to 3.5 dB by using the quantum-box structure.[45] The use of strained quantum wells can also improve the SLA performance. For example, MQW amplifiers can be made polarization insensitive by using barrier layers with tensile strain. In one set of experiments,[49] ten InGaAs quantum wells (lattice matched to InP) were separated by 11 tensile-strained barriers with a lattice mismatch of -1.7%. These amplifiers exhibited a polarization sensitivity of less than 0.5 dB while providing 27.5 dB signal gain near 1.55-μm together with a relatively high output saturation power (about 25 mW). In another set of experiments[50,51] polarization-insensitive MQW amplifiers were made by alternating multiple quantum wells with tensile and compressive strains. The use of strained quantum wells appears to improve the SLA performance considerably.

8.5. PULSE AMPLIFICATION

The large bandwidth of TW-type SLAs suggests that they are capable of amplifying ultrashort optical pulses (as short as a few picoseconds) without significant pulse distortion. However, when the pulse width τ_p becomes shorter than the carrier lifetime τ_c, gain dynamics play an important role, since both N and g in Eq. (8.24) become time dependent. Ultrashort pulse amplification in SLAs has been extensively studied[52-66] both theoretically and experimentally. This section discusses how pulse shape and spectrum are affected by the gain dynamics in SLAs.

8.5.1. Time-Dependent Amplifier Gain

In the slowly varying envelope approximation, the amplitude $A(z,t)$ of the pulse envelope in an SLA evolves as[55]

$$\frac{\partial A}{\partial z} + \frac{1}{v_g}\frac{\partial A}{\partial t} = \frac{1}{2}(1+i\beta_c)gA, \qquad (8.43)$$

where carrier-induced index changes are included through the linewidth enhancement factor β_c. The dispersive effects are not included in Eq. (8.43) because of negligible material dispersion and a short amplifier length (< 1 mm in most cases). The time dependence of the gain is governed by Eqs. (8.24) and (8.25). The two equations can be combined to yield

$$\frac{\partial g}{\partial t} = \frac{g_0 - g}{\tau_c} - \frac{gP}{E_s}, \qquad (8.44)$$

where the saturation energy E_s is defined as

$$E_s = h\nu(\sigma_m/\sigma_g), \qquad (8.45)$$

and g_0 is given by Eq. (8.27).

Equations (8.43) and (8.44) govern amplification of optical pulses in SLAs. They can be solved analytically for pulses whose duration is short compared with the carrier lifetime ($\tau_p \ll \tau_c$). The first term on the right side of Eq. (8.44) can then be neglected during pulse amplification. By introducing the reduced time $\tau = t - z/v_g$ together with $A = \sqrt{P}\exp(i\phi)$, Eqs. (8.43) and (8.44) can be written as[55]

$$\frac{\partial P}{\partial z} = g(z,\tau)P(z,\tau), \tag{8.46}$$

$$\frac{\partial \phi}{\partial z} = -\frac{1}{2}\beta_c g(z,\tau), \tag{8.47}$$

$$\frac{\partial g}{\partial \tau} = -\frac{1}{E_s}g(z,\tau)P(z,\tau). \tag{8.48}$$

Equation (8.46) is identical with Eq. (8.5) except that the gain g depends on both z and τ. It can easily be integrated over the amplifier length L to provide the output power as

$$P_{out}(\tau) = P_{in}(\tau)\exp[h(\tau)], \tag{8.49}$$

where $P_{in}(\tau)$ is the input power and $h(\tau)$ is the total integrated gain defined as

$$h(\tau) = \int_0^L g(z,\tau)dz. \tag{8.50}$$

If we integrate Eq. (8.48) over the amplifier length after replacing gP by $\partial P/\partial z$, we obtain[53,55]

$$\frac{dh}{d\tau} = -\frac{1}{E_s}[P_{out}(\tau) - P_{in}(\tau)] = -\frac{P_{in}(\tau)}{E_s}(e^h - 1), \tag{8.51}$$

where Eq. (8.49) was used to relate P_{out} to P_{in}. Equation (8.51) can be easily integrated to obtain $h(\tau)$. The quantity of practical interest is the amplification factor $G(\tau)$ related to $h(\tau)$ by $G = \exp(h)$. It is given by[6]

$$G(\tau) = \frac{G_0}{G_0 - (G_0 - 1)\exp[-E_0(\tau)/E_s]}, \tag{8.52}$$

where G_0 is the unsaturated amplifier gain and

$$E_0(\tau) = \int_{-\infty}^{\tau} P_{in}(\tau)d\tau \tag{8.53}$$

is the partial energy of the input pulse defined such that $E_0(\infty)$ equals the input pulse energy E_{in}.

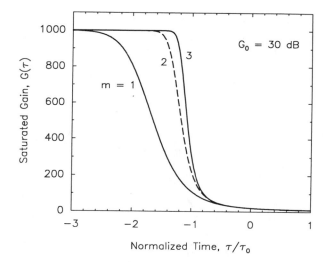

FIGURE 8.8. Time-dependent amplification factor for super-Gaussian input pulses of input energy such that $E_{in}/E_s = 0.1$. The unsaturated value G_0 is 30 dB in all cases. The input pulse is Gaussian for $m = 1$ but becomes nearly rectangular as m increases.

8.5.2. Shape and Spectrum of Amplified Pulse

Solution (8.52) shows that the amplifier gain is different for different parts of the pulse. The leading edge experiences the full gain G_0, as the amplifier is not yet saturated. The trailing edge experiences the least gain, since the whole pulse has saturated the amplifier gain. The final value G_f of $G(\tau)$ after passage of the pulse is obtained from Eq. (8.52) by replacing $E_0(\tau)$ by E_{in}. The intermediate values of the gain depend on the pulse shape. Figure 8.8 shows the shape dependence of $G(\tau)$ for the so-called super-Gaussian input pulses whose shape is governed by

$$P_{in}(\tau) = P_0 \exp[-(\tau/\tau_p)^{2m}], \tag{8.54}$$

where m is the shape parameter. The input pulse is Gaussian for $m = 1$ but becomes nearly rectangular as m increases. For comparison purposes the input energy is held constant for different pulse shapes by choosing $E_{in}/E_s = 0.1$. The shape dependence of the amplification factor $G(\tau)$ implies that the output pulse is distorted when compared with the input pulse, and distortion is itself shape dependent. Figure 8.9 shows the pulse distortion occurring when a Gaussian input pulse ($m = 1$) of energy such that $E_{in}/E_s = 0.1$ is amplified by an SLA with 30 dB small-signal gain. Different curves correspond to different pulse widths τ_p. Pulses shorter than the carrier lifetime τ_c are distorted most since $G(\tau)$ is most nonuniform for such pulses. Interestingly enough, the energy gain G_E, defined as $G_E = E_{out}/E_{in}$, does not depend on the pulse shape. It depends only on the initial gain G_0 and the final gain G_f through the relation[6,55]

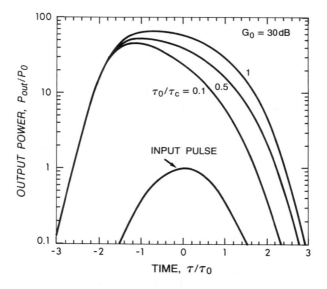

FIGURE 8.9. Amplified pulses for three values of τ_0/τ_c when a Gaussian input pulse of width τ_0 is amplified by a semiconductor laser amplifier with 30 dB small-signal gain.

$$G_E = \frac{\ln[(G_0-1)/(G_f-1)]}{\ln[(G_0-1)/(G_f-1)]-\ln(G_0/G_f)}. \tag{8.55}$$

The above discussion applies to any kind of amplifier. However, SLAs differ in one important aspect from other amplifiers. The difference is suggested by Eq. (8.47), which shows that gain saturation leads to a time-dependent phase shift across the pulse. The total phase shift is found by integrating Eq. (8.47) over the amplifier length and is given by

$$\phi(\tau) = -\frac{1}{2}\beta_c\int_0^L g(z,\tau)dz = -\frac{1}{2}\beta_c h(\tau) = -\frac{1}{2}\beta_c\ln[G(\tau)]. \tag{8.56}$$

Since the pulse modulates its own phase through gain saturation, this phenomenon is referred to as saturation-induced self-phase modulation.[55] Physically, gain saturation leads to temporal variations in the carrier population, which in turn change the refractive index through the linewidth enhancement factor β_c. Changes in the refractive index modify the optical phase. Pulses with a time-dependent phase indicate that the optical frequency varies with time along the pulse. Often such pulses are referred to as chirped. The frequency chirp, defined as the shift in the optical frequency at a certain time from the carrier frequency ω_0 of the incident pulse, is related to the phase derivative as

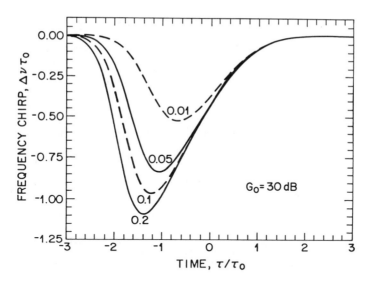

FIGURE 8.10. Frequency chirp imposed across the amplified pulse for several values of E_{in}/E_s. A Gaussian input pulse is assumed together with $G_0 = 30$ dB and $\beta_c = 5$ (after Ref. 52).

$$\Delta\nu_c = -\frac{1}{2\pi}\frac{d\phi}{d\tau} = \frac{\beta_c}{4\pi}\frac{dh}{d\tau} = -\frac{\beta_c P_{in}(\tau)}{4\pi E_s}[G(\tau)-1], \qquad (8.57)$$

where Eq. (8.5.9) was used. Figure 8.10 shows the chirp profiles for several input pulse energies when a Gaussian pulse is amplified in an SLA with 30 dB unsaturated gain. The frequency chirp is larger for more energetic pulses simply because gain saturation sets in earlier for such pulses.

Self-phase modulation and the associated frequency chirp can affect lightwave systems considerably. The spectrum of the amplified pulse broadens considerably and contains several peaks of different amplitudes.[53] The dominant peak is shifted toward longer wavelengths and is broader than the input spectrum. It is also accompanied by one or more satellite peaks. Figure 8.11 shows the expected shape and spectrum of amplified pulses when a Gaussian pulse of energy such that $E_{in}/E_s = 0.1$ is amplified by an SLA. The temporal and spectral changes depend on amplifier gain and are quite significant for $G_0 = 30$ dB. The experiments performed using picosecond pulses from mode-locked semiconductor lasers[52-55] confirm the expected qualitative behavior. In particular, the spectrum of amplified pulses is found to be shifted toward longer wavelengths by 0.5–1 nm (50–100 GHz) depending on the amplifier gain.[53] Spectral distortion in combination with the frequency chirp would affect the transmission characteristics when amplified pulses are propagated through optical amplifiers.

The frequency chirp imposed by the SLA is opposite in nature to that induced by directly modulated semiconductor lasers. If we also note that the chirp is nearly

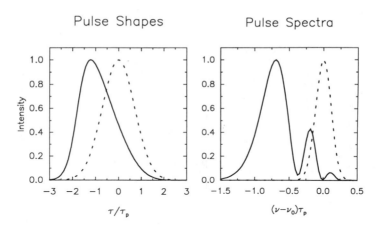

FIGURE 8.11. Shape and spectrum at the output of a semiconductor laser amplifier with $G_0 = 30$ dB and $\beta_c = 5$ for a Gaussian input pulse of energy $E_{in}/E_s = 0.1$. The dashed curves show for comparison input shape and spectrum.

linear over a considerable portion of the amplified pulse (see Fig. 8.10), it is easy to understand that the amplified pulse would pass through an initial compression stage when it propagates in the anomalous-dispersion region of optical fibers. Numerical simulations confirm this behavior. Figure 8.12 shows the input pulse, the amplified pulse, and the compressed pulse when a Gaussian pulse is first amplified in an SLA with 30 dB small-signal gain and then propagated through an optical fiber of length $0.3L_D$, where L_D is the dispersion length. The pulse is compressed by about a

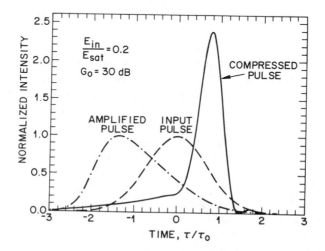

FIGURE 8.12. Input, amplified, and compressed pulses when a Gaussian input pulse of width τ_0 is amplified by a semiconductor laser amplifier with 30 dB small-signal gain and then propagated through an optical fiber of optimum length with anomalous dispersion (after Ref. 52).

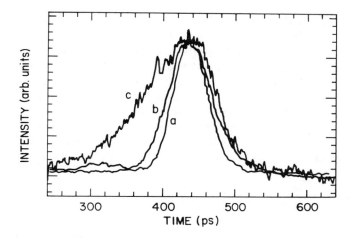

FIGURE 8.13. Streak-camera traces of the input pulse (trace a) and the amplified pulse after propagating in a 70-km-long fiber (trace b) for the case $G_0 = 30$ dB. Trace c shows the pulse shape when G_0 is reduced below 10 dB (after Ref. 55 © 1989 IEEE).

factor of 3 because of the frequency chirp imposed on the amplified pulse by the SLA. Such a compression was observed in an experiment[52] in which 40 ps optical pulses were first amplified in a 1.52-μm SLA and then propagated through 18 km of single-mode fiber with a dispersion of 15 ps/(km-nm). This compression mechanism can be used to design fiber-optic communication systems in which in-line SLAs are used to compensate simultaneously for both fiber loss and dispersion by operating SLAs in the saturation region so that they impose frequency chirp on the amplified pulse. The basic concept was demonstrated in 1989 in an experiment[54] in which the signal was transmitted over 70 km at 16 Gb/s by using an SLA. Figure 8.13 shows the pulse shapes measured by a streak camera at the fiber output with and without an SLA. In the absence of an SLA or when the SLA was operated in the unsaturated regime, the system was dispersion limited to the extent that the signal could not be transmitted over more than 20 km.

8.5.3. Subpicosecond Gain Dynamics

The preceding discussion applies quite well to the amplification of pulses whose width is ~ 10 ps. However, it needs to be modified for subpicosecond pulses since several new effects should be considered for such short optical pulses.[59–66] Pump-probe experiments have shown the existence of a fast gain-depletion process with a characteristic lifetime shorter than 1 ps attributed mainly to carrier heating.[59] They have also revealed fast phase changes occurring due to an almost instantaneous nonlinear contribution to the refractive index.[60–62] These effects become increasingly important as input pulses become shorter than a few picoseconds and should certainly be included for subpicosecond pulses.

Strictly speaking, the concepts of gain and refractive index themselves begin to break down when the pulse width becomes comparable to the dipole relaxation time governed by the intraband scattering processes occurring at a time scale ~ 100 fs. For such short optical pulses the rate-equation approximation, inherent in the use of Eq. (8.44), cannot be made since the SLA medium cannot be assumed to respond instantaneously.[67] In fact, even the concept of the linewidth enhancement factor looses its validity.[68] However, for pulses much longer than the intraband relaxation time ($\tau_p \gg 0.1$ ps) the rate-equation approach is approximately valid, and a phenomenological model that includes the fast dynamic processes such as carrier heating is found to be quite useful.[65,66] Equation (8.43) is then replaced by

$$\frac{\partial A}{\partial z} - \frac{i}{2}\beta_2 \frac{\partial^2 A}{\partial t^2} = \frac{1}{2}g(1+i\beta_c)A + \frac{1}{2}\Delta g_T(1+i\beta_T)A$$
$$- \frac{1}{2}\left(ig'A + \frac{1}{2}g''\frac{\partial^2 A}{\partial t^2} + \alpha A + (\alpha_2 + i\gamma)|A|^2 A\right),$$

$$(8.58)$$

where β_2 takes into account group-velocity dispersion, g' and g'' are the first and second derivatives of the gain spectrum at the carrier frequency, α is the linear loss, α_2 takes into account two-photon absorption, and $\gamma = (4\pi/\lambda)n_2$ represents instantaneous self-phase modulation governed by the material parameter n_2. Carrier heating is included through Δg_T and β_T such that Δg_T represents the gain change and β_T governs the corresponding index change.[66] Figure 8.14 shows the numerically simulated pulse shapes and spectra for an experiment in which 430 fs input pulses of 7.2 pJ energy are amplified in a tilted-stripe SLA (5° tilt) with 80-nm-thick, 5-μm-wide, and 500-μm-long active region. The pulse shape and the spectrum of the amplified pulse were in good agreement with the experimental data.[66]

8.6. MULTICHANNEL AMPLIFICATION

One of the advantages of using optical amplifiers is that they can be used to amplify several channels simultaneously as long as the carrier frequencies of the channels lie within the amplifier bandwidth. Ideally, the signal in each channel should be amplified by the same amount. In practice, several nonlinear phenomena in SLAs induce interchannel crosstalk, an undesirable feature that should be minimized for practical lightwave systems. Two such nonlinear phenomena are cross saturation and four-wave mixing (FWM).[69–95] They both are represented by the stimulated recombination term in the carrier rate equation. In the case of multichannel amplification, the power P in Eq. (8.25) corresponds to

$$P = \frac{1}{2}\left|\sum_{j=1}^{M} A_j \exp(-i\omega_j t) + \text{c.c.}\right|^2,$$

$$(8.59)$$

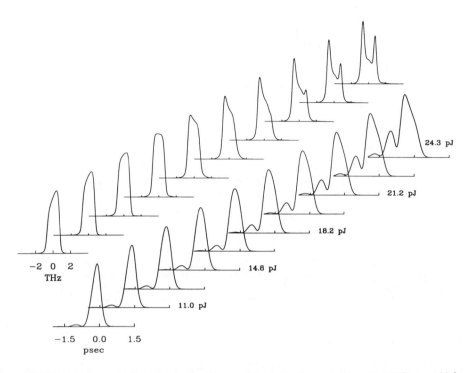

24.3 pJ

21.2 pJ

18.2 pJ

-2 0 2
THz

14.8 pJ

11.0 pJ

-1.5   0.0   1.5
psec

FIGURE 8.14. Numerically simulated evolution of the pulse shape and spectrum when a 430 fs input pulse with 7.2 pJ energy is amplified in a 500-μm-long SLA. Frames are separated by 50 μm (after Ref. 66 © 1994 IEEE).

where c.c. stands for complex conjugate, M is the number of channels, A_j is the (complex) amplitude, and ω_j is the carrier frequency for jth channel. Because of the coherent addition of individual channel fields, Eq. (8.59) contains a time-dependent term resulting from beating of the signal in different channels, that is,

$$P(t) = \sum_{j=1}^{M} P_j + \sum_{j \neq k}^{M} \sum^{M} 2\sqrt{P_j P_k} \cos(\Omega_{jk}t + \phi_j - \phi_k), \quad (8.60)$$

where $A_j = \sqrt{P_j} \exp(i\phi_j)$ was assumed together with $\Omega_{jk} = \omega_j - \omega_k$. The first term represents the total power in all channels and is responsible for cross saturation. The second term is the beating term and is responsible for FWM.[69] The following subsections discuss these phenomena separately.

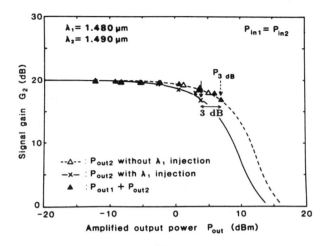

FIGURE 8.15. Signal gain of a 1.49 μm channel as a function of the output channel power in the presence of another 1.48 μm channel. Dashed curves show for comparison the channel gain in isolation. Crosses and triangles show the experimental data in the two cases whereas solid and dashed lines show the theoretical prediction based on cross saturation (after Ref. 71).

8.6.1. Cross Saturation

The origin of cross saturation is evident from Eq. (8.60). The first term on the right side shows that the power P in Eq. (8.26) should be replaced by the total power in all channels. Thus, the gain of a specific channel is saturated not only by its own power but also by the power of neighboring channels, a phenomenon known as cross saturation. In fact, Eq. (8.5) can be easily generalized to the multichannel case. The power growth along the amplifier length for each channel is found by solving

$$\frac{dP_j}{dz} = \frac{g_0 P_j}{1 + P_T/P_s},\tag{8.61}$$

where P_j is the channel power, g_0 is the small-signal gain, P_s is the saturation power, and P_T is the total power. In general, this set of coupled equations should be solved numerically to obtain the output channel powers which can then be used to obtain the channel gains. Cross saturation reduces the gain of a specific channel.

Cross-saturation-induced gain reduction has been observed in several experiments.[70-72] Figure 8.15 shows the gain of a 1.49 μm channel as a function of the output channel power in the presence of another channel at 1.48 μm for an SLA operating with about 20 dB of small-signal gain. The dashed curve shows for comparison the data obtained in the absence of the second channel. In the low-power region cross saturation is negligible as long as the total power remains well below the saturation power. However when the total power in all channels becomes comparable to the saturation power, channel gain is reduced significantly even though power in the specific channel is not large enough to saturate the SLA.

Cross saturation can lead to considerable interchannel crosstalk and is undesirable particularly for direct-detection or amplitude-shift keying (ASK) coherent systems, in which the channel power changes with time depending on the bit pattern. The signal gain of one channel then changes from bit to bit, and the change depends on the bit pattern of neighboring channels. The amplified signal appears to fluctuate more or less randomly. Such fluctuations degrade the effective SNR at the receiver. The cross-saturation-induced interchannel crosstalk occurs regardless of the extent of channel spacing. It can only be avoided by operating SLAs in the unsaturated regime. It is also absent for phase-shift keying (PSK) and frequency-shift keying (FSK) coherent systems, since the power in each channel, and hence the total power, remains constant with time.

8.6.2. Four-Wave Mixing

The time-dependent beating term in Eq. (8.60) accounts for FWM. Indeed, when Eq. (8.60) is substituted in Eq. (8.25), the carrier population is also found to oscillate at the beat frequency Ω_{jk}. Since the gain and the refractive index both depend on N, they are also modulated at the frequency Ω_{jk}. Such a modulation is referred to as the creation of gain and index gratings by the multichannel signal. These gratings generate FWM by scattering a part of the signal from one channel to another.[69] The mathematical description of FWM in SLAs is based on the solution of Eq. (8.25) with $P(t)$ given by Eq. (8.60). To simplify the analysis, consider the case of M equispaced channels with the channel spacing Ω. In that case, an approximate solution of Eq. (8.25) is given by[81]

$$N(t) = \bar{N} + \sum_{m=1}^{M-1} [\Delta N_m \exp(-im\Omega t) + \text{c.c.}],\qquad(8.62)$$

where

$$\bar{N} = N_0 + \frac{N_0(I/I_0 - 1)}{1 + P_T/P_s},\qquad(8.63)$$

with $I_0 = qVN_0/\tau_c$, $P_T = \displaystyle\sum_{m=1}^{M} |A_m|^2$ and

$$\Delta N_m = -\frac{C(\bar{N} - N_0)\Sigma_j A_j A_{j-m}^*}{1 - im\,\Omega\tau_c + P_T/P_s}.\qquad(8.64)$$

In Eq. (8.62) $\bar{N}$ represents the average value of the carrier population in the presence of both self and cross saturation, while ΔN_m governs the modulation of carrier population at the frequency $m\Omega$ as a result of beating between the channels separated by $m\Omega$. The parameter C in Eq. (8.64) is the overlap factor resulting from the

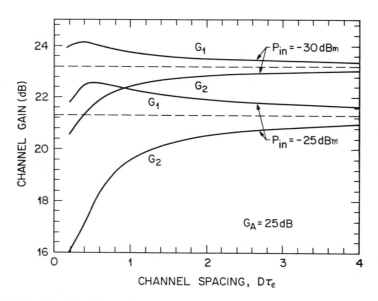

FIGURE 8.16. Calculated channel gains for an SLA used for simultaneous amplification of two channels and biased to provide 25 dB small-small gain for each channel. Solid curves show different channel gains for input channel powers of −10 and −25 dBm. Dashed lines show the equal channel gains expected in the absence of four-wave mixing (after Ref. 81 © 1990 IEEE).

nonplane wave nature of the fields propagating inside the SLA. The parameter C has typical values in the range 0.5–1.

To obtain the channel gains, one must solve the wave equation by taking into account changes induced in the gain medium by the carrier-density modulation. Although such an analysis can be carried out numerically for an arbitrary number of channels,[82] the main qualitative features can be seen by considering the much simpler two-channel case. The evolution of channel powers is governed by a set of two coupled differential equations[77]

$$\frac{dP_1}{dz} = \frac{g_0 P_1}{1+P_1+P_2}\left(1 - \frac{(1+P_1+P_2-\beta_c\Omega\tau_c)P_2}{(1+P_1+P_2)^2+(\Omega\tau_c)^2}\right), \tag{8.65}$$

$$\frac{dP_2}{dz} = \frac{g_0 P_2}{1+P_1+P_2}\left(1 - \frac{(1+P_1+P_2+\beta_c\Omega\tau_c)P_1}{(1+P_1+P_2)^2+(\Omega\tau_c)^2}\right), \tag{8.66}$$

where the channel powers P_1 and P_2 are normalized to the saturation power P_s. The term inside brackets represents the effect of FWM. This term involves the linewidth enhancement factor β_c because modulation of the carrier density at the beat frequency Ω affects both the gain and the refractive index of the active medium. Figure 8.16 shows the gains of two channels for an SLA with 25 dB small-signal gain when the input channel powers are −30 and −25 dBm (1 and 3.2

μW, respectively) by taking $\beta_c = 6$. Dashed lines show the equal channel gain expected in the absence of FWM. The gain of the low-frequency channel is enhanced while the gain of high-frequency channel is reduced as a result of energy transfer initiated by FWM. The channel gains can differ by more than 5 dB for large input powers and small channel spacing.

Interchannel crosstalk induced by FWM can occur for all multichannel coherent communication systems irrespective of the modulation format used.[73-85] However, in contrast with cross saturation, it occurs only when the channel spacing is not too large and becomes negligible when channel spacing exceeds 10 GHz simply because carrier population is not able to respond when $\Omega\tau_c \gg 1$. For closely spaced channels the channel gains can differ by more than a factor of 2. Such gain variations result in a crosstalk penalty and degrade the system performance considerably. The crosstalk penalty was measured in a system experiment[79] where three channels spaced about 2 GHz were used to transmit a coherent FSK signal through a 50-km-long fiber. The bit rate was 560 Mb/s. The bit-error rate (BER) degraded severely as a result of four-wave mixing with a large penalty (> 2 dB) at 10^{-9}. System performance could be improved by reducing the channel powers or by increasing the channel spacing. Since the cross-talk penalty depends on the linewidth enhancement factor, it is expected to be lower for MQW amplifiers. Indeed, in a two-channel experiment[83] intermodulation distortion was found to be reduced by 15 dB by the use of MQW amplifiers.

8.6.3. Highly Nondegenerate Four-Wave Mixing

The FWM phenomenon discussed in the preceding section is often referred to as nearly degenerate FWM since it dominates only when the channel spacing Ω is relatively small ($\Omega \leqslant 10$ GHz). The reason is that carrier density is unable to respond to temporal variations occurring on a time scale much shorter than the carrier lifetime τ_c. Mathematically, this feature is evident from Eq. (8.64), which shows that the carrier-density modulation ΔN_m becomes negligible for $\Omega\tau_c \gg 1$. However, it was pointed out[86] in 1987 that FWM can occur in semiconductor lasers and amplifiers even when $\Omega\tau_c \gg 1$ through nonlinear mechanisms involving intraband carrier dynamics occurring on a much shorter time scale (~ 100 fs). Such FWM is referred to as highly nondegenerate FWM to distinguish it from the phenomenon discussed in Sec. 8.6.2. It has been extensively studied[86-95] particularly because it can be used as a frequency-domain tool to investigate ultrafast dynamics of semiconductor lasers and amplifiers.

Several different intraband processes can lead to highly nondegenerate FWM; two most notable are spectral hole burning[69,86] and carrier heating.[59,60] The response time of spectral hole burning is related to the dipole relaxation time T_2 which in turn is governed by the intraband carrier scattering time τ_{in}. Since τ_{in} is ~ 50–100 fs, FWM induced by spectral hole burning can occur for a frequency spacing Ω in excess of 1 THz. By contrast, the response time of carrier heating is estimated to be in the range 0.6–1 ps. As a result, FWM induced by carrier heating can occur only for frequency spacing well below 1 THz. The dependence of FWM on the response time can be exploited to estimate the response time of different nonlinear phenomena contributing to the FWM process.[91-95] The basic idea is to perform a pump-probe type of experiment in which the frequency separation $\Omega = \omega_p - \omega_s$ between the pump and probe beams incident on an SLA is varied over a large range from 1 GHz to > 1 THz. The amplitude of the conjugate wave generated through FWM is measured as a function of Ω. Assuming that different nonlinear phenomena contribute to the FWM process independently, the corresponding response times can be inferred through a curve-fitting process. In one experiment[93] Ω was varied over a wide range from 10 GHz to 1.7 THz. The results show the contribution of three nonlinear phenomena to the FWM signal. At frequencies below 50 GHz, the dominant contribution comes from the carrier-density modulation discussed in Sec. 8.6.2. For frequencies in excess of 100 GHz both carrier heating and spectral hole-burning contribute to FWM. The data indicate 650 fs response time for carrier heating and 50 fs response time for spectral hole burning. The latter number is so close to the resolution limit of the experiment that it can be interpreted only as a rough estimate. A more accurate measurement would require FWM measurements over a wider frequency range. Indeed, the 50 fs response time was also measured in an experiment that extended the frequency range to 5 THz.[94] The data also indicated a smaller contribution (about 15%) from carrier heating with a response time of 0.6 ps.

The impact of highly nondegenerate FWM on interchannel cross talk is expected to be minimal. The reason is that intraband nonlinearities are much weaker compared with the carrier-density modulation effect discussed in Sec. 8.6.2. As a result, the energy transfer between the two channels separated by 100 GHz or more is relatively insignificant. This is good news for multichannel local-area networks making use of wavelength-division multiplexing whose channel spacing typically exceeds 100 GHz.

8.7. AMPLIFIER APPLICATIONS

The potential use of SLAs has been demonstrated for many system applications ranging from optical communications to photonic switching. Some of these applications are discussed in Chaps. 9 and 10. This section briefly reviews the present status of SLAs from the standpoint of technological applications. For simplicity of discussion, these applications are classified into three following subsections.

8.7.1. Optical Communication Systems

Optical amplifiers can serve several purposes in the design of fiber-optic communication systems; four such applications are shown schematically in Fig. 8.17. An important application for long-haul systems is in replacing electronic regenerators with optical amplifiers. Such a replacement can be carried out as long as the system performance is not limited by the cumulative effects of dispersion and spontaneous

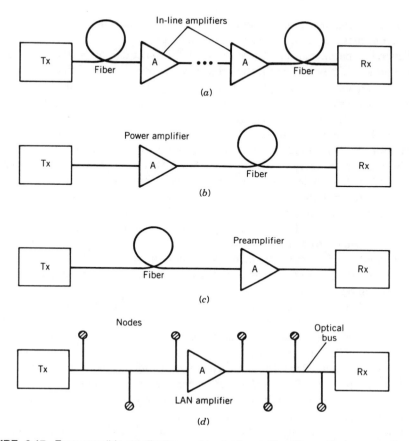

FIGURE 8.17. Four possible applications of optical amplifiers in optical communication systems: (a) as in-line amplifiers; (b) as booster of transmitter power; (c) as a preamplifier to the receiver; and (d) for compensation of distribution losses in local-area networks.

emission. The use of optical amplifiers is particularly attractive for multichannel lightwave systems, since electronic regeneration requires demultiplexing of channels before each channel signal is regenerated using separate receiver and transmitters, a rather costly procedure. Optical amplifiers can amplify all channels simultaneously. When optical amplifiers are used to replace electronic regenerators, they are called in-line amplifiers.

Another way to use optical amplifiers is to increase the transmitter power by placing an amplifier just after the transmitter. Such amplifiers are called power amplifiers or boosters, as their main purpose is to boost the transmitted power. A power amplifier increases the transmission distance by 10–100 km depending on the amplifier gain and the fiber loss. Transmission distance can also be increased by putting an amplifier just before the receiver to boost the received power. Such amplifiers are called preamplifiers. Another application of optical amplifiers is to use them for compensating distribution losses in local-area networks. They can also be used to make components such as optical taps.

Several experiments have demonstrated the potential of SLAs as an optical preamplifier.[96-98] In this application the signal is optically amplified before it falls on the receiver. The preamplifier boosts the signal to such a high level that the receiver performance is limited by shot noise rather than by thermal noise. The basic idea is similar to the case of avalanche photodiodes (APDs), which amplify the signal in the electrical domain. However, just as APDs add additional noise, preamplifiers also degrade the SNR through spontaneous-emission noise. The relatively large noise figure of SLAs ($F_n = 5-6$ dB) makes them less attractive as a preamplifier. Nonetheless, they can improve the receiver sensitivity considerably. In an 8-Gb/s transmission experiment, the use of an SLA as a preamplifier resulted in a 3.7 dB improvement over the best sensitivity achieved by using APD receivers.[98] SLAs can also be used as power amplifiers to boost the transmitter power. It is, however, difficult to achieve powers in excess of 10 mW because of a relatively small value ($\sim 5-7$ mW) of the output saturation power.

SLAs have been used as in-line amplifiers in several system experiments.[99-108] In one experiment,[101] signal at 1 Gb/s was transmitted over 313 km by using four cascaded SLAs. This experiment used the conventional IM/DD (intensity modulation with direct detection) technique for signal transmission. In a coherent transmission experiment[102] four cascaded SLAs were used to transmit 420-Mb/s FSK signal over 370 km. In a 140-Mb/s coherent experiment,[105] the transmission distance was extended to 546 km by using ten cascaded SLAs. Interest in the use of SLAs as in-line amplifiers declined after the advent of erbium-doped fiber amplifiers in 1989 in view of their superior performance compared with SLAs.[2-4]

Recently, SLAs have been employed to overcome distribution losses in the local-area network applications.[107,108] In one experiment[108] an SLA was used as a dual-function device. It amplified five FSK channels multiplexed by using the subcarrier multiplexing (SCM) technique, improving the power budget by 11 dB. At the same time, it was used to monitor the network performance through a baseband control channel. The 100 Mb/s baseband control signal modulated the carrier density of the amplifier, which in turn produced a corresponding electric signal which was used for monitoring. SLAs can also be used to make other components such as optical taps that are useful for local-area networks. In one demonstration an asymmetric flared twin-waveguide design was used to demonstrate an optical tap with as much as 30 dB gain.[109]

An important use of SLAs is in the design of photonic integrated circuits.[110] For example, an SLA can be integrated with a semiconductor laser to obtain a high-power, narrow-linewidth optical source useful for coherent communication systems. MQW amplifiers have been integrated with both distributed feedback[111] (DFB) and distributed Bragg reflector[112] (DBR) lasers to demonstrate such an optical source. Figure 8.18 shows schematically integration of a DBR laser with a MQW amplifier. This scheme can easily be generalized to amplify the output of several lasers simultaneously by using a single amplifier. In one demonstration, the output of four

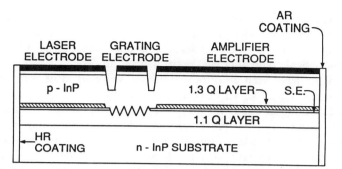

FIGURE 8.18. Schematic illustration of a distributed Bragg reflector laser integrated with a MQW amplifier.

DBR lasers, fabricated on a single chip, was channeled through passive waveguides to a MQW amplifier that was also a part of the same chip. The amplifier bandwidth was large enough that the output of all four DBR lasers could be amplified simultaneously without significant cross talk. The resulting photonic integrated circuit can act as a source in a fiber-optic communication system where several channels are multiplexed in the wavelength domain (wavelength-division multiplexing) and transmitted simultaneously over the same fiber.

8.7.2. Photonic Switching

Switching is an essential operation not only for telecommunication networks but also for digital computers and other signal processing systems. Currently this operation is performed in the electrical domain even for optical communication systems by converting an optical signal to the electrical signal before performing the switching operation. However, as the bit rates begin to exceed a few Gb/s, it becomes increasingly more difficult to perform electrical switching at such high speeds. The obvious solution is to resort to direct switching of optical signals, a technology referred to as photonic switching.[113] The potential of SLAs for photonic switching applications has been extensively studied and they are likely to play an important role in this novel technology.[114–121] Since Chap. 10 covers optical switching applications in detail, only brief description is included here.

Both FP-type and TW-type SLAs can be used in photonic switching applications. Optical bistability in FP-type SLAs was demonstrated in 1983.[115] Since then, SLAs have been shown to be capable of performing various logic operations.[116] They can also be used for the purpose of optical memory, differential amplifications, and optical limiting.[117,118] By contrast, TW-type SLAs play a different role in photonic switching systems. They are used as a gain element to overcome the losses associated with other switching elements such as a $LiNbO_3$ switch. In one experiment[119] SLAs were used to demonstrate a 128-line space-division photonic switching system by using 8×8 $LiNbO_3$ switches. In another experiment[120] 15 dB gain from a 1.53-μm SLA provided an InP directional coupler that could be used as a lossless optical switch over a wavelength range of 80 nm. The use of SLAs as a gain element is expected to find wide applications in many other areas such as optical interconnects, optical computing, and optical signal processing.

Recently SLAs have been used for shifting the wavelength of a channel while preserving the modulated signal carried by that channel.[121-126] This capability of SLAs is useful for applications requiring a change in the carrier frequency of an optical data stream. An example is provided by wavelength-routing photonic networks in which the signals are routed based on their wavelengths. The physical mechanism behind wavelength wavelength conversion can be cross saturation or four-wave mixing, both of which have been discussed in the context of multi-channel amplification in Sec. 8.6. In the case of cross saturation,[121-123] the modulated signal is launched into the SLA together with a cw probe beam at a different wavelength within the gain bandwidth of the SLA. The signal beam is intense enough to saturate the SLA. It modulates the carrier density through gain saturation in such a way that the carrier density nearly mimics the bit pattern imposed on the signal beam as long as the bit rate B is smaller than the inverse of the carrier lifetime ($B\tau_c < 1$). The cw probe beam sees this modulation of the carrier density during its amplification inside the SLA. Because of the time-dependent nature of the amplifier gain, the modulation pattern imposed on the signal beam is transferred to the probe beam at a different wavelength. In a system experiment wavelength conversion over 17 nm at 4 Gb/s was performed with only 1.5 dB power penalty.[123] In the case of four-wave mixing,[124-126] wavelength conversion over up to 65 nm has been demonstrated. The conversion efficiency was more than 14% for shifts as large as 24 nm.[126]

8.7.3. High-Power Applications

Many applications of semiconductor lasers require high power levels in excess of 1 W. Such power levels are difficult to obtain from a single semiconductor laser directly mainly because of a relatively small active-region volume. The simplest solution is to increase the active-region volume by using lateral dimensions ~ 100 μm or more. Such broad-area lasers have several problems related to filamentation and mode stability. These problems can be solved to some extent by adopting a master-oscillator power-amplifier (MOPA) design in which the output of a low-power semiconductor laser, acting as a master oscillator, is amplified in a TW-type SLA acting as a power amplifier. If the master oscillator and the SLA are integrated monolithically on the same chip, the resulting device can provide power levels in the range 1–10 W while maintaining single-mode operation and a good beam quality. SLAs have been used extensively in recent years to provide high-power lasers through a MOPA design.[127-140] Both edge-emitting and surface-emitting geometries have been pursued. This subsection reviews briefly the current progress for the two MOPA geometries.

Figure 8.19 shows the MOPA design for an edge-emitting high-power laser.[133] It

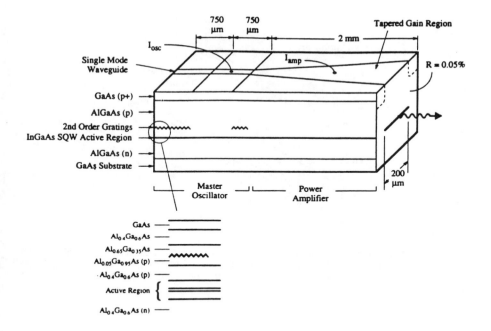

FIGURE 8.19. Schematic illustration of an edge-emitting MOPA with separate oscillator and amplifier contacts (after Ref. 128 © 1993 IEEE).

consists of a DBR master oscillator coupled to a tapered (or flared) power amplifier. An $In_{0.25}Ga_{0.75}As$ quantum-well active layer provides gain for both the master oscillator and the power amplifier through two separate electrical contacts. The DBR laser has a 750-μm-long active region. A second-order grating is etched on one of the cladding layers on both sides of the 750-μm-long active region to provide distributed feedback and select a single longitudinal mode. A 4-μm-wide waveguide controls the lateral mode through index guiding and forces the master oscillator to oscillate in a single transverse mode. The output of the DBR laser is injected into a 2-mm-long amplifier. The gain region in the amplifier is tapered in such a way that it increases from 4 to 250 μm over 2 mm length, allowing the optical mode to diffract freely along the amplifier length. The output facet reflectivity is reduced to below 0.1% through an antireflection coating. Such a device delivered 2 W of diffraction-limited output power under cw operation when the DBR laser was biased at 150 mA and 3.5 A current was injected into the amplifier. The device was bonded p-side down on a copper heat sink whose temperature was maintained at 5 °C. Further design improvements increased the maximum output power to 3 W under cw operation.[138] Even higher powers have been achieved in MOPAs where master oscillator and power amplifiers were not integrated monolithically. In one experiment,[132] 4.5 output power under cw operation was obtained by injecting 150

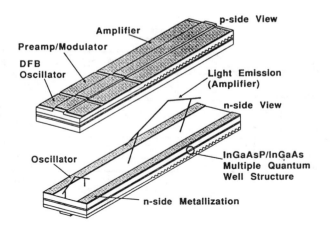

FIGURE 8.20. Schematic illustration of a surface-emitting MOPA consisting of a DFB master oscillator, a preamplifier, and an active-grating amplifier. Both *p*-side and *n*-side views are shown. The output is taken from the *n* side through the transparent substrate (after Ref. 130 © 1992 IEEE).

mW from a Ti:sapphire laser into a tapered 1.5-mm-long SLA. In another experiment the output power increased to 5.25 W when 200 mW of input power was injected from a Ti:sapphire laser.[139] Under pulsed operation broad-area SLAs have been used to provide output powers as high as 21 W,[129] indicating the ultimate potential of the MOPA design. Recently 39-W of power was obtained from an eight-element monolithic array of MOPAs.[141]

In the surface-emitting geometry of the MOPA design, the output of the amplifier is produced through distributed coupling from an off-resonant grating and appears in a direction nearly normal to the surface (off by about 10°). Figure 8.20 shows schematically the design of such a surface-emitting MOPA.[135–137] Similar to the case of edge-emitting MOPA shown in Fig. 8.19, a quantum-well active layer provides gain for both the master oscillator and the power amplifier. However, the gratings in the master oscillator and the power amplifiers are made different by changing the grating pitch. The grating in the master oscillator is designed such that it is in resonance with respect to the gain-peak wavelength. It helps to select a single longitudinal mode in a way similar to conventional DFB lasers. In contrast, the grating pitch in the amplifier region is made larger (by about 10%) so that the amplifier grating is off resonant with respect to the wavelength emitted by the master oscillator. Such a grating produces amplifier output in a direction nearly normal to the chip surface, typically at an angle $\sim 10°$ off surface normal. The grating can cover the entire amplifier length or a portion of it depending on the design. In the latter case the amplifier region without the grating serves as a preamplifier and can be used to modulate the output power. The optical mode is confined in the lateral direction through a ridge waveguide (width ~ 3–5 μm) that extends continuously through the master oscillator, the preamplifier, and the active-grating amplifier and is $\sim$ 1–3 mm depending upon its length. The angular beam divergence of such a device is $\sim$ 0.03°–0.05° because of a relatively large beam size. Even though the output power is not as large as in the case of edge-emitting MOPAs, it can exceed 1 W with a proper design.[134] An output power of 150 mW at 1.6 μm was obtained in a device with a 0.4-mm-long master oscillator, 0.5-mm-long preamplifier, and 3-mm-long active-grating amplifier.[135] In another device operating at 1.7 μm, the far-field beam divergence at 146 mW power was reduced to below 0.02° by using a 7-mm-long active-grating amplifier.[137] These results indicate that SLAs can be integrated with a low-power master oscillator to provide high optical powers.

8.8. FUTURE PROSPECTS

Although the potential of SLAs for lightwave system applications has been demonstrated, SLAs need to overcome several drawbacks before their use becomes practical. A few among them are polarization sensitivity, interchannel cross talk, and a large coupling loss. Even though SLAs can have chip gain as high as 30–35 dB, the usable gain is reduced by 8–10 dB because of a large coupling loss occurring at the input and output ends. Fiber amplifiers are preferable from this standpoint, since the coupling loss (due to fusion splices) is negligible for them. They are also nearly polarization insensitive and have negligible interchannel crosstalk. SLAs would nonetheless find some applications since they can be monolithically integrated within the transmitter or the receiver. Current trends indicate that SLAs are unlikely to be used as in-line amplifiers in optical communication systems.

Two most important characteristics of SLAs that make their use preferable in some applications over other kinds of amplifiers are their compact size and the ability of electrical pumping. Both of these characteristics allow SLAs to be integrated monolithically within a multicomponent optoelectronic or photonic integrated circuit. The future applications of SLAs are likely to explore this property of SLAs. As discussed in Sec. 8.7, SLAs have been integrated within optical transmitters, receivers, and high-power lasers. Their use in photonic switching applications including optical interconnects, optical computing, and optical data processing is likely to become more common as the technology involving the fabrication of opto-electronic integrated circuit becomes more mature.

ACKNOWLEDGMENTS

This work was partially supported by the U.S. Army Research Office through the University Research Initiative Program.

REFERENCES

1. G. P. Agrawal, *Fiber-Optic Communication Systems* (Wiley, New York, 1992).
2. M. J. F. Digonnet, Ed., *Rare Earth Doped Fiber Lasers and Amplifiers* (Dekker, New York, 1993).

3. A. Bjarklev, *Optical Fiber Amplifiers: Design and System Applications* (Artech House, Boston, 1993).
4. Special Issue on Optical Amplifiers, J. Lightwave Technol. **9**, 145 (1991); special Issue on Photonic Integrated Circuits, Opt. and Photon. News., March issue **4**, 8 (1993).
5. J. M. Senior, *Optical Fiber Communications*, 2nd ed. (Prentice–Hall, New York, 1992), Chap. 10.
6. A. E. Siegman, *Lasers* (University Science Books, Mill Valley, California, 1986).
7. P. W. Milonni and J. H. Eberly, *Lasers* (Wiley, New York, 1988).
8. M. J. Coupland, K. G. Mambleton, and C. Hilsum, Phys. Lett. **7**, 231 (1963).
9. J. W. Crowe and R. M. Graig, Jr., Appl. Phys. Lett. **4**, 57 (1964).
10. W. F. Kosonocky and R. H. Cornely, IEEE J. Quantum Electron. **QE-4**, 125 (1968).
11. D. Schicketanz and G. Zeidler, IEEE J. Quantum Electron. **QE-11**, 65 (1975).
12. S. Kobayashi and T. Kimura, Electron. Lett. **16**, 232 (1980).
13. Y. Yamamoto, IEEE J. Quantum Electron. **QE-16**, 1047 (1980).
14. V. N. Luk'yanov, A. T. Semenov, and S. D. Yakubovich, Sov. J. Quantum Electron. **10**, 1432 (1980).
15. F. Favre, L. Jeunhomme, I. Joindot, M. Monerie, and J. C. Simon, IEEE J. Quantum Electron. **QE-17**, 897 (1981).
16. J. C. Simon, Electron. Lett. **18**, 438 (1982).
17. T. Mukai, Y. Yamamoto, and T. Kimura, IEEE J. Quantum Electron. **QE-18**, 1560 (1982).
18. T. Saitoh and T. Mukai, Electron. Lett. **23**, 218 (1987); IEEE J. Quantum Electron. **QE-23**, 1010 (1987); J. Lightwave Technol. **6**, 1656 (1988).
19. J. C. Simon, J. Lightwave Technol. **LT-5**, 1286 (1987).
20. G. Eisenstein, B. C. Johnson, and G. Raybon, Electron. Lett. **23**, 1020 (1987).
21. M. G. Öberg and N. A. Olsson, Electron. Lett. **24**, 99 (1988).
22. M. J. O'Mahony, J. Lightwave Technol. **6**, 531 (1988).
23. N. A. Olsson, J. Lightwave Technol. **7**, 1071 (1989).
24. T. Saitoh and T. Mukai, in *Coherence, Amplification, and Quantum Effects in Semiconductor Lasers*, edited by Y. Yamamoto (Wiley, New York, 1991), Chap. 7.
25. G. P. Agrawal and N. K. Dutta, *Semiconductor Lasers*, 2nd ed. (Van Nostrand Reinhold, New York, 1993).
26. R. H. Clarke, Int. J. Electron. **53**, 495 (1982); G. Eisenstein, AT&T Bell Lab. Tech. J. **63**, 357 (1984).
27. D. R. Kaplan and P. P. Deimel, AT&T Bell Lab. Tech. J. **63**, 857 (1984).
28. T. Saitoh, T. Mukai, and O. Mikami, J. Lightwave Technol. **LT-3**, 288 (1985).
29. C. Vassallo, Electron. Lett. **21**, 333 (1985); **24**, 61 (1988).
30. C. E. Zah, J. S. Osinski, C. Caneau, S. G. Menocal, L. A. Reith, J. Salzman, F. K. Shokoohi, and T. P. Lee, Electron. Lett. **23**, 990 (1987).
31. J. T. K. Chang, and J. I. Vukusic, J. Mod. Opt. **35**, 355 (1988).
32. J. Salzman, R. J. Hawkins, and T. P. Lee, Opt. Lett. **13**, 455 (1988).
33. B. Mikkelsen, D. S. Olesen, K. E. Stubkjaer, Z. Wang, A. J. Collar, and G. D. Henshall, Electron. Lett. **25**, 357 (1989).
34. D. Marcuse, J. Lightwave Technol. **7**, 336 (1989).
35. N. A. Olsson, R. F. Kazarinov, W. A. Nordland, C. H. Henry, M. G. Oberg, H. G. White, P. A. Garbinski, and A. Savage, Electron. Lett. **25**, 1048 (1989).
36. I. Cha, M. Kitamura, H. Honmou, and I. Mito, Electron. Lett. **25**, 1241 (1989).
37. S. Cole, D. M. Cooper, W. J. Devlin, A. D. Ellis, D. J. Elton, J. J. Isaak, G. Sherlock, P. C. Spurdens, and W. A. Stallard, Electron. Lett. **25**, 314 (1989).
38. G. Großkopf, R. Ludwig, R. G. Waarts, and H. G. Weber, Electron. Lett. **23**, 1387 (1987).
39. N. A. Olsson, Electron. Lett. **24**, 1075 (1988).
40. M. Sumida, Electron. Lett. **26**, 1913 (1989).
41. M. Koga and T. Mutsumoto, IEEE Photon. Technol. Lett. **12**, 431 (1989); J. Lightwave Technol. **9**, 284 (1991).
42. G. Bendelli, K. Komori, and S. Arai, IEEE J. Quantum Electron. **28**, 447 (1992).
43. G. Eisenstein, U. Koren, G. Raybon, T. L. Koch, J. M. Wisenfield, M. Wegener, R. S. Tucker, and B. I. Miller, Appl. Phys.. Lett. **56**, 1201 (1990).
44. T. Saitoh, Y. Suzuki, and H. Tanaka, IEEE Photon. Technol. Lett. **2**, 794 (1990).
45. K. Komori, S. Arai, and Y. Suematsu, IEEE Photon. Technol. Lett. **3**, 39 (1991).
46. P. J. A. Thijs, L. F. Tiemeijer, P. I. Kuindersma, J. J. M. Binsma, and T. van Dongen, IEEE J. Quantum Electron. **27**, 1426 (1991).

47. D. Tauber, R. Nagar, A. Livne, G. Eisenstein, U. Koren, and G. Raybon, IEEE Photon. Technol. Lett. **4**, 238 (1992).
48. K. S. Jepsen, B. Mikkelsen, J. H. Povlsen, M. Yamaguchi, and K. E. Stubkjaer, IEEE Photon. Technol. Lett. **4**, 550 (1992).
49. K. Magari, M. Okamoto, H. Yasaka, K. Sato, Y. Noguchi, and O. Mikami, IEEE Photon. Technol. Lett. **2**, 556 (1990); K. Magari, M. Okamoto, and Y. Noguchi, IEEE Photon. Technol. Lett. **3**, 998 (1991); Y. Suzuki, K. Magari, M. Ueki, T. Amano, O. Mikami, and M. Yamamoto, IEEE Photon. Technol. Lett. **5**, 404 (1993).
50. L. F. Tiemeijer, P. J. A. Thijs, T. van Dongen, R. W. M. Slootweg, J. M. M. van der Heijden, J. J. M. Binsma, and P. C. M. Krijn, Appl. Phys. Lett. **62**, 826 (1993); M. A. Newkirk, B. I. Miller, U. Koren, M. G. Young, M. Chien, R. M. Jopson, and C. A. Burrus, IEEE Photon. Technol. Lett. **5**, 406 (1993).
51. S. Duboritsky, A. Mathur, W. H. Steir, and P. D. Dapkus, IEEE Photon. Technol. Lett. **6**, 176 (1994).
52. G. P. Agrawal and N. A. Olsson, Opt. Lett. **14**, 500 (1989).
53. N. A. Olsson and G. P. Agrawal, Appl. Phys. Lett. **55**, 13 (1989).
54. N. A. Olsson, G. P. Agrawal, and K. W. Wecht, Electron. Lett. **25**, 603 (1989).
55. G. P. Agrawal and N. A. Olsson, IEEE J. Quantum Electron. **25**, 2297 (1989).
56. I. W. Marshall, D. M. Spirit and M. J. O'Mahony, Electron. Lett. **23**, 818 (1987).
57. J. M. Wiesenfeld, G. Eisenstein, R. S. Tucker, G. Raybon, and P. B. Hansen, Appl. Phys. Lett. **53**, 1239 (1988).
58. T. Saitoh, H. Itoh, Y. Noguchi, S. Sudo, and T. Mukai, IEEE Photon. Technol. Lett. **1**, 297 (1989).
59. K. L. Hall, J. Mark, E. P. Ippen, and G. Eisenstein, Appl. Phys. Lett. **56**, 1740 (1990).
60. C. T. Hultgren and E. P. Ippen, Appl. Phys. Lett. **59**, 635 (1991); K. L. Hall, A. M. Darwish, E. P. Ippen, U. Koren, and G. Raybon, Appl. Phys. Lett. **62**, 1320 (1993).
61. R. S. Grant and W. Sibbett, Appl. Phys. Lett. **58**, 1119 (1991).
62. P. J. Delfyett, Y. Silberberg, and G. A. Alphonse, Appl. Phys. Lett. **59**, 10 (1991).
63. A. Uskov, J. Mark, and J. Mork, IEEE Photon. Technol. Lett. **4**, 443 (1992).
64. J. Mark and J. Mork, Appl. Phys. Lett. **61**, 2281 (1992).
65. A. Dienes, J. P. Heritage, M. Y. Hong, and Y. H. Chang, Opt. Lett. **17**, 1602 (1992).
66. M. Y. Hong, Y. H. Chang, A. Dienes, J. P. Heritage, and P. J. Delfyett, IEEE J. Quantum Electron. **30**, 1122 (1994).
67. C. M. Bowden and G. P. Agrawal, Opt. Commun. **100**, 147 (1993).
68. G. P. Agrawal and C. M. Bowden, IEEE Photon. Technol. Lett. **5**, 640 (1993).
69. G. P. Agrawal, Opt. Lett. **12**, 260 (1987); J. Opt. Soc. Am. B **5**, 147 (1988).
70. G. Großkopf, R. Ludwig, and H. G. Weber, Electron. Lett. **22**, 900 (1986).
71. T. Mukai, K. Inoue, and T. Saitoh, Electron. Lett. **23**, 396 (1987).
72. R. M. Jopson, K. L. Hall, G. Eisenstein, G. Raybon, and M. S. Whalen, Electron. Lett. **23**, 510 (1987).
73. G. P. Agrawal, Electron. Lett. **23**, 1175 (1987).
74. K. Inoue, Electron. Lett. **23**, 1293 (1987); K. Inoue, T. Mukai, and T. Saitoh, Appl. Phys. Lett. **51**, 1051 (1987).
75. R. M. Jopson, T. E. Darcie, K. T. Gaylard, R. T. Ku, R. E. Tench, T. C. Rice, and N. A. Olsson, Electron. Lett. **23**, 1394 (1987).
76. T. E. Darcie, R. M. Jopson, and R. W. Tkach, Electron. Lett. **24**, 31 (1988); T. E. Darcie and R. M. Jopson, Electron. Lett. **24**, 630 (1988).
77. I. M. I. Habbab and G. P. Agrawal, J. Lightwave Technol. **7**, 1351 (1989).
78. R. P. Webb and T. G. Hodgkinson, Electron. Lett. **25**, 491 (1989).
79. S. Ryu, K. Mochizuki, and H. Wakabayashi, J. Lightwave Technol. **7**, 1525 (1989).
80. F. Favre, D. LeGuen, J. C. Simon, and P. Doussiere, Electron. Lett. **25**, 272 (1989); F. Favre and D. LeGuen, IEEE J. Quantum Electron. **26**, 858 (1990).
81. G. P. Agrawal and I. M. I. Habbab, IEEE J. Quantum Electron. **26**, 501 (1990).
82. T. G. Hodgkinson and R. P. Webb, IEEE Photon. Technol. Lett. **2**, 69 (1990); J. Lightwave Technol. **9**, 605 (1991).
83. Y. C. Chung, J. M. Wiesenfeld, G. Raybon, U. Koren, and Y. Twu, IEEE Photon. Technol. Lett. **3**, 130 (1991).
84. R. Hui and A. Mecozzi, Appl. Phys. Lett. **60**, 2454 (1992); A. Mecozzi, A. D'Ottavi, and R. Hui, IEEE J. Quantum Electron. **29**, 1477 (1993).
85. S. Jiang and M. Dagenais, Appl. Phys. Lett. **62**, 2757 (1993); Opt. Lett. **18**, 1337 (1993).

86. G. P. Agrawal, Appl. Phys. Lett. **51**, 302 (1987).
87. J. G. Provost and R. Frey, Appl. Phys. Lett. **55**, 519 (1989).
88. S. Murata, A. Tomita, J. Shimizu, M. Kitamura, and A. Suzuki, Appl. Phys. Lett. **58**, 1458 (1991).
89. L. F. Tiemeijer, Appl. Phys. Lett. **59**, 499 (1991).
90. S. R. Chinn, Appl. Phys. Lett. **59**, 1673 (1991).
91. K. Kikuchi, M. Kakui, C. Zah, and T. P. Lee, IEEE J. Quantum Electron. **28**, 151 (1992).
92. J. Zhou, N. Park, J. W. Dawson, K. J. Vahala, M. A. Newkirk, U. Koren, and B. I. Miller, Appl. Phys. Lett. **62**, 2301 (1993).
93. J. Zhou, N. Park, J. W. Dawson, K. J. Vahala, M. A. Newkirk, and B. I. Miller, Appl. Phys. Lett. **63**, 1179 (1993).
94. C. B. Kim, E. T. Peng, C. B. Su, W. Rideout, and G. H. Cha, IEEE Photon. Technol. Lett. **4**, 969 (1992).
95. A. D'Ottavi, E. Iannone, A. Mecozzi, S. Scotti, and P. Spano, Digest Quantum Electron. and Laser Science Conf., Paper QThA5 (Optical Society of America, Washington, CD, 1993).
96. N. A. Olsson and P. Garbinski, Electron. Lett. **22**, 1114 (1986).
97. I. W. Marshall and M. J. O'Mahony, Electron. Lett. **23**, 1052 (1987).
98. R. M. Jopson, A. H. Gnauck, B. L. Kasper, R. E. Tench, N. A. Olsson, C. A. Burrus, and A. R. Chraplyvy, Electron. Lett. **25**, 233 (1989).
99. I. W. Marshall. D. M. Spirit, and M. J. O'Mahony, Electron. Lett. **22**, 253 (1986).
100. H. J. Westlake and M. J. O'Mahony, Electron. Lett. **23**, 649 (1987).
101. N. A. Olsson, M. G. Öberg, L. A. Koszi, and G. J. Przybylek, Electron. Lett. **24**, 36 (1988).
102. M. G. Öberg, N. A. Olsson, L. A. Koszi, and G. J. Przybylek, Electron. Lett. **24**, 38 (1988).
103. G. Großkopf, R. Ludwig, and H. G. Weber, Electron. Lett. **24**, 551 (1988).
104. D. J. Maylon, D. J. Elton, J. C. Regnault, S. J. McDonald, W. J. Devlin, K. H. Cameron, D. M. Bird, and W. A. Stallard, Electron. Lett. **25**, 235 (1989).
105. S. Ryu, H. Taga, S. Yamamoto, K. Mochizuki, and H. Wakabayashi, Electron. Lett. **25**, 1682 (1989).
106. P. M. Gabla, S. Gauchard, and I. Neubauer, IEEE Photon. Technol. Lett. **2**, 594 (1990).
107. W. I. Way, C. E. Zah, and T. P. Lee, IEEE Trans. Microwave Theory Tech. **38**, 534 (1990).
108. K. T. Koai, R. Olshansky, and P. M. Hill, IEEE Photon. Technol. Lett. **2**, 926 (1990).
109. P. S. Mudhar, J. Singh, and D. A. H. Mace, IEEE Photon. Technol. Lett. **4**, 574 (1992).
110. N. K. Dutta, J. Lopata, R. Logan, and T. Tanbun-Ek, Appl. Phys. Lett. **59**, 1676 (1991).
111. U. Koren, B. I. Miller, G. Raybon, M. Oron, M. G. Young, T. L. Koch, J. L. DeMiguel, M. Chien, B. Tell, K. Brown-Geobeler, and C. A. Burrus, Appl. Phys. Lett. **57**, 1375 (1990).
112. T. L. Koch and U. Koren, IEEE J. Quantum Electron. **27**, 641 (1991).
113. B. E. A. Saleh and M. C. Teich, *Fundamentals of Photonics* (Wiley, New York, 1991), Chap. 21; T. K. Gustafson and P. W. Smith, Eds., *Photonic Switching* (Springer, New York, 1988).
114. A. Ehrhardt, M. Eiselt, G. Großkopf, L. Kuller, R. Ludwig, W. Pieper, R. Schnabel, and H. G. Weber, J. Lightwave Technol. **8**, 1287 (1993).
115. N. Nakai, N. Ogasawara, and R. Ito, Jpn. J. Appl. Phys. **22**, L310 (1983); K. Otsuka and S. Kobayashi, Electron. Lett. **19**, 262 (1983).
116. W. F. Sharfin and M. Dagenais, Appl. Phys. Lett. **46**, 819 (1985); **48**, 321 (1986); **48**, 1510 (1986).
117. N. Ogasawara and R. Ito, Jpn. J. Appl. Phys. **25**, L739 (1986); H. Kawaguchi, IEE Proc. J. **140**, 3 (1993).
118. M. J. Adams, H. J. Westlake, and M. J. O'Mahony, in *Optical Nonlinearities and Instabilities in Semiconductors* (Academic, San Diego, 1988), Chap. 15.
119. M. Fujiwara, H. Nishimoto, T. Kajitani, M. Itoh, and S. Suziki, J. Lightwave Technol. **9**, 155 (1991).
120. R. van Roijen, J. M. M. van der Heijden, L. F. Tiemeijer, P. J. A. Thijs, T. van Dongen, J. J. M. Binsma, and B. H. Verbeek, IEEE Photon. Technol. Lett. **5**, 529 (1993).
121. B. Glance, J. M. Wiesenfeld, U. Koren, A. H. Gnauck, H. M. Presby, and A. Jourdan, Electron. Lett. **28**, 1714 (1992).
122. I. Valiente, J. C. Simon, and M. LeLigne, Electron. Lett. **29**, 502 (1993).
123. C. Joergensen, T. Durhus, C. Braggard, B. Mikelsen, and K. E. Stubkjaer, IEEE Photon. Technol. Lett. **5**, 660 (1993).
124. S. Murata, A. Tomita, J. Shimizu, and A. Suzuki, IEEE Photon. Technol. Lett. **3**, 1179 (1991).
125. J. Zhou, N. Park, J. W. Dawson, K. J. Vahala, M. A. Newkirk, and B. I. Miller, IEEE Photon. Technol. Lett. **6**, 50 (1994).
126. K. J. Vahala, J. Zhou, N. Park, M. A. Newkirk, and B. I. Miller, Conf. Lasers & Electro-Optics,

Anaheim, California, May 8–13, 1994, Paper CPD4.
127. D. Mehuys, R. Parke, R. G. Waarts, D. F. Welch, A. Hardy, W. Streifer, and D. R. Scifres, IEEE J. Quantum Electron. **27**, 1574 (1991).
128. J. Andrews and G. Schuster, Opt. Lett. **16**, 913 (1991).
129. L. Goldberg and D. Mehuys, Appl. Phys. Lett. **61**, 633 (1992).
130. L. Goldberg, D. Mehuys, and D. C. Hall, Electron. Lett. **28**, 1082 (1992).
131. D. Mehuys, D. F. Welch, and L. Goldberg, Electron. Lett. **28**, 1944 (1992).
132. D. Mehuys, L. Goldberg, R. Waarts, and D. F. Welch, Electron. Lett. **29**, 219 (1993).
133. R. Parke, D. F. Welch, A. Hardy, R. Lang, D. Mehuys, S. O'Brien, K. Dzurko, and D. Scifres, IEEE Photon. Technol. Lett. **5**, 297 (1993).
134. N. W. Carlson, IEEE J. Quantum Electron. **28**, 1884 (1992).
135. N. W. Carlson, P. Gardner, R. Menna, J. Andrews, R. Stolzenberger, A. Triano, E. Vangieson, D. Bour, G. A. Evans, S. K. Liew, J. Kirk, and W. Reichert, IEEE Photon. Technol. Lett. **4**, 988 (1992).
136. S. K. Liew, N. W. Carlson, and R. Amantea, IEEE Photon. Technol. Lett. **5**, 209 (1993).
137. N. W. Carlson, R. Menna, P. Gardner, S. K. Liew, J. Andrews, A. Triano, J. Kirk, and W. Reichart, Appl. Phys. Lett. **62**, 2006 (1993).
138. R. Parke, D. F. Welch, S. O'Brien, and R. Lang, Digest Conf. Lasers and Electro-Optics (Optical Society of America, Washington, DC, 1993), paper CTuI4.
139. D. Mehuys, L. Goldberg, and D. F. Welch, IEEE Photon. Technol. Lett. **5**, 1179 (1993).
140. D. Mehuys, R. S. Geels, W. E. Plano, and D. F. Welch, Electron. Lett. **30**, 961 (1994).
141. J. S. Osinski, D. Mehuys, D. F. Welch, K. M. Dzurko, and R. J. Lang, IEEE Photon. Technol. Lett. **6**, 1185 (1994).

CHAPTER 9

Applications of Semiconductor Lasers

G. R. Gray

Electrical Engineering Department, University of Utah,
Salt Lake City, Utah

9.1. INTRODUCTION

If one judges the importance of a laser type solely by production volume,[1] then the semiconductor laser (SL) is by far the single most important type of laser. There are two reasons for this massive production yield: Due to their small size, SLs are naturally grown in large numbers rather than one at a time; and just as importantly, the SL has found use in a plethora of applications that has fueled the tremendous amount of research and development devoted to SLs during the past 30 years. There are many properties of SLs that make them ideal for a variety of applications. As mentioned above SLs are extremely small devices: The active region volume of a conventional bulk SL is merely 10^{-10} cm^3 or 100 μm^3. The use of quantum wells in one or more dimensions lowers this value accordingly. Even when the confining and blocking layers are considered, the total chip volume ($\sim 5 \times 10^{-3}$ mm^3) is considerably less than one cubic millimeter. This small size coupled with the relatively high efficiency of SL devices leads to rather low threshold currents and small power requirements. Order-of-magnitude threshold and operating currents for "low-power" (< 30 mW) SLs used in optical recording and optical communication range from 20–80 mA threshold and 50–120 mA operating current. Such values are, of course, highly temperature dependent. The associated output powers, though small in comparison to those of high-power lasers producing many watts of power, are adequate for the optical communication and storage applications discussed in this chapter.

Aside from low power requirements, high efficiency, and small size, the SL possesses additional characteristics useful in applications. Information can be transmitted at GHz rates by the SL simply through direct modulation of the laser injection current. The intensity and frequency noise levels, although higher than those of certain other lasers, are generally low enough for applications. SLs have shown good reliability, with 10^5 h (> 10 years) of lifetime being typical. Many laser diodes have become available at very modest prices, as manufacturers have made improvements in production yields and because of increased demand, increased the supply. The SL can operate in many wavelength ranges from the visible to the infrared, and the number of output wavelengths continues to grow. Currently the two most common wavelength ranges are those used for optical communication ($\lambda = 1.3–1.6$ μm) and for optical data storage ($\lambda \sim 0.8$ μm). Recent research advances have led to the development of SL pumps ($\lambda \sim 0.98$ μm) for erbium-doped fiber amplifiers as well as visible red diode lasers ($\lambda \sim 0.63–0.67$ μm) for laser pointers, optical scanners, and other traditional helium–neon laser applications. Considerable effort is being devoted to further shorten the wavelength of SLs to the blue-green, either by frequency doubling near-infrared lasers or by using ZnSe-based semiconductors (see Chap. 7).

Earlier chapters have described recent advances in SLs, including quantum well and strained quantum well, high power, vertical cavity, narrow linewidth, short pulsed and visible output (both red and blue-green) devices. The potential use of these lasers in applications spurs the research into these devices. This chapter is dedicated to describing some of the major applications of semiconductor lasers. Usually research papers on diode lasers merely list a variety of suitable applications, without mentioning the details of how the laser is used in a particular application. This chapter represents an attempt to fill in some of those missing details. Since SLs are used in so many applications, a complete description would fill much more than a single chapter. Thus we concentrate on a few select applications: Fiber-optic communications, optical data storage, select medical applications, and some applications from atomic physics. In each section we describe how the laser is used in the particular application as well as both present and future requirements the laser must satisfy. Where appropriate, we discuss inherent problems the laser must face in a given application.

9.2. OPTICAL COMMUNICATION

It is no understatement to say that optical fiber based communication has revolutionized the industry. In fact the application of diode lasers to optical communication spurred much of the SL research in the 70's and 80's. Many other applications have benefited from the research aimed at communication lasers, which operate primarily at wavelengths of 1.3 and 1.55 μm. This section concentrates on optical-fiber communication, and most of the remarks are valid regardless of whether the fiber is carrying voice or data or even video. To be specific, we now describe the use of SLs in long-distance voice communication. The ability of the laser to be directly modulated is a key feature here, since an electrical signal representing a voice (actually thousands of voices) modulates the laser intensity. The most common modulation and detection scheme is referred to as IM/DD, for intensity modulation and direct detection. The laser intensity is modulated between high and low values representing the 1's and 0's of a digital stream. Direct detection refers to the fact that the photodetector responds only to the total intensity incident upon it; all phase information is lost. Other digital modulation schemes rely on modulation of the laser frequency (FSK—frequency-shift keying) or phase (PSK—phase-shift keying) to encode the information. Communication utilizing FSK or PSK is referred to as coherent, and the detection becomes more complex than for intensity modulation. We will discuss briefly the laser requirements for coherent communication in Sec. 9.2.3; first we consider the simpler case of IM/DD.

One of the primary reasons for pursuing optical-fiber communications in the first place was the bandwidth increase available over microwave communications. The increase is essentially due to the difference in source frequencies: $\sim 10^{10}$ Hz for microwaves and 10^{14} Hz for light. The effective bandwidth of the optical fiber can easily exceed 1 THz. The efficient use of such a large bandwidth requires sophisticated *multiplexing* techniques that transform low bandwidth single voice signals into high-bandwidth communication signals. Specifically, if an ordinary phone conversation is digitized using 8 bits for a SNR (signal-to-noise ratio) of 30 dB, then the resulting bandwidth B required of a single phone call is $B > 31$ kb/s.[2] Allowing for a margin of error, the bandwidth used in practice is 64 kb/s. A single telephone call using 64 kb/s would certainly not make efficient use of an optical-fiber bandwidth, which exceeds 1 THz; therefore different conversations are multiplexed together in order to increase the overall bandwidth. Time-division multiplexing (TDM) in the electrical domain is typically performed first, boosting the bit stream by a factor of 1000 to about 51 Mb/s. One of the advantages of TDM is that it can be continued to further enhance the system bandwidth. Since 1990, commercial systems have been available that operate at 2.2 Gb/s.

The electrical signal at 2.2 Gb/s represents a composite of over 30 000 telephone conversations. As such it is an extremely complex electrical signal, which is then used to modulate a single laser diode. There are basically three major requirements that such a laser must satisfy in this application; these are related to laser speed, spectral content, and output power.

9.2.1. Speed Requirements

If we consider an electrical bit stream at $B = 2.2$ Gb/s, then each 1 or 0 occupies its own bit slot of width $T_B = 1/B = 450$ ps. Thus the laser must be able to be turned on and off in a time somewhat shorter than T_B. A common experimental and analytical method used to estimate the speed of a laser involves the small-signal modulation response, which measures how well the laser responds to a small-amplitude sinusoidal modulation. A typical modulation response is shown in Fig. 9.1.[3] At low frequencies, the laser follows the modulation perfectly, and so the response is unity. The response is enhanced when the modulation frequency nears the laser relaxation-oscillation frequency ν_R, which is simply the natural resonance frequency of this nonlinear oscillator. Beyond ν_R the response drops rapidly because the laser is no longer able to follow the modulation current. On the basis of the

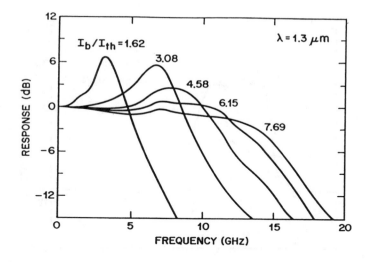

FIGURE 9.1. Measure small-signal modulation response as a function of modulation frequency for a 1.3 μm DFB laser. Several different bias levels are shown (after Ref. 3).

small-signal modulation response, then, it is desirable to have ν_R as high as possible. The relaxation-oscillation frequency scales as[4]

$$\nu_R = \frac{1}{2\pi} \sqrt{\frac{G_N(I-I_{th})}{q}}, \tag{9.1}$$

where G_N is the differential gain coefficient, I and I_{th} are the bias current and the threshold value, respectively, and q is the magnitude of the electron charge. Therefore, high-speed lasers require a large differential gain, and the speed increases as the injection current is raised. As discussed in Chap. 1, the differential gain coefficient can be a factor of 2 larger for quantum well (QW) lasers compared to bulk devices, making QW lasers excellent candidates for high-speed operation. Equation (9.1) holds as long as the relaxation-oscillation damping rate Γ_R is not too large; i.e., $\Gamma_R^2 \ll (2\pi\nu_R)^2$. Actually, Γ_R depends mainly on the strength of nonlinear gain, which is also increased for QW lasers. In that case, Eq. (9.1) may hold only approximately at high operating powers where nonlinear effects become important.

The small-signal modulation response represents a good starting point for the evaluation of the speed of a SL, but in practice the modulation applied to the laser is far from "small." Typically the modulation current brings the laser from below threshold to far above threshold, so that an analytic treatment of modulation response is not possible. The response can still be measured experimentally, however, and computer simulations can also be used to model the large-signal response. Figure 9.2 shows the calculated large-signal modulation response of a SL to a 500 ps rectangular current pulse,[5] indicating that the output power is able to follow the input current reasonably well at the assumed 2 Gb/s bit rate. The dashed

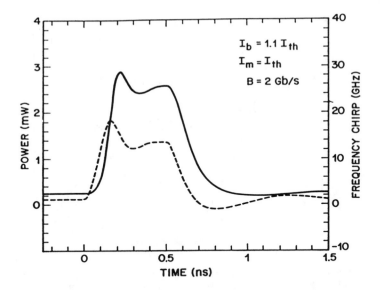

FIGURE 9.2. Large-signal modulation response calculated by computer simulation for a semi-conductor laser in response to a 500 ps current pulse. The solid curve shows the shape of the emitted optical pulse; the dashed curve indicates that the laser frequency also changes with modulation. Bias current is 10% above threshold, modulation current equals the threshold current, and the linewidth-enhancement factor $\alpha = 5$ (after Ref. 5).

line in Fig. 9.2 indicates that frequency modulation accompanies the intensity modulation, a fact which is always true for directly modulated SLs. The origin of the intensity-frequency coupling lies with the linewidth-enhancement factor, which has a value of 2–8 for most SLs. Because of this coupling, the laser frequency shifts to higher values near the leading edge of the pulse and toward lower values at the trailing edge of the pulse. Such a time-dependent frequency shift is referred to as chirp. Pulses emitted from directly modulated SLs are always negatively chirped due to the positive value of the linewidth-enhancement factor, commonly denoted as α.

The inherent speed requirement of SLs or light-emitting diodes (LEDs) in digital communication systems, therefore, is essentially determined by the need to turn the laser on and off faster than the bit slot. Additional electrical parasitics can sometimes limit the modulation speed to a value less than the inherent laser value, particularly for speeds in excess of 10 GHz.

9.2.2. Spectral and Power Requirements

A common figure of merit used to judge a communication link is the product of bit rate B and repeater spacing L. An efficient system will operate at a B-L value very close to the system limit. The power requirement is obviously related to the repeater spacing: Losses along the fiber, including coupling and splicing losses, require that more power be launched into the fiber for a given repeater distance, to achieve the same SNR at the receiver. The spectral requirement stems from the fact that an optical pulse is comprised of multiple frequencies that travel at different speeds down the fiber because of the dispersion of the group velocity. The broadening of a pulse by group velocity dispersion (GVD) has two effects: The pulse tends to smear out and enter adjacent bit slots, leading to intersymbol interference; and the energy contained in the bit slot gets reduced with a corresponding reduction in SNR. Power would need to be increased to overcome this SNR reduction. Thus, spectral and power requirements are actually intimately related. As the bit rate is increased to obtain higher capacity, the bit slot correspondingly decreases making dispersion more of a problem.

The B-L limit due to fiber losses can be obtained from the expression for repeater spacing L:

$$L = \frac{10}{\alpha_f} \log_{10}\left(\frac{P_{tr}}{P_{rec}}\right), \qquad (9.2)$$

where P_{tr} and P_{rec} are the transmitted and received optical powers, respectively, and α_f is the fiber loss in dB/km, which should include all sources of loss. The received power is related to the bit rate by $P_{rec} = N_p h \nu B$, where N_p is the number of received photons, $h\nu$ is the photon energy, and B is the bit rate. Substitution of P_{rec} into Eq. (9.2) yields the maximum B-L product if only fiber losses are considered. The actual B-L value is best determined graphically, however, as shown in Fig. 9.3, where repeater spacing is plotted vs bit rate.[6] The solid lines represent the maximum values limited only by loss for different generations of communications systems. The slope of the solid lines is simply $10/\alpha_f$, and depends only on the total fiber loss. Although the fiber loss decreases significantly as the wavelength increases from 0.85 to 1.55 μm, the different slopes appear nearly parallel since repeater distance is plotted on a logarithmic scale. The maximum (loss-limited) B-L product has increased from about 3 Gb/s km for the first generation system operating at 0.85 μm (using multimode fiber) to about 2000 Gb/s km for third generation systems operating in the low-loss window of the fiber near 1.55 μm, where the loss is merely 0.2 dB/km. For multimode fibers, intermodal dispersion is the main source of pulse broadening and limits their use to bit rates below 100 Mb/s, as seen in Fig. 9.3.

In addition to fiber losses, fiber dispersion represents a fundamental limitation in fiber-communication systems. With respect to dispersion-induced pulse broadening in single-mode fibers, considerable effort has been devoted to modeling this nontrivial problem. A particularly useful model,[7] although it assumes a Gaussian-shaped input pulse, calculates the amount of broadening as a function of input pulse width, input spectral width, input pulse chirp, and third-order dispersion, which can become important when operating at the zero-dispersion point. This model can be used to estimate the source spectral width and chirp that is tolerable for a given B-L product. In general, the spectral requirements that a communication source must satisfy become more stringent as the system capacity increases. The limits imposed by dispersion for the different system generations are also shown in Fig. 9.3 as the

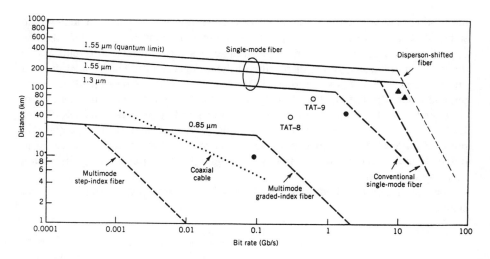

FIGURE 9.3. Repeater spacing vs bit rate for the various generations of communication systems. The solid lines indicate loss-limited behavior, and the dashed curves represent dispersion-limited behavior. Filled circles denote actual terrestrial systems, and open circles denote undersea Transatlantic (TAT) systems. The solid triangles represent laboratory experiments using dispersion-shifted fibers. The dotted line shows a coaxial cable system for comparison (after Ref. 6).

dashed lines. We consider briefly the different limits given by this model for the various generations employing single-mode fiber.

For LED or multimode laser-based systems, the source spectral width is large and is the main source of dispersion. For such systems, in order to have each pulse remain substantially in its bit slot, the dispersion-limited B-L product is bounded by[8]

$$ BL < \frac{1}{4|D|\sigma_\lambda}, \tag{9.3} $$

where D is the dispersion in ps/(km-nm), and σ_λ is the root-mean-square (rms) source spectral width in nanometers. The factor of 4 arises from the criterion that the pulse width should remain less than $\frac{1}{4}$ of the bit slot. The wavelength dependence of dispersion for standard silica fiber as well as dispersion-shifted and dispersion-flattened fiber is shown in Fig. 9.4. Operation near the zero-dispersion wavelength can result in values of $|D| \sim 1$–2 ps/(km-nm). For a spectral bandwidth of 1 nm, which is typical for multimode lasers, the maximum dispersion-limited B-L product is B-$L \sim 100$ Gb/s km. As seen in Fig. 9.3, such a system generally operates loss limited for bit rates below 1 Gb/s but operates dispersion limited at higher bit rates. Note that such a system could operate at rather high bit rates over short distances, such as in a local-area network (LAN) environment.

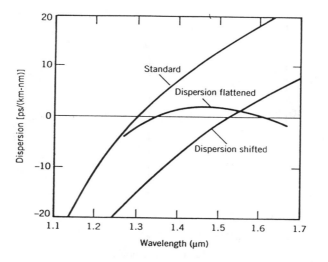

FIGURE 9.4. Dispersion vs wavelength characteristics of standard, dispersion flattened, and dispersion-shifted fibers (after Ref. 5, p. 44).

To increase the B-L product further, single-mode lasers can be used. Single-mode lasers operate with a source bandwidth orders of magnitude lower than a multimode laser, at least when operated continuously. Actually, the third generation lightwave systems employ single-mode distributed feedback (DFB) lasers, but operating near 1.55 μm, to take advantage of the lowest fiber loss. From Fig. 9.3, this system operates loss limited up to bit rates of several Gb/s, but then becomes dispersion-limited beyond that. Because dispersion is so much larger at 1.55 μm [$D \sim 16$ ps/(km-nm)], the primary source of broadening tends to be chirp induced. As mentioned above and shown in Fig. 9.2, a SL pulse is always negatively chirped (leading edge is blue shifted, trailing edge is red shifted), when the laser is directly modulated. A frequency chirp of 10 GHz is already two orders of magnitude larger than the laser linewidth, which is typically below 100 MHz. The effect of the B-L product as a function of chirp is shown in Fig. 9.5,[9] which is calculated numerically by solving the nonlinear Schrödinger equation for pulse propagation in fiber. For both Gaussian and super-Gaussian input pulses, the B-L product remains limited to below 100 Gb/s km for chirp parameter $C(= \alpha)$ of 6. Interestingly, when the pulse chirp is positive, the B-L product is greatly increased: Because the pulse is initially compressed in this regime (positive chirp and normal dispersion), much longer propagation occurs before dispersion-induced broadening sets in. Physically, a positively chirped pulse has its leading edge red-shifted compared to its trailing edge. In the normal dispersion regime the red-shifted components travel more slowly than the blue-shifted parts, allowing the trailing edge to "catch up" with the leading edge, and pulse compression occurs. After a certain distance, however, the pulse chirp effectively changes sign, and the pulse once again broadens with further

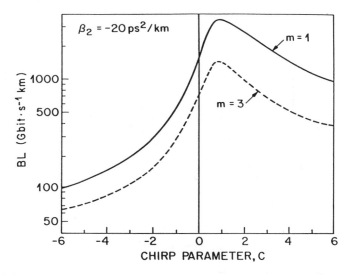

FIGURE 9.5. B-L product vs chirp parameter in the dispersion limit for Gaussian (solid curve) and super-Gaussian (dashed curve) input pulses. The assumed dispersion is 16 ps/(km-nm) (after Ref. 9).

propagation. It is possible, by using the nonlinear-refractive index, to cancel the GVD-induced broadening (which is always present due to the finite source spectral width) and produce soliton pulses that travel without changing shape. Communication systems have been designed and tested using this principle.[10]

In order to reduce the chirp-induced dispersion, one can either reduce the chirp or reduce the magnitude of dispersion. Reducing the chirp can be accomplished by biasing the laser close to threshold.[11] Although such biasing reduces the contrast between a 1 and a 0, and therefore the SNR at the receiver, a compromise can be reached, as shown in Fig. 9.6.[12] The dashed line in Fig. 9.6 represents the power penalty associated with decreasing the contrast (i.e., extinction ratio). For a given SNR at the receiver, power must be increased as the contrast is reduced. The dotted line represents the chirp-induced penalty, showing that lowering the extinction ratio can reduce the chirp penalty. The lowest overall penalty due to both these effects can be achieved for an extinction ratio of -10 dB. The chirp can be further reduced by lowering the value of the linewidth-enhancement factor α, since the chirp is directly proportional to α. Quantum-well lasers, which operate at higher speeds than bulk lasers due to their larger gain derivative, also possess lower values of α, and so are desirable for reducing the chirp-induced broadening. Another way of reducing or actually eliminating chirp is to modulate the laser externally. Rather than reducing the chirp, one can instead reduce the magnitude of dispersion by using either dispersion-shifted fibers or dispersion-flattened fibers, whose properties are shown in Fig. 9.4.

When a single-longitudinal mode laser is used in a telecommunication link, it can

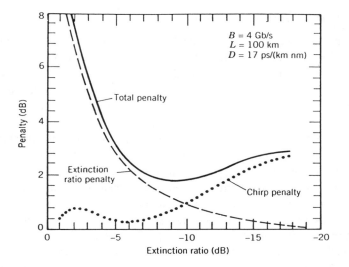

FIGURE 9.6. Power penalty vs extinction ratio (after Ref. 12 © 1987 IEEE).

also be limited by mode-partition effects. That is, if any side modes are excited during modulation, they will travel at different speeds down the fiber due to GVD and lead to a desynchronization of the pulse and a reduction in SNR. As long as the side modes are suppressed by more than a factor of 100, then this effect is basically negligible.[13]

As noted above, problems associated with dispersion broadening or fiber losses are often discussed in terms of a power penalty, i.e., a necessary increase in laser power in order to overcome the detrimental effect. It is important to realize that laser power cannot be increased indefinitely. Operating at higher power degrades the reliability of the laser by shortening its lifetime. Also, the electrical power requirements are greater if higher operating currents are used, leading to increased system cost. In addition, there are fundamental effects due to fiber nonlinearities that limit the amount of laser power launched into the fiber. Such nonlinear effects include stimulated Raman scattering (SRS), stimulated Brilliouin scattering (SBS), and self-phase modulation (SPM). Since in-depth coverage of fiber nonlinearities can be found elsewhere,[14] we discuss them only briefly here. In practical terms, SRS and SBS represent additional loss and broadening mechanisms for single-channel communication systems. In both cases, certain frequency components of a pulse get scattered either to other parts of the pulse (a source of broadening) or out of the pulse altogether (a source of loss). The threshold for SRS is rather high (> 100 mW) for a single-channel system but the threshold for SBS, which depends on both the bit rate and the modulation format,[15] is only about 10 mW. This low SBS threshold represents a fundamental upper limit for the fiber launch power.

9.2.3. Coherent Communications

As mentioned above, systems that employ modulation of the laser frequency or phase are termed coherent, because the coherence of the laser phase must be very stable. Direct detection, which destroys all coherence information, cannot be used in systems employing such modulation schemes; instead, either homodyne or heterodyne detection must be used. Homodyne and heterodyne systems enjoy two basic advantages over an IM/DD system: Coherent detection allows a 20 dB enhancement in receiver sensitivity, which can translate into a longer allowable repeater spacing; and more efficient use of fiber bandwidth can be made since many closely spaced channels can be sent down the fiber simultaneously. Because of these advantages research into coherent communication received considerable attention during the 1980s.[16–19] However, the introduction of the erbium-doped optical fiber amplifier has perhaps delayed the commercialization of coherent systems because the amplifier effectively takes away the first advantage. That is, IM/DD systems can use an erbium amplifier to achieve the same receiver sensitivity as coherent detection provides. The second advantage is still intact, however, and coherent techniques will likely play an important role in future lightwave systems. We briefly discuss in this section the requirements of SLs used in coherent communication.

In a coherent system utilizing heterodyne detection, the output from the fiber is mixed on a detector with light from a stable local oscillator. In this case the detector current depends explicitly on the phases of the transmitter laser and the local oscillator laser as well as on the frequency difference $\Delta\omega$ between them. If the phase of either laser varies much during the duration of a bit, then the information will be partially washed out and errors will occur. The time over which a laser remains coherent can be approximated by the coherence time, which is just $1/\Delta\nu$, where $\Delta\nu$ is the laser linewidth. Thus a requirement is for $\Delta\nu < B$, the exact criterion depending greatly on the type of modulation and demodulation the system utilizes.[16] The most strict linewidth requirement occurs for homodyne detection ($\Delta\omega = 0$), for which $\Delta\nu/B < 5\times10^{-4}$ in order to keep the power penalty below 1 dB.[17] In contrast, a system using FSK and asynchronous heterodyne detection only requires $\Delta\nu/B \sim 0.1$.[18,19] An interesting consequence of this relationship is that the linewidth requirement becomes more stringent as the bit rate is lowered. Physically, this is due to the fact that the time during which the phase must remain coherent, given by the bit slot, increases as the bit rate is reduced. Hence, high data rate systems are better suited for coherent communications since the linewidth requirement is less stringent.

In addition to the narrow-linewidth requirement, coherent systems also require a SL that is tunable over at least a few nanometers.[20] This is because the local oscillator laser must be able to match the transmission laser frequency either exactly (for homodyne detection) or within the required intermediate frequency (for heterodyne detection). Multisection distributed-Bragg reflector (DBR) lasers[21,22] and multiquantum well DFB lasers[23] have been developed for this purpose.

9.2.4. MultiChannel Systems

As shown in Fig. 9.3, single-channel lightwave systems are limited by dispersion to bit rate-length products of B-L < 1 Tb/s km. In order to make more efficient use of the available fiber bandwidth, which exceeds 12 THz or 100 nm in the low-loss window at λ = 1.5 μm, multiple channels should be sent down the fiber simultaneously. The procedure is known as frequency-division multiplexing (FDM) when it is performed optically, or subcarrier multiplexing (SCM) when it is performed electrically. In practice these techniques can be combined to achieve the most efficient use of the fiber bandwidth. In this section we describe these two techniques, paying particular attention to the additional requirements placed on the semiconductor lasers used in multichannel systems.

9.2.4.1. Frequency-Division Multiplexing

In frequency-division multiplexing (FDM), multiple high bit rate streams, which may have been produced by TDM or SCM, modulate separate SLs operating at slightly different wavelengths. The output of all of the lasers is combined and sent down a single optical fiber. This scheme is sometimes referred to as wavelength-division multiplexing (WDM) when the channel spacing is much larger than the bit rate. An IM/DD system operating at bit rate B could be enhanced by a factor of N simply by modulating N separate semiconductor lasers with different bit streams and combining the light output from all lasers into the same fiber. Such a system was demonstrated in 1985 using ten lasers spaced by 1.35 nm at $\lambda \approx 1.5$ μm.[24] The system performed with a repeater spacing of 68.3 km for an effective B-L product of 1.37 Tb/s km. The fiber bandwidth used by such a system is 13.5 nm, which is still far less than the total fiber bandwidth. Although more than 70 such channels could in principle be combined into a single fiber, in practice multiplexers and demultiplexers tend to limit the channel numbers to much smaller values. The limitations stem primarily from input coupling losses. Many different types of multiplexers have been studied, including those based on gratings[25,27-29] and those based on interferometers.[26,30] Since a grating is an angularly dispersive element, light containing multiple wavelengths incident on the grating is dispersed into its component parts, which can be focused to separate detectors for demultiplexing. In practice the grating structure can be integrated onto a waveguide structure to provide a compact, low loss design.[27,28] Such grating-based multiplexers have been used for 32 channels but are limited to channel spacings of about 1 nm for operation near 1.55 μm.[29]

A schematic of an eight-channel multiplexer based on three Mach–Zhender interferometers is shown in Fig. 9.7.[26] In practice, the entire structure can be fabricated onto a SiO$_2$ substrate. For each interferometer, different arm lengths produce wavelength-dependent phase shifts such that the two input powers appear at only one output port. The structure has been extended to produce a 128 channel demultiplexer employing seven serially connected interferometers.[30] Because of the coupling losses that exist when light enters the multiplexer, it is desirable to inte-

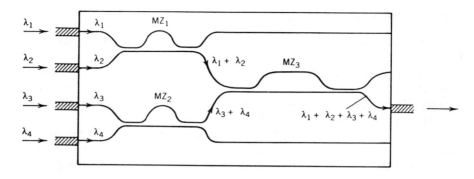

FIGURE 9.7. Schematic layout of a four-channel multiplexer constructed from three Mach–Zhender waveguide interferometers (after Ref. 26 © 1988 IEEE).

grate the transmitter lasers and the multiplexer onto the same chip. Such integration has been achieved recently with InP-based optoelectronic integrated circuit (OEIC) technology. Several examples of such photonic integrated circuits (PICs) have been fabricated.[31] In particular, four completely independent tunable MQW-DBR lasers are multiplexed through passive waveguides, and then the light is sent through an amplifier section on the same chip to compensate for the coupling losses.[32] Such photonic integrated circuits are likely to play an important role in future communications systems, particularly in the high bandwidth systems needed for simultaneous transmission of voice, data, FAX, and video.

9.2.4.2. SubCarrier Multiplexing

When the data signals are separated in frequency in the electrical domain, prior to modulation of the optical carrier, the multiplexing scheme is known as subcarrier multiplexing (SCM). SCM has received considerable attention lately[36] because it is compatible with both analog and digital transmission and because it makes use of microwave components, which are part of a relatively mature technology. As shown schematically in Fig. 9.8, each bit stream modulates a different microwave carrier, each with its own base frequency. Several such microwave carriers are combined in a microwave power combiner, and the output is used to modulate a single semiconductor laser. Upon detection, the original bit stream signals can be recovered using standard microwave filtering techniques.

SCM has found particular application in the transmission of video signals.[35–39] For instance, in the cable television (CATV) industry, a set number of video channels (~ 100) needs to be sent to a much larger set of subscribers (thousands or even millions). Since each television signal requires a bandwidth of only 4 MHz, 100 channels could be subcarrier multiplexed for a composite bandwidth of less than 1 GHz. A single SL can be used therefore to transmit a few hundred cable television signals. However, the SCM laser is subject to stricter requirements than a typical

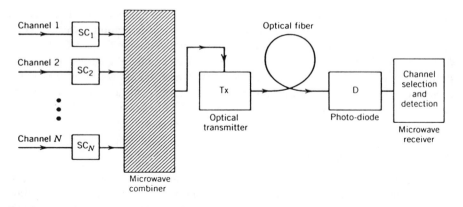

FIGURE 9.8. Schematic of a multichannel SCM lightwave system. Multiple microwave carriers (SC_1, SC_2,...,SC_N), representing N separate data channels are combined in the microwave combiner. The resulting composite stream modulates a single optical transmitter (after Ref. 36).

IM/DD laser, at least in terms of intensity-noise level and linearity. The reason stems from the CATV industry's use of *analog* techniques for transmission of the video signals, rather than the digital ones used nearly exclusively for lightwave systems. Analog modulation implies that the laser power must follow the modulation signal exactly, in order to maintain the video waveform. Any nonlinearity present in the light-current (L-I) curve would alter the transmitted waveform. Since an L-I curve can become nonlinear at elevated temperatures and near the end of the SL life, such conditions must be avoided when using analog modulation. Note further that the modulating current must never fall below threshold, since the linearity would be immediately ruined. Linearity can also be affected by transmission of the signal through the optical fiber, which is an intrinsically nonlinear medium. The subcarrier multiplexed signal contains many different frequency components that can interact inside the fiber producing higher-order products.[33] Considerable progress has been made in controlling the amount of distortion resulting from these higher-order products.[34]

The requirement of the intensity noise level of the laser used in an analog system employing SCM is particularly severe. For the analog amplitude-modulation (AM) scheme used for cable television,[35] the SNR needs to be about 50 dB. For a modulation index of 0.1 and other typical values,[36] this translates into a nominal relative-intensity noise (RIN; see Sec. 9.3.3.2) value of -150 dB/Hz, which is a fairly stringent requirement. Since the RIN decreases inversely with the power cubed, such low values of RIN can be obtained at high power ($>$ 5 mW). High values of laser power, however, strain the transmitter power budget and limit the transmission distance due to nonlinear effects in the fiber. The RIN requirement is less severe if frequency modulation (FM) is used instead of AM.[38,39] Although the channel bandwidth increases from 4–30 MHz, the required SNR decreases by about 30 dB, allowing the RIN requirement to be relaxed to $<$ -135 dB/Hz. Both AM and FM transmission has been demonstrated. An AM system demonstrated the transmission of 35 channels by using erbium amplifiers to overcome distribution losses.[37] In one FM experiment,[38] 120 channels occupied the frequency range 2.7–7.5 GHz with a 40 MHz spacing. One of the benefits of the SCM technique, the ability to transmit simultaneous analog and digital signals, was demonstrated when 60 FM-video channels were combined with a 100 Mb/s digital signal and used to modulate the same laser.[39] Commercial CATV systems using SCM became available in 1992.

9.2.5. Additional Considerations

Semiconductor-laser reliability[40] is an issue that is important for virtually all applications employing these lasers. It can be particularly crucial in undersea or satellite communication systems because repair in these situations becomes extremely expensive.[41] Although some SLs fail catastrophically and abruptly, a typical laser degrades rapidly during the first thousand hours or so of operation, but then stabilizes and degrades much more slowly during the rest of its lifetime. This initial degradation period is normally used to identify weak lasers, by exposing the lasers to an accelerated aging period of high-power, high-temperature operation. The assumption is that weak lasers will fail catastrophically during this initial high-stress period and can then be discarded for applications. In fiber communications, not only does the reliability of the laser pose a concern but the reliability of the laser-to-fiber coupling must also be considered. The transmitter package must be carefully designed such that the coupling efficiency is extremely stable and does not degrade with age.[42] Related to the question of reliability is the sensitivity of the laser-to-fiber coupling to various environmental effects, including humidity, shock, temperature, pressure, etc. The temperature variations are particularly important because the transmitter may be exposed to a wide range of operating temperatures, and because the laser is known to be extremely sensitive to temperature variations. The emitting wavelength, the output efficiency, and lifetime are all dependent upon the operating temperature. For this reason the laser temperature may be stabilized by use of a thermoelectric cooler.

Optical feedback (OFB) is another effect that can degrade the laser performance.[43,44] OFB can occur because of parasitic reflections from the near and far ends of the optical fiber and because of scattering from fiber imperfections and splices. The exact effect of OFB depends on many parameters, including distance to the reflector, operating power, and feedback strength, but generally the laser intensity noise and phase noise get increased. Even extremely weak amounts of feedback can affect the laser linewidth, either broadening it or narrowing it depending on the phase of the returned light.[43] This effect is particularly detrimental in coherent systems, which rely on a stable and narrow linewidth. For IM/DD systems, on the other hand, the main problem with OFB is the increased intensity noise, which leads directly to reduced SNR at the receiver. For IM/DD systems, it has been found that most of the detrimental effects of OFB can be avoided if the feedback ratio is maintained below 0.1%.[45] The OFB problems in the transmitter can therefore be solved by using an optical isolator, which generally reduces feedback to below 30 dB. High-quality optical isolators, based on the Faraday effect, are a somewhat expensive option on a per-transmitter basis, but the relatively small number of transmitters required makes it an affordable option for the system as a whole. In optical-recording heads, in contrast, the use of such an optical isolator in a commercial product would render it too expensive. For this and related reasons, OFB may be a greater problem in optical-recording applications, as discussed in Sec. 9.3.4.

9.3. OPTICAL DATA STORAGE

Of all the applications that use semiconductor lasers, the optical recording industry utilizes the largest number of lasers. Although the first commercial product for optical storage was the videodisc player, it was the (continuing) success of audio compact disk technology that has been responsible for entrenching the importance of optical recording. In terms of storage capability, optical storage has always invoked comparison with its magnetic counterpart. Some of the advantages that the optical technology holds are higher data density, due to extremely small track pitch (~ 1 μm compared to 10 μm for magnetic devices); greater reliability: head crashes do not occur since the optical head is not in close contact with the disk; and (read-only) disks allow for efficient mass replication. Optical drives also suffer disadvantages compared to magnetic drives: Due to the relatively massive optical head, disk access times are slower; servo systems (e.g., for tracking) are more complex because of the smaller track pitch; and most optical-recording media are nonerasable. The nonerasability may be the most significant disadvantage of optical-storage devices. Although erasable optical systems are commercially available, they are significantly more expensive than their magnetic counterparts; certainly until this is reversed, magnetic devices will retain a significant share of the market.

This section considers the use and requirements of semiconductor lasers in optical data recording applications. There are three basic types of optical-recording systems, known as read-only, write-once, and erasable. In read-only systems, a high-power argon-ion or helium-cadmium laser is used to record a "master" disk. From this master, virtually unlimited quantities of read-only disks can be *replicated*. Since these disks are mass produced by replication, their media is never designed for optical recording, and hence the name "read-only." Audio CD's and CDROM's fit into this category. For write-once media each disk is encoded individually by a laser, often a semiconductor laser. The recording process is irreversible, meaning that the data is very stable and reliable; however, no erasure is possible. The information on such a disk can be read a virtually unlimited number of times, so these systems are often designated as WORM, for Write-Once, Read-Many times. Erasable types of recording systems either use media that allow reversible phase changes or use magneto-optical recording.

Semiconductor lasers can be used to read and write data onto an optical disk, the requirements being somewhat different in each case. First we consider the SLs used

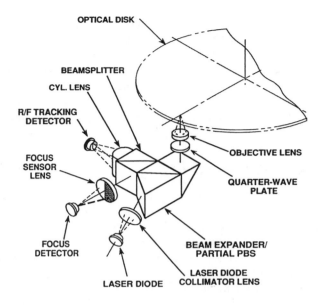

FIGURE 9.9. Schematic of a typical write-once optical head (after Ref. 54).

for reading from an optical disk. Since all CD and CD-ROM players contain a SL, these lasers represents a large fraction of lasers used.

9.3.1. Reading Data from an Optical Disk

In all recording systems, reading and writing is performed by tightly focusing the light from a laser onto a tiny portion of the disk. Such a focused beam is often termed the optical stylus, in analogy with LP recording. Figure 9.9 is a schematic of a typical optical head for a write-once system. Some of the elements in Fig. 9.9 are related to the detection system, and the rest will be discussed below. The primary difference between reading and writing is the power level of the optical stylus. Read power must be low enough that the light does not alter the disk surface. In mass-replicated read-only disks, pits (or marks) of a certain depth are made on a highly reflective substrate (or land). The depth of the pits is chosen such that light reflected from the pit is 180° out of phase with light reflected from the substrate. Specifically, as the disk is being scanned beneath the optical stylus, when the stylus moves from a land area to a mark, the reflected light intensity drops due to the destructive interference between the two portions of the beam. Clearly, a relatively high degree of spatial coherence must be present across the beam, dictating the use of a laser for readout. For complete destructive interference between land and mark reflections, the mark depth d can be found from $4\pi nd/\lambda \approx \pi$. For a refractive index $n = 1.5$ and $\lambda = 800$ nm, the required mark depth is 133 nm.

Reading from write-once media and erasable phase-change media is simpler than read-only, because the 1's and 0's have a substantially different reflectivity, making the phase difference of only minor importance. For large SNR this difference should be as great as possible. Whether the 1 or 0 has a larger reflectivity depends on the type of medium.[46] When light from a continuous, low power (~ 1 mW) SL is focused onto the rotating disk, the detector of the reflected light records an intensity modulation corresponding to the presence of 1's and 0's.

A substantially different readout scheme is utilized for magneto-optic (M-O) recording. In this case a 1 or 0 is encoded as a magnetization of the medium, either oriented "up" (away from the disk surface) or "down" (into the disk surface). Readout of these two magnetizations is accomplished by the optical Kerr effect, which induces a slight rotation of the incident linear polarization. The optical Kerr effect involves a combination of circular birefringence and circular dichroism. Circular birefringence exists if right- and left-handed circular polarized light waves experience different indices of refraction when incident on the medium. Circular dichroism exists when the two circular polarizations experience different absorptions by the medium. These relationships can be written compactly in terms of the complex index of refraction, $\tilde{n}$:

$$\tilde{n} = n'_j + i n''_j, \tag{9.4}$$

where the subscript $j = R$ or L stands for right and left circular polarizations, and the $'$ and $''$ indicate real and imaginary parts, respectively. Thus, circular birefringence corresponds to the case $n'_R \neq n'_L$, whereas circular dichroism exists when $n''_R \neq n''_L$. When linear polarization, which can be decomposed into equal amounts of right- and left-handed circular polarization, is focused onto the medium, the reflected beam is returned as elliptically polarized light (due to circular dichroism) with the major axis of the ellipse rotated slightly (due to circular birefringence) from the direction of initial linear polarization. The sense of the rotation, which is typically less than 1°, depends on the direction of the surface magnetization. In order to sense this tiny rotation, a polarizer can be placed before the detector. For highest contrast of the signal, the polarizer angle can be set nearly orthogonal to the laser polarization, although this orientation results in a small absolute signal level. The best polarizer angle to use depends on whether the system is limited by laser noise, electronic noise, or shot noise.[47] An alternative detection scheme, which yields higher contrast and larger signal, is based on differential detection. The two polarization components of the reflected light are separated (for example, by a polarization beam splitter) and detected by separate detectors and then subtracted electronically. One of the advantages of this method is that noise sources common to both polarizations, such as laser noise, can be nearly eliminated.

9.3.2. Semiconductor Lasers Used for Optical Writing

Semiconductor lasers are used for writing onto write-once media as well as onto erasable media. Whenever any laser is used to record optically, the primary writing mechanism is generally a thermal effect. For write-once media, two common methods used for making marks are laser ablation and irreversible phase changes. In the former case, the heat from the laser essentially burns a hole in the disk surface. The reflectivity of the hole is usually much lower ($\sim$ 4%) than that of the substrate ($\sim$ 35%). In the case of phase-change media, the material which is prepared in the amorphous state, crystallizes irreversibly when heated abruptly by the laser.

Some phase-change materials can be used for erasable recording. The material is prepared in the crystalline state, and rapid heating produces a mark by changing the material to the amorphous state. The media is erasable because a slower heating of the mark can recrystallize (i.e., anneal) the surface back to the original state.[48]

Writing onto a magneto-optic disk is accomplished by first heating the medium above the Curie temperature T_c (typically 200 °C) at which the medium becomes disordered and loses its magnetization. The medium is then cooled and takes on whatever residual magnetization is present, usually imposed by a "bias" magnetic field. At low read powers, the marks are stable because below T_c the medium strongly resists any change in its magnetization from applied magnetic fields (high coercivity).

9.3.3. Requirements for Semiconductor Lasers in Optical Recording

Many of the advantages of SLs listed above in Sec. 9.1 are particularly important for optical-recording applications. The size of the laser is a minuscule part of the standard hermetically sealed can into which the laser is mounted. Even the can accounts for only a fraction of the relatively massive optical head, as seen in Fig. 9.9. The cost of a CD readout laser is below $10 in high volume quantities, primarily because very large quantities are manufactured for recording applications. The price of higher power SLs used for writing is dropping for the same reason. The efficiency and low power requirements generally keep the laser's power consumption at only a small fraction of the total optical device power. The predominantly linear polarization of the laser relaxes the need for a polarizer in some optical heads. Finally, as is the case in optical communications, the ability of the laser to be directly modulated allows simple transfer of information, in this case to the optical disk rather than through an optical fiber. What requirements the laser must satisfy in typical optical-recording applications is the subject of this section.

Many of the requirements that an optical-recording laser must satisfy are related either to spatial beam quality and coherence or to laser intensity and wavelength fluctuations. First we consider those requirements related to the spatial quality of the beam.

9.3.3.1. Spatial Coherence

The reason a laser can be used so effectively for optical recording is that its supreme directionality allows the light to be focused to a very small spot. The details of focusing can be very complicated, requiring scalar and even vector diffraction theory for an accurate analysis. From a simpler analysis based on Gaussian-beam propagation,[49] one can show the focused spot diameter to be approximately given by $0.6\lambda/NA$, where NA is the numerical aperture of the focusing lens. For $\lambda \sim 800$ nm, and NA = 0.4, the focused diameter is 1.2 μm. This extremely small diameter leads to a high potential data density and one of the advantages of optical recording. However, the SL beam differs from a more typical laser beam in two important respects: The beam profile is quite elliptical, and there can be astigmatism between the two orthogonal transverse-beam directions. Both of these features are related to the asymmetric emitting area of typical SLs. The width of the emitting region is generally at least an order of magnitude larger than the active-region thickness. The smaller thickness causes greater diffraction so that after collimation the beam size in the direction perpendicular to the junction will be larger than the size in the parallel direction by a ratio of 3:1 or 4:1. The focused spot size will also be elliptical, although it can been shown that the ellipticity is reduced by focusing.[50] For example, an ellipticity of 3:1 might be reduced to 1.6:1 assuming the beam fills the focusing lens properly. A more detrimental effect of the elliptical beam is that the effective focused spot size is larger, so that the irradiance (power per area) is only about 70% of the irradiance obtained from an equal power circular beam. The laser power must be increased above that required for a circular-beam laser to achieve a given irradiance.

Because of the above power penalty, it is often desirable to circularize the beam, particularly in write-once heads where power is at a premium. Circularization can be accomplished in three ways: By apodization of the beam, by use of a cylindrical telescope, or by use of an anamorphic prism pair. Apodization simply implies placing an aperture in the beam so that only the central circular part is passed. Obviously this procedure is somewhat wasteful of power. A cylindrical telescope can be used to stretch the short axis of the ellipse or to compress the long axis so that the beam is circular. Usually for efficient focusing of the beam onto the disk, the short axis part of the beam would be stretched so that the resultant beam diameter before focusing is maximum. The disadvantage of this technique is that cylindrical lenses are not as common as their spherical counterparts; hence they tend to be more expensive. The anamorphic prism pair also stretches the short axis of the beam, and has the added benefit of being reasonably achromatic. The planar faces of the prisms, however, can lead to aberrations if the input beam is not well collimated.

A reasonable compromise among these approaches is to use a single prism instead of two. Since the beam exits the prism in a different direction from that of the input beam, the prism is also used to turn the beam, thereby making the head assembly more compact (see Fig. 9.9). Although true circularization cannot be accomplished by a single prism, it is possible to reduce the ellipticity from 3:1 to as low as 1.5:1. Then after focusing, the spot ellipticity will be only about 1.3:1 and the irradiance nearly 90% of the maximum value. Such an arrangement is often used in write-once heads and represents a suitable compromise between perfect circularization and the mass and expense of one or two additional optic elements.

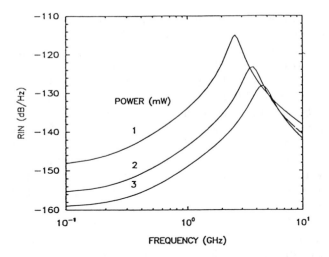

FIGURE 9.10. Typical RIN spectra for three different power levels (after Ref. 4).

Aside from emitting elliptical beams, SLs sometimes emit beams that are slightly astigmatic, such that the two transverse directions appear to diverge from different points inside the laser. This implies that, when focused, the two directions will have two different minimum spot sizes, the separation of which is often used as a measure of the astigmatism. Since the focused spot size as well as the typical mark length are about 1 μm, clearly an astigmatism of more than 1 μm is unacceptable. Fortunately, most index-guided lasers (at least in the near-infrared) are designed to emit a beam with a fundamental transverse mode and negligible astigmatism. This may not automatically be the case for shorter-wavelength lasers.

9.3.3.2. Intensity and Wavelength Fluctuations

Additional requirements that the optical-recording laser must satisfy relate to intensity and wavelength fluctuations. Compared to some lasers, semiconductor lasers are relatively noisy, in part because of their high rate of spontaneous emission ($\sim 10^{12}$ photons per second). Although, the intensity-noise level is generally low enough for many applications, operation in the presence of random longitudinal mode hopping and/or optical feedback (OFB) can raise the intensity-noise level by 20 dB or more. In the absence of mode hopping and OFB, the laser relative intensity noise (RIN) behaves schematically as shown in Fig. 9.10. The noise is typically below -120 dB/Hz in the low-frequency regime, peaks at the relaxation-oscillation frequency, and decreases rapidly for higher frequencies. In optical-recording applications, only the intensity noise well below 1 GHz is of interest. Laser noise should be low near the data frequency (~ 5 MHz) and low at frequencies where various servo controls operate (~ 20 kHz).

The spectral behavior of a typical "Fabry–Perot" laser diode depends sensitively

Regions of Single Mode Operation

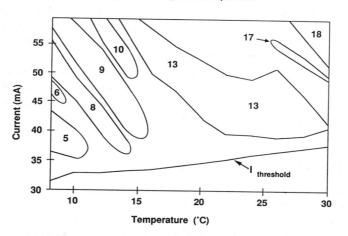

FIGURE 9.11. Regions of single-mode operation (after Ref. 51).

on laser temperature and operating current.[51,52] A representative example is shown in Fig. 9.11, for a 3 mW TJS (transverse junction stripe) laser operating at 780 nm. The "islands" seen in the figure represent operating points for which the laser is operating in a predominantly single-longitudinal mode; i.e., all side modes remain at least 10 dB below the main mode. The mode numbers are arbitrary, but higher numbers imply longer wavelength modes. When the laser temperature and current are tuned outside an island, the laser emits in multiple longitudinal modes. Two distinct types of multimode behavior can be distinguished: Simultaneous operation of several modes (mode partition), and unstable competition between two or more modes (mode hopping). The noise level exhibited by the total intensity is greatly dependent on the character of the multimode behavior. In the case of mode partition, the modes fluctuate in an anticorrelated fashion so that the total power remains relatively constant and the RIN remains low. The mode hopping regime, in contrast, is marked by a violent exchange of power between modes.[51] In this regime, usually only two modes contain most of the laser power, but the power is shared between them in a random, bistable manner: Only one mode may be "on" at any given time. The low-frequency RIN is greatly increased during mode-hopping behavior. Mode hopping can generally be avoided either by temperature stabilizing the laser diode or by using a single-mode laser such as a DFB laser. Both approaches are rather expensive. An alternative approach utilizes self-pulsing SLs. such lasers emit pulses with a repetition rate near the laser relaxation-oscillation frequency (~ 1 GHz). Self- pulsing lasers operate with many longitudinal modes, reducing the problem of mode hopping. In addition, the pulsing behavior reduces the coherence length, making optical feedback (Sec. 9.3.4) less of a problem. Such lasers have found use in certain read-only heads.

Even when mode hopping can be avoided, there are other detrimental effects caused by temperature variations. One is that the laser efficiency worsens as the temperature increases, leading to a power penalty associated with temperature rise. In general, the laser's power supply must be designed to supply additional power to account for both temperature-rise effects and lifetime effects. Another effect, which is evident in Fig. 9.11, is that the laser wavelength shifts to higher values as either the temperature or the current rises. This shift means that the laser wavelength at low read power will differ from that at high write power. It has been shown that such wavelength shifts can lead to either a compression or an elongation of the mark length depending on the direction of disk rotation.[53]

9.3.4. Effects of Optical Feedback

Optical feedback (OFB) can be a severe problem in any type of optical-storage head, because the technique used to focus the laser light onto the disk simultaneously images the reflected light back to the laser. Semiconductor lasers are known to be extremely sensitive to even small amounts of OFB.[43] Although OFB can be beneficial to the laser, precise control of the external-cavity length is required. In an optical-recording system, where the disk serves as the external reflector, both the reflectivity and length of the external cavity fluctuate, and the OFB can be detrimental to laser performance. From a practical standpoint, OFB leads to a tremendous increase in the low-frequency RIN. Increased noise below 100 MHz can affect mark-length stability, while noise below 100 kHz causes problems with the servos, which operate below 20 kHz. In this section we consider some of the consequences of OFB in optical recording and examine ways to reduce the deleterious effects.

Although most of the research into OFB effects on SLs was performed in connection with optical communication, OFB is probably a greater problem for optical recording. The reason is that, as explained in Sec. 9.2.5, Faraday optical isolators can be used in optical communication transmitters to keep the OFB below -30 dB. The use of such an isolator in optical heads is typically prohibitively expensive. Actually, many write-once heads and some read-only heads do employ an optical isolator based on a quarter-wave plate (QWP), as shown in Fig. 9.9. In such an isolator, linearly polarized light incident at 45° with respect to the fast axis of the QWP is transformed into circularly polarized light. Upon reflection at the disk, the "handedness" of the circular polarization is reversed, so that the polarizer transmits none of the reflected light, effectively isolating the laser. Such an isolator works far less than ideally for a number of reasons: The polarizer is never perfect, the QWP is only a true QWP for a single wavelength (the light from the laser is often comprised of several wavelengths), and perhaps most importantly, the protective cover sheet of the optical disk is often birefringent. Such birefringence severely affects the performance of the isolator, since the reflected light is no longer perfectly circularly polarized. The result is that OFB in the range of 0.1 to a few % OFB (ratio of reflected power to emitted power) will affect the laser dynamics and noise.

Depending upon the exact strength of OFB, the precise distance of the laser to the disk, the laser power, and many other factors, OFB affects the laser in a variety of

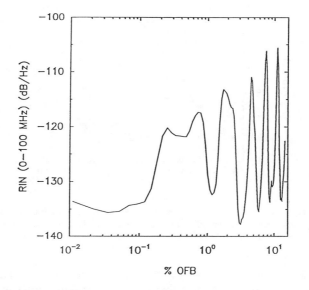

FIGURE 9.12. Calculated low-frequency RIN vs %OFB for a single-mode laser.

ways. As in IM/DD optical communication, the effect that is most degrading in optical-recording systems is the large increase in RIN that sometimes accompanies OFB. An example of this RIN-increase problem is shown in Fig. 9.12, the result of computer simulation of single-mode SL rate equations with OFB. Low-frequency RIN averaged over the range 0–100 MHz is plotted as a function of % OFB. When the feedback level is below 0.1%, the RIN is essentially unaffected by feedback. But for feedback levels greater than 0.1%, the RIN is increased by more than 20 dB. Numerical simulations of the rate equations with the spontaneous emission noise sources *neglected* indicate that the RIN increase is actually a manifestation of optical chaos.[54] Note that the regions of high RIN are interrupted many times by stable regions with low RIN. Such regions correspond to a locking of the laser to one of the modes of the external cavity. The narrow frequency-locked states are not of any practical value in typical optical heads because the external-cavity length as well as the feedback strength are not constant but vary slightly due to unavoidable system variations.

Given that a certain amount of OFB is inevitable in optical-recording heads, what can be done to keep the RIN at a low value? One technique, known as high-frequency injection (HFI) has been used experimentally[55] and studied theoretically[54,56] for its ability to suppress the OFB-induced high RIN values. HFI consists of sinusoidal (or square-wave) modulation of the laser current at a frequency f_m. HFI can be used to delay the onset of chaos, and associated high RIN values, to feedback levels not accounted in practice. For example, Fig. 9.13 shows the variation of low-frequency RIN with the OFB level for a five-mode laser for three different modulation frequencies: 0 (no modulation), 275, and 500 MHz.[57]

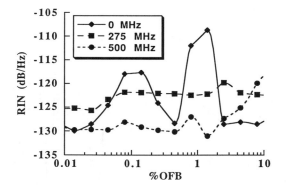

FIGURE 9.13. Calculated low-frequency RIN vs %OFB for a multimode laser for three different HFI modulation frequencies: 0 (no modulation), 275, and 500 MHz. Modulation current is 20 mA, peak to peak (after Ref. 57 © 1994 IEEE).

The unmodulated case is very similar to the single-mode case; namely, the intensity noise increases rapidly as the OFB increases. The remaining curves in Fig. 9.13 are for 20 mA peak-to-peak modulation. Although RIN is reduced for $f_m = 275$ MHz, the choice of 500 MHz is much better since the RIN is close to the solitary-laser value over a large range of OFB. As observed also experimentally, the ability to suppress the RIN enhancement is sensitive to the choice of the modulation frequency.

9.3.5. Summary of Optical-Recording Requirements

It is difficult to identify a single requirement that was mentioned in the preceding sections as being "most important" to optical reading and writing. Rather, the success of the optical storage technology is due to the combination of many SL features. The ability of the laser light because of its spatial coherence to be focused to a tight spot, is the heart of optical data storage. However, focusing already produces spots close to the diffraction limit, and smaller spots and associated higher recording densities will depend on the availability of shorter wavelength sources. Short-wavelength lasers is an area of vigorous research effort, since halving of the wavelength should result in a quadrupling of the data density. Even visible red laser diodes would mean increased data density. Such laser diodes have been commercially available in low power versions and can be found in laser pointers, scanners, and other traditional helium–neon laser applications. High power SLs emitting near and below 680 nm have become available recently and are being planned for use in write-once heads. Certainly methods for producing shorter wavelength laser diodes will continue to be a vigorous area of research.

Another important requirement concerns the laser's intensity-noise level, which must be kept sufficiently low during reading or writing so that a low error rate can be maintained. This requirement usually implies that the mode-hopping regime, which can exist under certain operating conditions or in the presence of optical feedback, must be avoided. Optical feedback often leads to increased intensity noise even when as in a single longitudinal mode laser, it does not initiate longitudinal mode hopping. In such systems, some type of optical isolation becomes a necessity, unless HFI can alleviate the problems.

A highly polarized output is particularly useful in magneto-optic heads, since linear polarization is required at the disk surface. It can also obviate the need for a bulk polarizer in the optical isolator used in some heads. Finally, the ellipticity of the beam is generally more of a problem for a laser used for writing, since beam ellipticity leads to a lower power density at the focus. For this reason, certain vertical cavity, surface-emitting lasers are attracting attention for use in optical recording, because they emit a circular beam. Power levels currently available from such devices are too low for use in optical recording, however. Perhaps the future optical recording head will consist of an entirely integrated structure, with laser, detectors, beamsplitter, and focusing elements all contained on the same chip. Such a design would be extremely lightweight, and it would allow optical recording to overcome its supreme disadvantage relative to magnetic recording: The massive optical head. Nevertheless, optical recording represents a vital application of laser diodes, an optoelectronic application that is here to stay.

9.4. DIODE LASERS IN MEDICAL APPLICATIONS

The use of lasers for medical applications has excited researchers since the invention of the first laser in the early 1960's. The high power and supreme directionality of the laser gave promise for a much "cleaner" tool for surgery than traditional surgical cutting tools. The narrow spectral output would allow absorption only by specific tissues, allowing new noninvasive procedures based on light. Today, although lasers are used in a variety of medical settings, few would argue that the presence of lasers has revolutionized the medical profession to the degree it has revolutionized the field of communications. One reason is certainly that the majority of lasers used in medical applications today are large and bulky, not very user friendly or reliable, and very expensive. The attributes of diode lasers stand in stark contrast to those of traditional lasers, and by 1984 the use of semiconductor lasers in a variety of medical applications was beginning to be explored.[58] In this section we describe the use of SLs in various medical applications, including uses in ophthalmology and the optical imaging of tissues.

9.4.1. Ophthalmology

It is fitting that ophthalmic applications be discussed first, since ophthalmology was one of the first medical fields to embrace the new technology of lasers in the 1960's. As research continued into the 1970's, the argon-ion laser became the standard for most ophthalmic procedures. Although the argon laser can be used with considerable success to treat a variety of retinal problems, the laser suffers from several

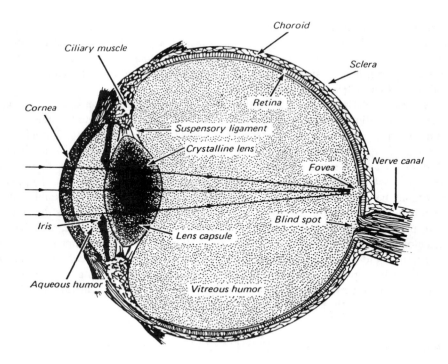

FIGURE 9.14. Schematic of a human eye (after Ref. 61).

disadvantages that make it far from ideal. Most high-power ion lasers require special electrical hookups that are not commonly available. They also require vigorous water or air cooling to maintain tube life. Even with good cooling they typically last only for a few years before tube replacement, at up to 20% the original laser cost, becomes necessary. Because of these problems, combined with the original expense of the laser, only specialized facilities are able to maintain such medical lasers. The portability, low cost, and high reliability of SLs make them of high interest in ophthalmic and other medical laser applications. But the research that resulted in the success of the communications and recording fields discussed above concentrated mostly on devices with powers that were generally too low to be used in medicine. This situation changed, however, as research into high-power SLs and SL arrays resulted by the mid-eighties in output powers exceeding 500 mW. SL output exceeding many tens of watts is commercially available today.[59]

By 1987 SL output powers were large enough that SLs began to be explored for use in photocoagulation of the retina in animals.[60] This procedure is a mainstay of clinical argon laser treatments and consists of making several lesions (i.e., burns) on or near the retina of the eye in order to inhibit the creation of new blood vessels that could cause retinal bleeding. The burns are produced because light gets absorbed by the retinal pigment ephythelium and by the choroid. See Fig. 9.14 for a schematic of a human eye.[61] This first animal study utilized laser powers (emitted from an optical fiber) of 50–170 mW and exposure times of 0.2–2 s. For an optical power of 130 mW, a power density at the surface of 34 W/cm^2 resulted and, for an exposure of 0.2 s, a dose of 10 J/cm^2. The lesions that were produced were as effective as those produced by other lasers, notably argon and krypton lasers, and the power levels required were similar. The optical fiber was inserted into a needle, which was subsequently inserted into the sclera of the eye, rather than the pupil. The small size and portability of the SL allowed the device to be mounted directly on a standard ophthalmic slit lamp microscope, so that later clinical studies made use of transpupillary (through the pupil of the eye) delivery of the SL light.[62]

In addition to the aforementioned SL advantages over argon lasers in terms of size and electrical power service requirements, there are wavelength advantages. The near infrared (NIR) wavelength (~ 810 nm) allows greater penetration (less scattering and absorption) through the cornea and lens of the eye. This greater penetration is particularly important for older patients, whose eyes sometimes develop a yellowish tint causing transmission of blue-green argon light to be reduced by 20% compared to only 7% for transmission of SL light.[62] The transmission of NIR light is therefore less age dependent. Even through ordinary, nondiscolored eyes, the scattering and reflection of laser light by the eye is greatly reduced by using light in the NIR rather than at argon laser wavelengths. The scattering of argon light requires a shutter system to be used to protect the ophthalmologist. No shuttering system is needed for SL devices, and the retinal lesions can thus be viewed continuously during the exposure for greater accuracy.[62] One downside of the greater transmission of NIR light is that so much light gets concentrated at the back of the eye on the retina and choroid, that local temperature increases can be quite high, with deeper lesions and some patient discomfort resulting.[64]

Glaucoma and related diseases also benefit from laser technology. The high intraocular pressure characteristic of glaucoma can be reduced by a laser treatment known as trabeculoplasty.[63] As in photocoagulation, the argon laser is the standard by which new lasers are judged. Recently diode lasers have been shown to be essentially as effective as argon in treating glaucoma.[64] The treatment consists of focusing the laser light to a small spot at the junction between the pigmented and nonpigmented trabecular meshwork of the eye. The area is blanched by the laser, and the process is repeated for the complete 360° circumference of the eye. About 20–25 applications are made per quadrant. The performance of the SL in glaucoma treatment is comparable to that of the argon laser with minor differences. First, in tests, the power level required for similar effectiveness was 675 mW for argon but 1135 mW for the diode laser. The different requirements are probably due to the differences in absorption at the two wavelengths; also, since the SL lesions tend to be deeper, more power was necessary to create the same visual blanch at the surface. The second difference between the two lasers in this comparison study was the spot size used: 50 μm for the argon, whereas the minimum spot size available for the SL was 100 μm. Although this spot size difference leads to an irradiance

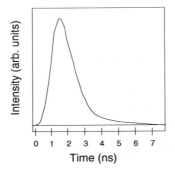

FIGURE 9.15. Typical detection intensity vs time for light scattered by tissue.

(power per area) difference of a factor of 4, no treatment dissimilarities were noted due to the unequal spot sizes.

9.4.2. Transillumination Imaging

In the past few years diode lasers have been explored for use in imaging techniques for the detection of tumors and cancer diagnosis.[65-69] Although such techniques as the use of x rays have traditionally been utilized for tumor detection, concern over the hypothetical risk of cancer induction by the ionizing x rays has prompted interest in nonionizing, optical imaging techniques. Red and near-infrared light show great promise in this regard, as it travels with minimal absorption through most tissues and the brain. However, unlike x rays, which travel in linear paths, light is multiply scattered as it travels through tissue. Since the mean-free path of a photon due to scattering is merely 100 μm, essentially no photons remain unscattered when traveling through tissues of centimeter thickness. The scattering significantly reduces the contrast between healthy and malignant tissue. Two techniques that show particular promise in overcoming the scattering problem are time-domain transillumination (TDT)[66] and coherence-domain transillumination (CDT).[67,68] Both techniques can be performed with semiconductor-laser technology, which will be required in any event for a cost-effective clinical instrument. A hybrid system has also been developed that combines TDT and CDT by using an interferometer and a short-pulsed source.[69]

When a pulse of light traverses a tissue or other scattering material of thickness on the order of a few centimeters, the light detected along the optic axis appears schematically as shown in Fig. 9.15. The rising part of the pulse represents the earliest arriving photons, ones that have been scattered the least. Note that photons arriving after 1 ns have experienced a travel distance of nearly 30 cm, which is much wider than the material, indicating that even the early arriving photons have been scattered a multitude of times. The long-time exponential decay of the pulse is representative of the absorption of the material. Although spatial information is lost

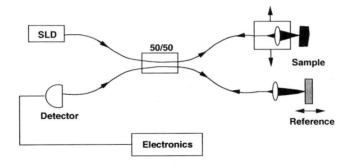

FIGURE 9.16. Schematic of fiber-optic Michelson interferometer used for low-coherence imaging.

due to the multiple scattering, the absorption tail can be used to determine the composition of the material, for example, the amount of oxygen present in hemoglobin.[66] To obtain spatial (imaging) information, the TDT approach makes use of only the early arriving photons, the ballistic component, since they have traveled most closely to the unscattered optical path. Since the scattering of photons is less in a tumor,[70] more photons arrive in the detection window allowing the tumor to be detected. One system applying this technique used a pulsed diode laser operating at 815 nm with a 30 ps pulse length. The sample was moved with computer-controlled x-y translation stages allowing one- or two-dimensional imaging of the sample.[70] The technique was also used *in vivo* for imaging of a human hand, which always has greater scattering due to blood flow, for example.[71]

Whereas the above technique utilizes a fixed time-gated detection window, a slightly different approach allows a variable window but detects a fixed percentage of total number of photons. The method makes use of three pulsed SLs at different wavelengths, 780, 850, and 904 nm.[72] The pulse widths are 100 ps, but because only the early arriving photons are used to construct the image, the effective time resolution is significantly better than this. The technique has been used to successfully image all of the major organs of a rat.

Transillumination imaging has also been performed using interferometric means. In the coherence-domain approach, an interferometer (e.g., Mach–Zhender or Michelson) is constructed with the sample inserted into one arm. The basic principle is straightforward: an appreciable output (i.e., constructive interference) is obtained only when the two path lengths are equal, to within the coherence length of the source. Therefore, in this application a low-coherence length source is desirable, so that the best depth resolution may be obtained. In one realization of this technique,[67] a super-luminescent diode with 20 μm coherence length was used in a fiber optic Michelson interferometer, as shown in Fig. 9.16. The two scanning directions allowed two-dimensional imaging to be performed. For instance, for a given reference mirror location, the sample beam can be scanned laterally across the sample. The interferometric signal is detected only when the two path delays are equal to

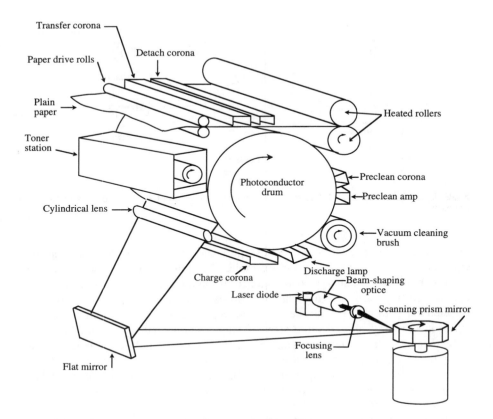

FIGURE 9.17. Schematic of a laser printer.

within 20 μm. Since the source coherence is low, the detector output falls off rapidly when the relative delay exceeds the coherence length, allowing for sharp contrast. Subsequent movement of the reference mirror and repeated sample beam scanning allows the depth of the sample to imaged. Heterodyne detection allows a sensitivity of 10 fW for an input power of only 20 μW.

Using heterodyne detection implies the need for both high- and low-coherence laser diodes. The low-coherence requirement of the light source in a CDT system is relatively rare. In communication applications, in particular for coherent communications, and in the atomic-physics applications discussed below in Sec. 9.5.2, extremely narrow linewidth lasers are required.

9.5. ADDITIONAL APPLICATIONS

As mentioned in the Introduction, diode lasers are used in such a large number of different applications that entire books could be filled in discussing them. Therefore, the previous sections have discussed in detail only three important applications:

Dot Row	Bit Map
01	00000000 11111100 00001111 1100000
02	00000111 11111111 00011111 1110000
03	00001111 11111111 10011111 1100000
04	00111111 11010111 11011110 0000000
05	00111110 00000001 11111110 0000000
06	01111100 00000000 01111110 0000000
07	01111000 00000000 01111110 0000000
08	11110000 00000000 00111110 0000000
09	11110000 00000000 00111110 0000000
10	11110000 00000000 00011110 0000000
11	11110000 00000000 00011110 0000000
12	11110000 00000000 00011110 0000000
13	11110000 00000000 00011110 0000000
14	11110000 00000000 00111110 0000000
15	01111000 00000000 00111110 0000000
16	01111000 00000000 01111110 0000000
17	01111100 00000000 11111110 0000000
18	00111110 00000001 11111110 0000000
19	00011111 11010111 11011110 0000000
20	00001111 11111111 10011110 0000000
21	00000111 11111111 00011110 0000000
22	00000000 10101000 00011110 0000000
23	00000000 00000000 00011110 0000000
24	00000000 00000000 00011110 0000000
25	00000000 00000000 00011110 0000000
26	00000000 00000000 00111110 0000000
27	00000000 00000000 00111100 0000000
28	00000000 00000000 01111100 0000000
29	00000000 00000001 11111000 0000000
30	00000001 11111111 11110000 0000000
31	00000011 11111111 11100000 0000000
32	00000001 11111111 10000000 0000000

FIGURE 9.18. Low-resolution bit-map image of lower-case letter "*g*" (after Ref. 74).

optical communication, optical recording, and medical applications. In this section, two additional applications will be briefly described. One of the applications is so mundane (the laser printer) that it appears in virtually every business office. The other application is rather esoteric and is taken from the realm of fundamental atomic physics.

9.5.1. Laser Printing

Like optical recording, laser printing represents a relatively mature application of laser diodes emitting near 800 nm. A schematic of a typical laser printer is shown in Fig. 9.17.[73] The laser printer operates as follows.[74] The photoconductor drum gets positively charged by the charge corona. Laser pulses, representing the information to be printed, are focused onto the drum by the scanning mirror and cylindrical lens assemblies. Because of the material properties of the photoconductor drum, when light strikes the drum, the charge is stripped away. The rotating drum next contacts the toner station, where the negatively charged toner particles are attracted to the residual positively charged drum material, the part that was not exposed to laser light.[75] Further rotation of the drum brings it into contact with the paper, which has been positively charged by the transfer and detach coronas. The toner particles jump to the paper, which is more positively charged than the drum. The heated rollers fuse the toner particles permanently to the paper. As the drum continues to rotate, the preclean corona and lamp loosen any remaining toner particles, which are cleared away by the vacuum brush, and the discharge lamp neutralizes the drum before the charge corona initiates the process all over again.

The key to the information transfer process achieved by the laser diode is the representation of text and graphics by bit-mapped images. Figure 9.18 shows a rather low-resolution bit map of a lower-case letter "g," where the dark and light areas are represented by 1's and 0's, respectively. The printer has access to all of these bit-mapped images, either in its own memory or stored on the host computer. Further printing details depend somewhat on the particular printing language (e.g., postscript) that is being used. The resolution with which the printer operates determines how tightly the laser must be focused onto the drum surface. For a 300 dots per inch (DPI) resolution, which is considered good letter quality resolution, the laser light must be focused to about 80 μm spots. This is certainly a far less stringent requirement than the 1 μm spots needed in optical recording.

9.5.2. Atomic Physics

Another area that has found use for laser diodes is the study of atomic physics. Some of the atomic-physics community have embraced SLs for the same reasons as the medical community: Laser diodes are low-cost, compact and reliable, high efficiency light sources. Several excellent reviews have been written that describe some of the problems of atomic physics that can be studied using diode lasers.[76] These reviews also discuss the various requirements that a SL must satisfy to be an effective research tool. Some of the areas of research include optical pumping,[77] in which large numbers of atoms can be prepared in a given state by the absorption of laser light; laser cooling and trapping,[78] in which atoms again absorb laser photons causing the atoms to slow and nearly stop; and spectroscopy, for which various atomic transitions are probed and studied. For many of the atomic-physics applications, the laser must be tunable, have a narrow linewidth, and possess good frequency and amplitude stability. Before SLs were available, dye lasers provided all of these attributes. However dye lasers are expensive because they are usually pumped by another laser, typically argon ion, and the liquid dyes themselves are frequently hazardous materials. Fortunately, by placing a SL into an external cavity where a grating takes the place of the external mirror, the laser can be made tunable (~ 30 nm tuning range) and its linewidth reduced to < 1 MHz (~ 100 kHz).[79] The amplitude and frequency of the SL can be stabilized in a variety of additional ways, including coupling to a high-Q resonator,[80] employing electronic feedback,[81] and making use of optical phase-locked loops.[82]

The cooling and trapping of atoms is a particularly interesting application of laser diodes, since use is made of one of the peculiarities of SLs, namely, the dependence of the laser frequency on injection current (i.e., chirp; see Sec. 9.2.1). For example, in order to slow an atomic beam of Cesium atoms, laser diodes emitting at 852 nm are required, since this frequency corresponds to one of the common transition frequencies of the Cs atom. When collimated laser light is directed at oncoming Cs atoms, each time an atom absorbs a photon, it consequently receives a momentum "kick" in the direction opposite to the atomic velocity and with magnitude $h\nu/c$, where h is Plank's constant, ν is the light frequency, and c is the speed of light. When the atom relaxes by spontaneous emission, it will undergo another momentum kick, but since spontaneous emission is equally likely in all directions, these kicks average to zero. For a typical thermal beam of Cs atoms with velocity of 2.7×10^4 cm/s, each absorption reduces the velocity by 0.35 cm/s, so that about 76 000 absorptions must occur for an atom to be brought to rest. The deceleration is governed by the spontaneous lifetime, and for a value of 31 ns, the atoms can be stopped in about 5 ms time and 62 cm distance.[78] The situation is somewhat more complicated, however, because after an atom has absorbed about 100 photons it becomes Doppler-shifted out of resonance with the laser. However, by applying a suitable ramp to the driving current of the laser, the frequency of the laser can be made to change with time (and therefore distance) so that the Doppler shift can be exactly compensated. The ramp that is required in this case provides 310 MHz frequency change in 5 ms, or 67 MHz/ms, an amount easily attainable with a SL. The resulting effective temperature of the stopped atoms is only 1 K.[78] Three-dimensional trapping and cooling of Cs atoms has also been accomplished using diode lasers.[83]

9.5.3. Conclusions and Future Directions

Semiconductor lasers are already used in a large number and wide variety of applications. This situation will only improve in the future as laser prices continue to drop, new wavelengths and structures (e.g., vertical-cavity lasers) continue to be developed, and higher powers become more common. This chapter has attempted to provide the reader with a flavor for the role the SL plays in a few of these applications. In particular, the use of the laser in optical communications and optical recording was described in some detail. The more recent interest in SLs for medical applications was also explored, and devices using SLs for ophthalmic treatments are now commercially available. Laser printing, a more common application, was seen to be an interesting combination of electrostatic effects and the information-carrying ability of laser diodes. SLs are also used in basic and fundamental research, and specific examples were given from the realm of atomic physics. It was seen that the coupled frequency and intensity modulation of SLs, which hinders their use in optical-fiber communication due to fiber dispersion, can be exploited in order to compensate for Doppler shifts in the laser cooling of atomic beams.

What future developments are likely to play the greatest role in the realization of new applications? There are several likely candidates. Certainly the commercialization of reliable and efficient blue laser diode sources will be a major development. The optical-recording industry would experience a minor revolution in that case, as data densities should increase fourfold. Another development that could revolutionize optical recording would be the monolithically integrated optical head, which would significantly reduce the optical head mass, the one feature making optical recording inferior to magnetic recording. The vertical-cavity laser, which emits in a circular beam, could also play an important role in optical recording and other applications where the usual elliptical profile of SLs yields complications. Perhaps the continued progress in high power devices has the greatest possibility of revolutionizing the laser industry, as high power laser arrays continue to substitute for more traditional lasers, for example, as a convenient clinical light source for medical applications or as optical pumps for compact solid-state lasers.

REFERENCES

1. Laser Focus, Jan. 1993, p. 72.
2. G. P. Agrawal, *Fiber-Optic Communication Systems* (Wiley, New York, 1992), p. 9.
3. H. Ishikawa, H. Soda, K. Wakao, K. Kihara, K. Kamite, Y. Kotaki, M. Matsuda, H. Sudo, S. Yamakoshi, S. Ioszumi, and H. Imai, J. Lightwave Technol. **LT-5**, 848 (1987).
4. G. P. Agrawal and N. K. Dutta, *Semiconductor Lasers*, 2nd ed. (Van Nostrand Reinhold, New York, 1993), Chap. 6.
5. G. P. Agrawal, *Fiber-Optic Communication Systems* (Wiley, New York, 1992), p. 117.
6. P. S. Henry, R. A. Linke, and A. H. Gnauck, in *Optical Fiber Telecommunications II*, S. E. Miller and I. P. Kaminow, Eds. (Academic, San Diego, CA, 1988), Chap. 21.
7. D. Marcuse, Appl. Opt. **19**, 1653 (1980); **20**, 3573 (1981).
8. G. P. Agrawal, *Fiber-Optic Communication Systems* (Wiley, New York, 1992), p. 52.
9. G. P. Agrawal and M. J. Potasek, Opt. Lett. **11**, 318 (1986).
10. G. P. Agrawal, *Fiber-Optic Communication Systems* (Wiley, New York, 1992), Chap. 9.
11. A. H. Gnauck et al., J. Lightwave Technol. **LT-3**, 1032 (1985).
12. P. J. Corvini and T. L. Koch, J. Lightwave Technol. **LT-5**, 1591 (1987).
13. R. A. Linke, B. L. Kasper, C. A. Burrus, I. P. Kaminow, J. S. Ko, and T. P. Lee, J. Lightwave Technol. **LT-3**, 706 (1985).
14. G. P. Agrawal, *Fiber-Optic Communication Systems* (Wiley, New York, 1992), pp. 61–65, 297–301; *Nonlinear Fiber Optics* (Academic, San Diego, 1989).
15. Y. Aoki, K. Tajima, and I. Mito, J. Lightwave Technol. **6**, 710 (1988).
16. L. G. Kazovsky, O. K. Tonguz, and D. G. Daut, Electron. Lett. **25**, 908 (1989).
17. L. G. Kazovsky, J. Lightwave Technol. **LT-3**, 1238 (1985); J. Opt. Comm. **7**, 66 (1986); J. Lightwave Tech. **LT-4**, 415 (1986).
18. G. J. Foschini, L. J. Greenstein, and G. Vannuchi, IEEE Trans. Commun. **36**, 306 (1988); L. J. Greenstein, G. Vannuchi, and G. J. Foschini, IEEE Trans. Commun. **37**, 405 (1989).
19. L. G. Kazovsky and O. K. Tonguz, J. Lightwave Technol. **8**, 338 (1990); I. Garret, D. J. Bond, J. B. Waite, D. S. L. Lettis, and G. Jacobsen, J. Lightwave Technol. **8**, 329 (1990).
20. Y. Kotaki, M. Matsuda, M. Yano, H. Ishikawa, and H. Imai, Electron. Lett. **23**, 325 (1987).
21. S. Murata, I. Mito, and K. Kobayashi, IEEE J. Quantum Electron. **QE-23**, 835 (1987); Electron. Lett. **23**, 403 (1987); K. Kobayashi and I. Mito, J. Lightwave Technol. **6**, 1623 (1988).
22. T. L. Koch, J. Lightwave Technol. **8**, 275 (1990).
23. M. Aoki, K. Uomi, T. Tsuchiya, S. Sasaki, M. Okai, and N. Chinone, IEEE J. Quantum Electron. **27**, 1782 (1991).
24. N. A. Olsson, J. Hegarty, R. A. Logan, L. F. Johnson, K. L. Walker, L. G. Cohen, B. L. Kasper, and J. C. Campbell, Electron. Lett. **21**, 105 (1985).
25. R. Watanabe, K. Nosu, and Y. Fujii, Electron. Lett. **16**, 108 (1980); J. Lipson, W. J. Minford, E. J. Murphy, T. C. Rice, R. A. Linke, and G. T. Harvey, J. Lightwave Technol. **LT-3**, 1159 (1985).
26. B. H. Verbeek, C. H. Henry, N. A. Olsson, K. J. Orlowsky, R. F. Kazarinov, and B. H. Johnson, J. Lightwave Technol. **6**, 1011 (1988).
27. K. Fujii and J. Minowa, Appl. Opt. **22**, 974 (1983); C. H. Henry, R. F. Kazarinov, Y. Shani, R. C.

Kistler, V. Pol, and K. J. Orlowsky, J. Lightwave Technol. **8**, 748 (1990).

28. S. Valette, J. Mod. Opt. **35**, 993 (1988); C. Cremer, G. Ebbinghaus, G. Heise, R. Muller-Navrath, M. Schienle, and L. Stoll, Appl. Phys. Lett. **59**, 627 (1991).

29. D. R. Wisely, Electron. Lett. **27**, 520 (1991).

30. K. Oda, N. Tokato, T. Kominato, and H. Toba, IEEE Photon. Technol. Lett. **1**, 137 (1989); N. Takato, T. Kominato, A. Sugita, K. Jinguji, H. Toba, and M. Kawachi, IEEE J. Sel. Areas Commun. **8**, 1120 (1990).

31. T. L. Koch and U. Koren, IEEE J. Quantum Electron. **27**, 641 (1991).

32. U. Koren, T. L. Koch, B. I. Miller, G. Eisenstein, and R. H. Bosworth, Appl. Phys. Lett. **54**, 2056 (1989).

33. T. E. Darcie, R. S. Tucker, and G. J. Sullivan, Electron. Lett. **21**, 665 (1985); **22**, 619 (1986); P. P. Iannone and T. E. Darcie, Electron. Lett. **23**, 1361 (1987).

34. J. H. Angenent, Electron. Lett. **26**, 2049 (1990).

35. W. I. Way, J. Lightwave Technol. **7**, 1806 (1989); J. A. Chiddix, H. Laor, D. M. Pangrac, L. D. Williamson, and R. W. Wolfe, IEEE J. Sel. Areas Commun. **8**, 1229 (1990).

36. G. P. Agrawal, Fiber-Optic Communication Systems (Wiley, New York, 1992), Sec. 7.5.

37. P. M. Gabla, V. Lamaire, H. Krimmel, J. Otterbach, J. Augé, and Al Dursin, IEEE Photon. Technol. Lett. **3**, 56 (1991).

38. R. Olshansky, V. A. Lanzisera, and P. M. Hill, J. Lightwave Technol. **7**, 1329 (1989).

39. R. Olshansky, V. A. Lanzisera, and P. M. Hill, Electron. Lett. **24**, 1234 (1988).

40. P. G. Eliseev, *Reliability Problems of Semiconductor Lasers* (Nova Science, New York, 1991).

41. F. R. Nash, W. B. Joyce, R. L. Hartman, E. I. Gordon, and R. W. Dixon, AT&T Tech. J. **64**, 671 (1985).

42. D. S. Alles and K. J. Brady, AT&T Tech. J. **68**, 183 (1989).

43. R. W. Tkach and A. R. Chraplyvy, J. Lightwave Technol. **LT-4**, 1655 (1986).

44. G. P. Agrawal and N. K. Dutta, *Semiconductor Lasers*, 2nd Ed. (Van Nostrand Reinhold, New York, 1993); K. Petermann, *Laser Diode Modulation And Noise* (Kluwer Academic, Boston, 1991), Chap. 9.

45. N. A. Olsson, W. T. Tsang, H. Temkin, N. K. Dutta, and R. A. Logan, J. Lightwave Technol. **LT-3**, 215 (1985).

46. A. B. Marchant, *Optical Recording* (Addison-Wesley, Reading, MA, 1990), pp. 42–43.

47. *ibid.*, pp. 211–215.

48. *ibid.*, Sec. 4.3.

49. A. Siegman, *Lasers* (University Science Books, Mill Valley, CA, 1986), Sec. 17.2.

50. A. B. Marchant, *Optical Recording* (Addison-Wesley, Reading, MA, 1990), p. 142.

51. G. R. Gray and R. Roy, J. Opt. Soc. Am. B **8**, 632 (1991).

52. M. Ohtsu, Y. Teramachi, and T. Miyazaki, IEEE J. Quantum Electron. **24**, 716 (1988); M. Ohtsu, Y. Teramachi, Y. Otsuka, and A. Osaki, *ibid.* **QE-22**, 535 (1986); H. Ishikawa, H. Imai, T. Tanahashi, and M. Takusagawa, Appl. Phys. Lett. **38**, 962 (1981).

53. E. C. Gage and B. J. Bartholomeusz, J. Appl. Phys. **69**, 569 (1991).

54. G. R. Gray, A. T. Ryan, G. P. Agrawal, and E. C. Gage, Opt. Eng. **32**, 739 (1993).

55. K. Stubkjaer and M. B. Small, Electron. Lett. **19**, 388 (1983); A. Arimoto, M. Ojima, N. Chinone, A. Oishi, T. Gotoh, and N. Ohnuki, Appl. Opt. **25**, 1398 (1986); E. C. Gage and S. Beckens, SPIE Optical Data Storage **1316**, 199 (1990).

56. T. Kanada, Trans. IECE Jpn. **68**, 180 (1985); M. Yamada and T. Higashi, IEEE J. Quantum Electron. **27**, 380 (1991).

57. A. T. Ryan, G. P. Agrawal, E. C. Gage, and G. R. Gray, IEEE J. Quantum Electron. **30**, 668 (1994).

58. R. Pratesi, IEEE J. Quantum Electron. **QE-20**, 1433 (1984).

59. Spectra Diode Labs catalog.

60. C. A. Puliafito, T. F. Deutsch, J. Boll, and K. To, Arch. Ophthalmol. **105**, 424 (1987).

61. F. A. Jenkins and H. E. White, *Fundamentals of Optics*, 4th ed. (McGraw–Hill, New York, 1976).

62. J. D. A. McHugh, J. Marchall, T. J. Ffytche, A. M. Hamilton, A. Raven, and C. R. Keeler, Eye **3**, 516 (1989).

63. J. B. Wise and S. L. Witter, Arch. Ophthalmol. **97**, 319 (1979).

64. R. Brancato, R. Caressa, and G. Trabucchi, Am. J. Ophthal. **112**, 50 (1991).

65. Y. Cho, T. Shiota, K. Wada, T. Umeda, and M. Osawa, CLEO 1993, paper CTuK7, p. 128.

66. L. Wang, P. P. Ho, C. Liu, G. Zhang, R. R. Alfano, Science **253**, 769 (1991); J. C. Hebden, R. A. Kruger, and K. S. Wong, Appl. Opt. **30**, 788 (1991); D. A. Benaron, M. A. Lenox, and D. K. Stevenson, SPIE **1641**, 35 (1992).

67. D. Huang, E. A. Swanson, and C. P. Lin, Science **254**, 1178 (1991).
68. R. P. Salathe, CLEO Baltimore, MD, paper CTuK1, 122 (1993); J. M. Schmitt, A. Knuttel, A. Grandjbakhche, and R. F. Bonner, SPIE.
69. J. A. Izatt, M. R. Hee, E. A. Swanson, J. M. Jacobsen, and J. G. Fujimoto, CLEO 1993, paper CTuG1, p. 94.
70. S. Svanberg, Optics and Photonics News **3**, 31 (1992).
71. S. Andersson-Engels, R. Berg, S. Svanberg, and O. Jarlman, Opt. Lett. **15**, 1179 (1990).
72. D. A. Benaron and D. K. Stevenson, Science **259**, 1463 (1993).
73. B. Smith, Byte, p. 186 (October, 1991).
74. S. J. Bennet and P. G. Randal, *The Laser Jet Handbook* (Simon & Schuster, New York, 1990).
75. Some laser printers, for example, some made by Ricoh, are arranged so that the toner particles are attracted to the exposed areas of the drum.
76. C. E. Wieman and L. Hollberg, Rev. Sci. Instrum. **62**, 1 (1991); C. C. Bradley, J. Chen, and R. G. Hulet, Rev. Sci. Instrum. **61**, 2097 (1990).
77. J. C. Camparo, Contenp. Phys. **26**, 443 (1985); J. C. Comparo and R. P. Frueholz, J. Appl. Phys. **59**, 3313 (1986); A. D. Streater, J. Mooibroek, and J. P. Woerdman, Appl. Phys. Lett. **52**, 602 (1988).
78. R. N. Watts and C. E. Wieman, Opt. Lett. **11**, 291 (1986); W. D. Phillips, P. L. Gould, and P. D. Lett, Science **239**, 877 (1988).
79. K. B. MacAdam, A. Steinbach, and C. Wieman, Am. J. Phys. **60**, 1098 (1992).
80. B. Dahmani, L. Hollberg, and R. Drullinger, Opt. Lett. **12**, 876 (1987); L. Hollberg and M. Ohtsu, Appl. Phys. Lett. **53**, 944 (1988); H. Li and H. R. Telle, IEEE J. Quantum Electron. **25**, 257 (1989).
81. M. Ohtsu, *Highly Coherent Semiconductor Lasers* (Artech House, Boston 1992), Chap. 4.
82. *ibid.* Chap. 5.
83. K. Sesko, C. G. Fan, and C. E. Wieman, J. Opt. Soc. Am. B **5**, 1225 (1988).

CHAPTER 10

Semiconductor Lasers and Amplifiers for Optical Switching

Guang-Hua Duan

Department Communications, Ecole Nationale Superieure des Telecommunications, 75634 Paris Cedex 13

10.1. INTRODUCTION

Switching is one of the most important functions in telecommunications and computing. Today, nearly all operational network switching systems are realized with electronics. The bit rate in modern systems for video, image, and data transmission keeps increasing. It is difficult for electronics alone to satisfy the demand of increasingly higher bit rates. In the last decade, the use of optical fibers as a transmission medium has become common in local networks and long-haul terrestrial or transoceanic systems. An optoelectronic conversion is necessary in these systems, as the switching function is realized with electronics while the transmission is optical. The introduction of optics in switching systems would increase the capacity and flexibility of these systems.

Optical transmission through silica optical fibers is now a mature technology, while photonic switching is in an early stage of development. Important progress in optoelectronic devices such as laser diodes and photodiodes has been made in the development of optical fiber transmission systems, making possible the use of these devices and technologies for switching applications. Semiconductor lasers (SLs) and amplifiers (SLAs) are used traditionally as light sources and linear amplifiers in transmission systems. Their characteristics have been extensively discussed.[1,2] In this chapter, we review their switching functions and their possible applications in photonic switching systems.

Photonic switching can be basically classified into three categories: Space division (SD), time division (TD), and wavelength division (WD), even though more complicated switching systems use a combination of two or three of these divisions.[3] The first two types use the same principles as those in electronic switching. In SD systems, there exists a specific physical path from any input channel to any output channel and SLAs are used as optical on-off gates. In TD systems, switching from an input channel to an output channel shares the same physical link but uses time-division multiplexing. In this case, SLAs and SLs can act as optical memory using their bistable characteristics. The self-pulsating phenomenon in semiconductor lasers can be used to extract the clock signal. A specific point associated with optics is the possibility to realize photonic switching in the wavelength division, in which the wavelength is used to represent the destination address of information. In WD systems, switching is achieved by using tunable sources, wavelength converters (WCs), and tunable filters, which can also be built using SLs and SLAs.

The performance of each switching system is largely conditioned by the components used. In this chapter, we discuss in particular the characteristics of SLs and SLAs as switching elements in those systems. In Sec. 10.2, the logic gate function of SLAs and their applications in SD switching systems are presented. In Sec. 10.3, the bistable function of SLAs and SLs are studied, including their applications as optical memory in TD systems. New functions such as clock recovery using self-pulsation semiconductor lasers are also discussed. Section 10.4 is devoted to wavelength converters, tunable filters, and their applications in WD systems. The results are summarized in Sec. 10.5.

10.2. LOGIC GATES USING SEMICONDUCTOR LASER AMPLIFIERS

Traditionally, SD switching systems are made using lithium-niobate ($LiNbO_3$) switch modules. $LiNbO_3$ switches are in a relatively mature stage of development and have already been used in some demonstration systems. However, $LiNbO_3$ switches present problems related to insertion loss and cross talk. To compensate for distribution and insertion losses, those systems require the use of optical amplifiers. In a 128×128 demonstration experiment using $LiNbO_3$ switches reported by Burke *et al.*,[4] the number of SLAs used was as large as 1024. Another limiting factor is the difficulty to integrate $LiNbO_3$ switches with electronics on the same wafer. These problems could be overcome by using SLAs as optical gates.

The main characteristics of a SLA for SD switching application are switching speed, contrast, and cross talk. The contrast is defined as the ratio between the output power at the "1" state and that at the "0" state when the amplifier is in the "pass" state. The cross talk is defined as the ratio of the output power at the "passing" state to that at the "blocking" state.

10.2.1. Basic Characteristics

SLAs were initially developed for applications as in-line amplifiers in the early 1980's.[5–8] Even though they are in competition with erbium-doped fiber amplifiers for applications in optical transmission, they are very well adapted for many applications in optical switching.

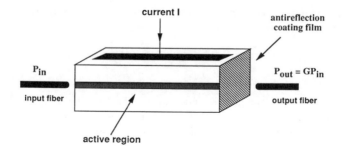

FIGURE 10.1. Schematic illustration of a semiconductor laser amplifier.

A SLA is schematically shown in Fig. 10.1. It has the same structure as a Fabry–Perot cavity laser, except that the facets of SLAs are usually antireflection coated in order to avoid gain modulation by the Fabry–Perot cavity. The residual reflectivity should be as low as 10^{-4} in traveling wave amplifiers.[6] The device is driven by injecting current into the active layer. In a SLA, the increase of injection current results in an increasing carrier density and a higher gain. The rate equation for the carrier density N is commonly written as[1]

$$\frac{dN}{dt} = \frac{I}{eV} - R(N) - v_g g P(z),$$ (10.1)

where I is the injection current, e the electron charge, V the volume of the active layer, v_g the group velocity in the active layer, and $P(z)$ the photon density. $R(N)$ is the carrier recombination rate given by[1]

$$R(N) = AN + BN^2 + CN^3,$$ (10.2)

where A is the recombination coefficient due to surface and defects, B is the radiative recombination coefficient, and C is the Auger recombination coefficient.[1] The modal gain g is an important material parameter. In the first-order approximation, it is related to the carrier density by[1]

$$g = g_d(N - N_0),$$ (10.3)

where $g_d = \partial g / \partial N$ is the differential gain, and N_0 is the carrier density at transparency.

A specific characteristic of semiconductor materials is that the refractive index changes with the carrier density. In fact, the refractive index change is related to the gain change through Kramers–Kronig relation. In practice, a linear expansion is usually valid:[2]

$$n = n_0 + \frac{\partial n}{\partial N} \Delta N,$$ (10.4)

with $\partial n/\partial N$ the derivative of the refractive index with respect to the carrier density. It is usual to define a linewidth enhancement factor α_H as[1]

$$\alpha_H = -2k_0 \frac{\partial n/\partial N}{\partial g/\partial N} \qquad (10.5)$$

with $k_0 = \omega/c$, ω the optical frequency and c the light velocity in vacuum. The factor α_H represents the variation of the refractive index relative to the gain change due to carrier density variation. It has some important consequences on the characteristics of SLs and SLAs.

By taking into account the variation of gain and index with carrier density, the equation governing the temporal evolution of the field's slowly varying part $E(z,t)$ is given by[2]

$$\frac{1}{v_g} \frac{\partial E(z,t)}{\partial t} + \frac{\partial E(z,t)}{\partial z} = \frac{1}{2}(\Gamma g - \alpha_L)(1 + j\alpha_H)E(z,t) \qquad (10.6)$$

where $j = \sqrt{-1}$, Γ is the confinement factor, and α_L is the internal loss due to scattering or intervalence band absorption of materials in the active layer.

The gain of a SLA is defined as a ratio between output power and input power. For a SLA with facet reflectivities R_1, R_2, it is given by:[5]

$$G = \frac{P_{\text{out}}}{P_{\text{in}}} = \frac{C_0 G_s(1-R_1)(1-R_2)}{(1-G_s\sqrt{R_1 R_2})^2 + 4G_s\sqrt{R_1 R_2}\,\sin^2(k_0 nL)}, \qquad (10.7)$$

where C_0 is the coupling efficiency, L the length, and G_s the single pass gain given by

$$G_s = \exp[(\Gamma g - \alpha_L)L]. \qquad (10.8)$$

The gain G achieves its maximum when the resonance condition is satisfied:

$$k_0 nL = m\pi, \qquad (10.9)$$

where m is an integer. The gain is a function of injection current through its dependency on carrier density. For traditional rectangular waveguide structure, the gain is different for TE and TM inputs due to the asymmetric nature of the waveguide. Advanced waveguide structures have been proposed to obtain polarization-independent gain.[9]

The performance of a typical traveling wave SLA can be summarized as follows. The single-pass gain changes more than 25 dB for an injection current variation from 0 to 100 mA.[8] The single pass gain is also a function of wavelength. The gain bandwidth broadens for increasing injection currents. The 3 dB gain bandwidth is as high as 70 nm for an injection current of 80 mA.[8] The saturation output power is about 5–10 dBm. The high variation of single pass gain with injection current allows us to obtain a low crosstalk by switching the injection current.

As any optical amplifier, SLAs will introduce noise through spontaneous emission,[7] which limits the contrast ratio and introduces a penalty on the bit-error rate (BER). The total noise of a SLA consists of signal-spontaneous emission beat noise, spontaneous-spontaneous emission beat noise, amplified signal shot noise, and spontaneous emission shot noise. The noise figure of a traveling wave SLA varies in a range of 4–8 dB depending on the injection current and the input wavelength.[6]

10.2.2. Dynamic Gating Characteristics

It is assumed at first that a SLA is modulated by a small sinusoidal signal. The analysis can be made by linearizing Eqs. (10.1) and (10.6) and by using Fourier transform. For a traveling wave SLA, the 3 dB modulation bandwidth (the frequency at which the response is decreased to half of the value at zero frequency) is approximately given by[10]

$$f_{3dB} = \frac{\sqrt{3}}{2\pi}\left(A + 2BN + 3CN^2 + \frac{P_{in}\lambda_{in}\Gamma g_d G_s}{wdch}\right),\qquad(10.10)$$

where w and d are the width and the thickness of the active layer respectively, h is Planck's constant, P_{in} is the input power, and λ_{in} is the input wavelength. In deriving Eq. (10.10), it has been assumed that the internal loss α_L is zero and the gain is independent of wavelength.[10] It is clear that the small-signal modulation bandwidth is determined by the carrier recombination rate and the stimulated emission rate.

To discuss the gating speed when a strong modulation current ΔI is applied, the rate equation for the carrier density variation ΔN is formally written as

$$\frac{d\Delta N}{dt} = \frac{\Delta I}{eV} - \frac{\Delta N}{\tau_R},\qquad(10.11)$$

where τ_R is the effective carrier lifetime given by

$$1/\tau_R = \partial R(N)/\partial N + v_g g_d P(z),\qquad(10.12)$$

which is a decreasing function of the carrier density and the photon density.

It is assumed that a switching current $I(t)$ is applied to the SLA:

$$I(t) = \begin{cases} I_0 & t < 0 \\ I_1 & t > 0 \end{cases}.\qquad(10.13)$$

As the effective carrier lifetime is a function of carrier density and photon density, there is no simple analytical solution for Eq. (10.11). In order to have a physical insight into the dynamics of a SLA, it is assumed that the effective carrier lifetime is nearly constant during switching. This assumption is only valid for some special cases but allows to obtain analytical solution. The solution of the carrier rate equation can then be written as

$$N(t) = N(\infty) + [N(0) - N(\infty)]\exp(-t/\tau_R), \tag{10.14}$$

with $N(0)$ the initial value and $N(\infty)$ the final value of the carrier density. In the case where the carrier density is very different for each state [$N(\infty) \gg N(0)$ in switch-on and $N(\infty) \ll N(0)$ in switch off] and the output power follows instantaneously the variation of the carrier density, the photon density variation is given by

$$P(t) = P(\infty)[1 - \exp(-t/\tau_R)], \quad \text{switch on}, \tag{10.15a}$$

$$P(t) = P(0)\exp(-t/\tau_R), \quad \text{switch off}, \tag{10.15b}$$

with $P(0)$ the initial value in switch off and $P(\infty)$ the final value in switch on. In this chapter, the rise (fall) time is defined as the time interval in which the optical power varies from 10% to 90% (from 90% to 10%) of the steady value in the "pass" state. By using Eqs. (10.15a) and (10.15b), the rise time τ_r and fall time τ_f are related to the carrier lifetime by

$$\tau_r = \tau_f = 2.2\tau_R. \tag{10.16}$$

The switching speed is thus mainly limited by the effective carrier lifetime in the active layer. Some important remarks can be made from Eqs. (10.10), (10.12), and (10.16):

(i) The carrier lifetime decreases with the carrier density. A high injection current leads to a high carrier density and to a short carrier lifetime.

(ii) It has been found that MQW structures improve the operation speed and the contrast ratio,[11] probably through different carrier density values in the active layer and the different values of material parameters such as A, B, C, and g_d.

(iii) A high input power would be favorable for a high-speed operation through stimulated carrier recombination.

Numerical simulation based on Eqs. (10.1) and (10.6) has been performed by Gillner.[10] It has been found that the rise time is slightly longer than the fall time. The rise and fall times become shorter when the modulation current ΔI increases. This is caused by the reduction of the carrier lifetime. Experimentally, the rise and the fall times are of the order of several hundreds of picoseconds. An example of the gated optical output of a SLA is shown in Fig. 10.2 for a cw light input.[12] The measured rise and fall times are less than 1 ns.

The contrast ratio of such a switch is limited by the spontaneous emission and the gain saturation. Contrast ratio as high as 20 dB has been measured. Cross talk is mainly limited by the amplitude of the switching signal applied to the amplifier. A cross talk of 40 dB has been reported by Fortenberry et al.[12]

10.2.3. Applications in Space-Division Switching Systems

Several switching experiments have been realized with SLAs. Figure 10.3 shows a 1×2 asynchronous transfer mode (ATM) package switching experiment.[12,15] In

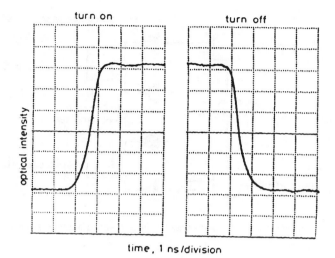

FIGURE 10.2. Gated optical output of a semiconductor amplifier gate with cw light input (after Ref. 12).

ATM systems, input signals are of the form of packets, which consist of header, representing the destination address of information, and subsequently information to transmit. In the configuration shown in Fig. 10.3, incoming optical packets enter the switch and approximately 10% of the input power is split off by an optical coupler and detected using an optoelectronic receiver. The receiver detects the address header of the incoming packet and feeds the detected signal to the address decoder. The address decoder in turn controls the electronic circuits which drive the SLAs.

Figure 10.4 shows a 2×2 directional coupler built with SLAs. It can be used for

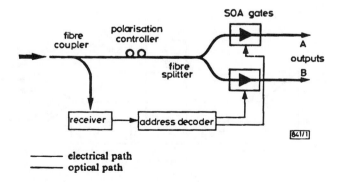

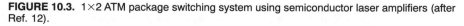

FIGURE 10.3. 1×2 ATM package switching system using semiconductor laser amplifiers (after Ref. 12).

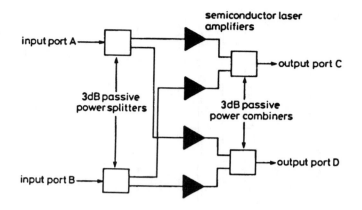

FIGURE 10.4. 2×2 directional coupler constructed with semiconductor laser amplifiers (after Ref. 3).

synchronous or asynchronous switching applications. Based on this type of directional coupler, large dimension multistage systems can be built.[13] Solutions for constructing Clos networks, in which nonblocking should occur, have also been proposed.[16] The number of cascaded amplifiers is limited by the spontaneous emission power which could saturate amplifiers. The channel selection function has also been examined experimentally with four inputs and one output selector.[17]

10.3. BISTABLE AND SELF-PULSATING SEMICONDUCTOR LASERS AND AMPLIFIERS

The principle of a TD switching system is shown in Fig. 10.5.[3] In this system, input signals from different channels are multiplexed in the time division into a single channel. The information is written into a memory matrix in the arrival order. The memory is read depending on the address of the channel. At the output, the signal

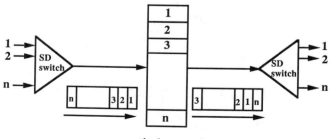

optical memories

FIGURE 10.5. Principle of a time-division switching system.

is demultiplexed. In this system, the multiplexer and demultiplexer can be realized with SD switches. One of the new key functions is the optical memory.

Optical memories can be realized with fiber delay line or optical bistable devices, such as self-electro-optic-effect devices and nonlinear Fabry–Perot etalons. The consideration of these devices is beyond the scope of the present chapter. We shall concentrate our attention on bistable semiconductor lasers and amplifiers. Excellent review papers have been published by Kawaguchi on this subject.[18,19] The objective of the present chapter is to explain the basic principles of operation and to review some more recent results.

Another important function used in TD systems is the clock recovery. Present systems use electronic circuits to extract the clock signal, which are usually designed for a specific bit rate. Direct optical clock recovery, allowing a flexible input bit rate, is very important for the development of future bit-rate transparent optical network.

10.3.1. Bistable Semiconductor Laser Amplifiers

Resonant type amplifiers, such as Fabry–Perot or DFB-type amplifiers, could have bistable behavior between input and output powers. The bistable behavior is due to the carrier density depletion by stimulated amplification of the input power, which in turn results in a power-dependent refractive index. In fact, the carrier density in the steady state in a SLA is related to the photon density $P(z)$ by using Eq. (10.1):

$$N(z) = N_0 + \frac{I\tau_e/(eV) - N_0}{1 + P(z)/P_s}, \quad P_s = \frac{1}{v_g g_d \tau_e}. \tag{10.17}$$

Here it is assumed that $R(N) = N/\tau_e$, with τ_e the carrier lifetime due to carrier recombination. P_s is the saturation photon density inside the amplifier. It is obvious that the carrier density decreases with the increasing photon density. As a result of this carrier density depletion, the gain and index will be changed simultaneously by the incident power through their dependency on the carrier density as described in Eqs. (10.3) and (10.4). The change of refractive index will result in a change of the resonance condition given by (10.9). In some conditions, the resonance condition is enhanced, leading to a more important photon density inside the cavity. That forms a delayed positive feedback loop as shown here:

$$P_{in}\uparrow \rightarrow P\uparrow \rightarrow N\downarrow \rightarrow n\uparrow \rightarrow \text{resonance condition more satisfied} \rightarrow$$

$$\tag{10.18}$$

This delayed positive feedback loop leads to the existence of a regime of instability. This instability will result in the bistability between input and output powers. Figure 10.6 shows the calculated results for a Fabry–Perot cavity SLA by using Eq. (10.7).[18] In this figure, $\phi_0 = k_0 n_0 L$ is the phase delay without injection. Other parameters used are $\Gamma = 0.5$, $\alpha_L = 25$ cm^{-1}, $\alpha_H = 3$, and $g = 0.95 g_{th}$ with g_{th} the threshold gain of lasing. The measured bistable curve has been reported by

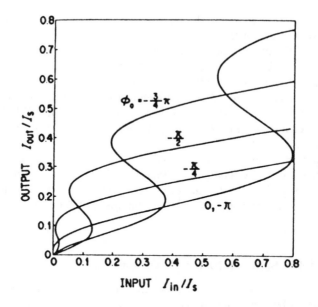

FIGURE 10.6. Calculated bistable curve of a resonant-type Fabry–Perot semiconductor laser amplifier (after Ref. 18).

Kawaguchi.[18] Multiple bistability and multistability have also been predicted in Fabry–Perot SLAs.[20]

The similar bistable behavior has also been predicted for distributed-feedback (DFB) or distributed Bragg-reflector (DBR) laser amplifiers.[21] As bistable behavior occurs only in the vicinity of threshold, a very small input power is sufficient to switch on bistable SLAs.

To study the dynamic response, the linearized rate equation (10.1) is transformed to the frequency domain. The resulted carrier density change due to input optical power variation is given by

$$\Delta \tilde{N} = -\frac{v_g g_0 \Delta \tilde{P}}{j\Omega + 1/\tau_R},\tag{10.19}$$

where g_0 is the average gain. The symbol $\tilde{}$ denotes the representation in the frequency domain. Thus the carrier density response has a bandwidth limited by the effective carrier lifetime τ_R. As the SLA is biased below threshold, the photon density is not very high inside the SLA. The operation speed is usually limited to the order of 1 GHz by the carrier lifetime.

Bistable SLAs can be thought of as dispersive bistable devices as the nonlinear refractive index is the origin of the bistability.[18] This type of device is very sensitive to the input wavelength as it needs the matching of the input wavelength to those corresponding to cavity resonance peaks.

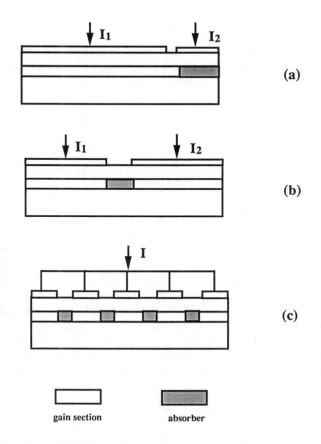

FIGURE 10.7. Structures of bistable semiconductor laser: (a) two-section (b) three-section, and (c) multisegment.

10.3.2. Bistable Semiconductor Lasers

Bistable semiconductor lasers (BSL) have been studied since the early stage of semiconductor laser development.[22-31] The first studied BSL structure is shown in Fig. 10.7(a). It consists of two sections, one gain section and one absorber section. Subsequently, other configurations have also been proposed and fabricated. These include

(1) A three-section Fabry–Perot cavity laser incorporating an absorber region shown in Fig. 10.7(b).[30] Two side gain sections are electrically pumped and the absorber section is not electrically pumped.

(2) A multisegment Fabry–Perot cavity laser, in which a large number of gain and absorber sections alternate [Fig. 10.7(c)].[19]

For all types of BSL structure, the active layer can be bulk or multiple quantum well (MQW).

10.3.2.1. Bistable and Self-Pulsating Regimes

For most models of BSL, it has been assumed that the photon density is constant along the cavity length and the carrier density changes from section to section. The rate equations governing the dynamics of such a laser are given by[23-25]

$$\frac{dP}{dt} = \left[\sum_i \Gamma f_i v_g g_{di}(N_i - N_0) - 1/\tau_p \right] P + \beta \sum_i f_i B_i N_i^2 + F_P(t) \quad (10.20)$$

$$\frac{dN_i}{dt} = \frac{I_i}{eV_i} - R_i(N_i) - v_g g_i P + F_{N_i}(t), \quad (10.21)$$

where P is the photon density inside the cavity, the subindex $i = 1,2,...$ represents different sections, $f_i = L_i/L$ is the fraction of section length L_i to the total cavity length L ($\Sigma_i f_i = 1$), τ_p the photon lifetime in the cavity, β the coupling efficiency of the spontaneous emission to the lasing mode, I_i the injection current into the ith section, and V_i the volume of the ith section. $F_P(t)$ and $F_{N_i}(t)$ represent the Langevin forces corresponding to the spontaneous emission and carrier shot noise, respectively. This is the standard rate-equation model for a multisection laser and has been widely used for the study of static and dynamic responses.[23-25]

Let us take an example of a two-section Fabry–Perot BSL. For small fluctuations, the rate equations can be linearized around a steady operation point and analyzed by using the Laplace transform. The results can be written in matrix form as

$$(sU - M) \begin{pmatrix} \Delta \tilde{N}_1 \\ \Delta \tilde{N}_2 \\ \Delta \tilde{P} \end{pmatrix} = \begin{pmatrix} \tilde{F}_{N_1} \\ \tilde{F}_{N_2} \\ \tilde{F}_P \end{pmatrix}, \quad (10.22a)$$

$$M = \begin{pmatrix} -\dfrac{1}{\tau'_{e1}} & 0 & -v_g \bar{g}_1 \\ 0 & -\dfrac{1}{\tau'_{e2}} & -v_g \bar{g}_2 \\ A_1 & A_2 & A_3 \end{pmatrix}, \quad (10.22b)$$

with U the unit matrix and τ'_{ei}, A_i given by

$$\frac{1}{\tau'_{ei}} = A + 2B\bar{N}_i + 3C\bar{N}_i^2 + v_g g_{di}\bar{P} \quad (10.23a)$$

$$A_i = f_i(\Gamma v_g g_{di}\bar{P} + 2\beta B\bar{N}_i), \quad i = 1,2, \quad (10.23b)$$

$$A_3 = \sum_i \Gamma f_i v_g g_{di}(\bar{N}_i - N_0) - 1/\tau_p. \quad (10.23c)$$

The symbols ~ and ‾ denote, respectively, the representation in the Laplace domain and the average value. The determinant of the matrix $sU - M$ is a polynomial of the third-order written as

$$\det(sU - M) = s^3 + C_1 s^2 + C_2 s + C_3, \qquad (10.24)$$

where $C_1, C_2,$ and C_3 are real constants depending on the laser's parameters.[40] The roots of the determinant are:

$$s_1, \quad s_2 + j\Omega_P, \quad s_2 - j\Omega_P, \qquad (10.25)$$

where $s_1, s_2,$ and Ω_P are real parameters. Negative values of s_1 and s_2 correspond to the stable regime, in which a laser after perturbation can return to its initial state through damped relaxation oscillations. The self-pulsating (SP) regime corresponds to $s_2 < 0$ and $|s_2| \ll \Omega_P$ and the bistable regime corresponds to s_1 or $s_2 > 0$.

The stability based on the above principle has been analyzed by Ueno,[23] Kawaguchi et al.[18] The results obtained by Kawaguchi for a two-section laser show that self-pulsation occurs only for small values of nonradiative carrier lifetime in the absorber section, which is usually achieved by ion implantation or reverse biasing of the absorber section.[18] In contrast, bistability, corresponding to the regime where there is an instability, could be easily obtained if there is a saturable absorber inside the laser structure.

10.3.2.2. Static Bistable Characteristics

As an example, the measured output power versus injection current of a three-section BSL is shown in Fig. 10.8(a).[30] Typically, a hysteresis loop is observed for a low current I_2. A contrast ratio, defined as the ratio between the output power at the "on" state and that at the "off" state, of higher that 10 dB can be obtained for a hysteresis width of several mA. The bistable behavior can be optimized by separately controling the injected current in both active sections. Figure 10.8(b) shows the three different regimes of operation in the plane of injection currents (I_1, I_2). The bistable loop of I_1 decreases with increasing current I_2. Modeling results show similar power-current relation and similar regimes of operation.[30]

Being biased under the bistable regime, BSLs can be switched on by an injected optical signal of photon energy higher than the energy band gap of the active material in the absorber section. The injected optical power plays the same role as electrical pumping. Consequently, there is also a bistable behavior between the output power and the input pumping power.

10.3.2.3. Dynamic Properties by Optical Triggering

Dynamic properties of BSLs by electrical or optical triggering have been studied by Liu and Ohlander et al.[25,27] Theoretical and experimental results have been reviewed by Kawaguchi.[19]

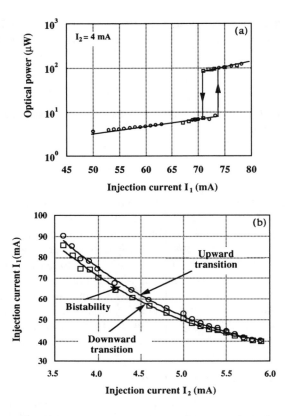

FIGURE 10.8. (a) Measured typical power-current characteristic and (b) different regimes of operation of the bistable semiconductor laser.

In the previous studies on dynamic response of BSLs, optical triggering is taken into account by adding a term of pumping in Eqs. (10.20) or (10.21) depending on whether the input signal is of the same wavelength as that of the BSL or not.[18] Recently it has been shown that an equivalent current of the absorber section can be introduced to represent the effect of optical pumping. For a three-section BSL as shown in Fig. 10.7(b), the expression of the equivalent current is given by[30]

$$I_{3 \text{ eff}} = \frac{\Gamma \alpha_{\text{in}}}{\alpha_T} \, \eta G_1 [1 - \exp(-\alpha_T L_3)] \, \frac{e P_{\text{in}}}{h \nu_{\text{in}}}, \qquad (10.26)$$

where G_1 is the single pass gain in the first section, η the quantum efficiency of the absorber material, P_{in} the input power, and ν_{in} the input frequency. This equivalent current shows clearly how the optical pumping depends on the injection power, the absorption coefficient, and the laser structure. It is important to note that the absorption coefficients α_{in} and α_T decrease with the carrier density N_3 inside the absorber.

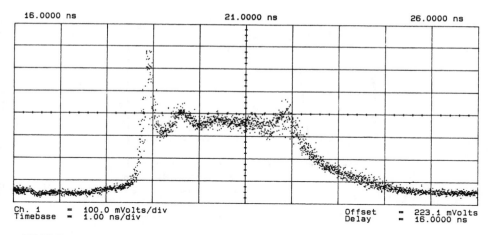

16.0000 ns   21.0000 ns   26.0000 ns

Ch. 1   = 100.0 mVolts/div   Offset = 223.1 mVolts
Timebase   = 1.00 ns/div   Delay = 16.0000 ns

FIGURE 10.9. Measured dynamic response of a bistable semiconductor laser under optical triggering.

As a consequence, the carrier density in the absorber should be always smaller than that at transparency, as above this value no absorption takes place.

Figure 10.9 shows the typical response of the optically triggered bistable laser for an input power at the on-state $P_{in} = 0.3$ mW.[28] The calculated result is also shown in Fig. 10.10. One can note that the laser has a fast rise time (300 ps) and undergoes relaxation oscillations before being stabilized at a steady-state value. When the input power vanishes, the laser's response decreases with a rather important fall time of the order of a nanosecond.

Measurement results on the transition time for the BSL are presented in Fig. 10.11. Figure 10.11(a) shows the rise time as a function of the injection currents. When the current in an active section is kept constant, the rise time decreases with increasing current in the other section. Also, it is observed that for a higher value of I_1, the carrier density at transparency in the absorber section can be reached more rapidly due to a higher amplification of the pump signal in the first gain section. Rise times as low as 300 ps have been measured. Figure 10.11(b) shows that the fall time increases with increasing injection currents. This behavior is due to the fact that a high electrical pumping in the active sections reduces the difference between the gain and the cavity overall loss (threshold gain), resulting in a slow decay of the output power. Figure 10.11(c) shows the overall transition time as a function of I_1. The opposite behavior of the rise and fall times leads to a minimum in these curves. An optimal value of I_1 for each value of I_2 minimizes the overall transition time. The minimum overall transition time decreases with increasing current I_1, indicating that a high current I_1 would be favorable for high-speed operation.

The spontaneous emission introduces an uncertainty on the delay time of the laser's response. This uncertainty is quantified as the turn-on jitter, which is defined

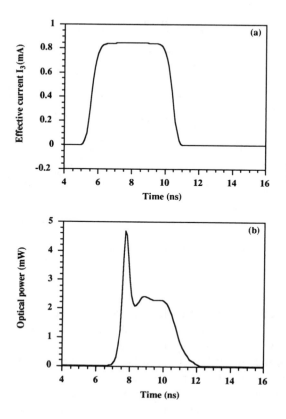

FIGURE 10.10. (a) Calculated effective current, and (b) calculated dynamic response of a bistable semiconductor laser under optical triggering (after Ref. 30).

as the standard deviation calculated at half the steady value in the laser's response. It has been found that the turn-on jitter decreases with increasing values of I_1,[28] due to the increasing stimulated emission rate with respect to the spontaneous emission rate.

In the above discussion, the BSL is biased below the downward threshold of the bistable regime. For optical memory applications, BSLs are usually biased inside the bistable hysteresis loop just below the upward threshold. An electrical reset signal is used to force BSLs to return to the below threshold state. In this case, the rise and fall times become extremely low.[27] Optical flip-flop operation up to 4 GHz has been obtained by Odagawa et al.[31]

The performance of discrete BSLs is very interesting for application as optical memories. However, applications in real TD systems need the integration of a certain number of BSLs in a single chip.

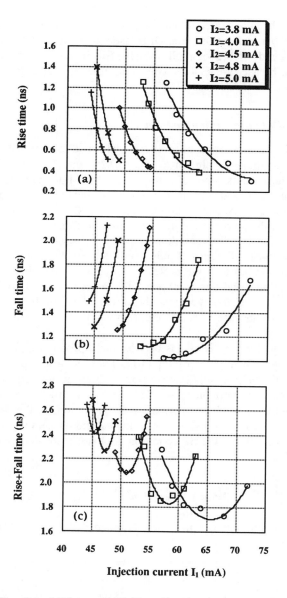

FIGURE 10.11. Rise time, fall time, and total transition time in the case of optical triggering of a bistable semiconductor laser (after Ref. 28).

10.3.3. Self-Pulsating Semiconductor Lasers and Clock Recovery

Self-pulsation (SP) is a phenomenon indicating that the optical power emitted from a laser diode varies periodically while the injection current is kept constant. This phenomenon has been observed in both multisection Fabry–Perot cavity lasers with a saturable absorber and multisection DFB lasers without a saturable absorber.[32-39] Self-pulsation is undesirable for applications of SLs in optical transmission systems. However, it can be used for clock recovery in order to realize bit-rate transparent optical networks.

10.3.3.1. Self-Pulsation Phenomenon

Theoretically the self-pulsation phenomenon was analyzed by Ueno and Lang by using a set of rate equations similar to Eqs. (10.20) and (10.21).[23] They concluded that self-pulsating occurs when the ratio τ_g/τ_a is high, with τ_g, τ_a the carrier lifetime in the gain and absorber section, respectively. In long-wavelength InGaAsP/InP lasers, the Auger recombination effect is important, which reduces the carrier lifetime in the gain section. It is thus more difficult to obtain self-pulsation in such lasers. One efficient solution proposed consists in doping the absorber section with zinc ions.[32] Stable self-pulsation has been observed in such lasers with oscillation frequency up to 5 GHz. Recently, repetitive SP phenomenon has also been observed in undoped two-electrode DFB lasers which do not include a saturable absorber.[33] It was believed that the origin of self-pulsation in DFB lasers without saturable absorbers is different from that in lasers with saturable absorbers.[33]

A theoretical analysis of SPLs has been recently proposed.[40] It has been shown that when the determinant given by Eq. (10.24) has two nearly pure imaginary zero points $(\pm j\Omega_P)$, the laser's output power will oscillate at the frequency Ω_P. Subsequently, higher-order harmonics at $2\Omega_P$, $3\Omega_P$,... will be generated through nonlinear phenomenon such as interband gain saturation. In the time domain, the corresponding signal is thus a series of periodical pulses with repetition frequency Ω_P. By using Eq. (10.24), the SP condition is obtained for double-section Fabry–Perot cavity SPLs:

$$\left(\frac{s_2}{\Omega_P}\right)^2 \approx 0 \rightarrow v_g g_1 A_1 \tau'_{e2} + v_g g_2 A_2 \tau'_{e1} + \left(\frac{1}{\tau'_{e1}} + \frac{1}{\tau'_{e2}}\right) = 0. \quad (10.27)$$

In obtaining this expression, it has been assumed that $A_3 = 0$. The self-pulsation frequency is given by:

$$\Omega_P = \sqrt{v_g g_1 A_1 + v_g g_2 A_2 + \frac{1}{\tau'_{e1}\tau'_{e2}} - A_3\left(\frac{1}{\tau'_{e1}} + \frac{1}{\tau'_{e2}}\right)}. \quad (10.28a)$$

In high output powers, this expression is reduced to the classical one:[1]

$$\Omega_P = \sqrt{v_g g_d \bar{P}/(\tau_P)}, \quad (10.28b)$$

which is proportional to the squared root of the output power. Here it is assumed that $A_i = f_i v_g g_{di} P$, $i = 1,2$, and $A_3 = 0$. Figure 10.12 gives the SP frequency as a function of the injection current in the gain section for the zinc ions implanted

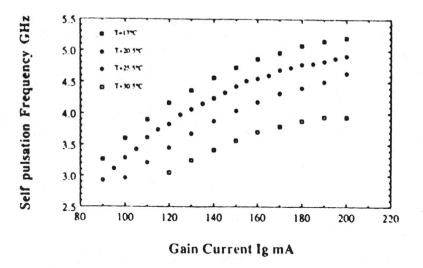

FIGURE 10.12. Self-pulsating frequency as a function of biasing current in the gain section (after Ref. 32).

laser, in which the absorber is unbiased.[32] One observes roughly a $\sqrt{I}$ variation law, in agreement with Eq. (10.28). The SP frequency can vary continuously from 2.5 to 5 GHz by changing the injection current into the gain section.

10.3.3.2. Frequency Locking Range

As any oscillator, self-pulsation can be locked to that of an input intensity modulated periodic signal under some conditions. This property can be used to extract the clock in an optical transmission system by synchronizing the SP frequency to the clock frequency. The frequency locking range has been found to increase with increasing input power at the clock frequency.[34,35] Figure 10.13 shows the locking range variation against input peak power for different return to zero input bit patterns.[34] It can be seen that the pattern "11111111" gives the largest loking range as its component at the clock frequency is the most important.[34,36] Furthermore, one can see that the locking range increases linearly with the input peak power for each pattern. This dependence law has been theoretically confirmed, as it is the optical power pulsation which is locked to the input power.[40] In classical injection locking of lasers or microwave oscillators, it is the electrical field which is locked to the input field. The corresponding locking range is proportional to the input field amplitude, or the squared root of the input power.

It has been found that the locking range is closely related to the spectral linewidth of the self-pulsating signal.[37] It has also been shown that SPLs can be locked to components of one half or one third of the input frequency.[38]

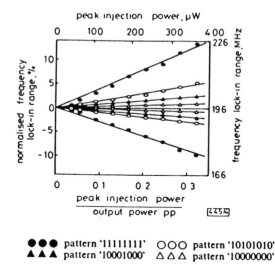

peak injection power, µW

●●● pattern '11111111'   OOO pattern '10101010'
▲▲▲ pattern '10001000'   △△△ pattern '10000000'

FIGURE 10.13. Injection locking range against input peak power of a self-pulsating multielectrode DFB laser for various RZ patterns (after Ref. 34).

10.3.3.3. Applications in Clock Recovery

All-optical clock extraction from 5 Gbit/s RZ data has been demonstrated using a two-electrode InGaAsP/InP SPL by Barnsley et al.[32] Injection of 10 µW optical data signal at a wavelength close to one of the Fabry–Perot wavelengths was sufficient to synchronize the self-pulsation to the incoming data stream. More recently, all-optical clock extraction at 18 Gbit/s was demonstrated by using a self-pulsating two-section DFB laser.[39]

The main advantages of such a clock recovery method are its simplicity and its adaptability to different bit rates.

10.3.4. Applications in Time-Division Switching Systems

The most remarkable TD switching experiment was realized by Suzuki et al.[41] The experimental configuration is shown in Fig. 10.14. The optical switch system contains a two 1×4 space switch and four BSLs as optical memory. The bit rate in the experiment is 256 Mbit/s and has been extended to 512 Mbit/s by using high speed BSLs.

10.4. WAVELENGTH CONVERTERS AND TUNABLE FILTERS

Silica optical fibers present low losses over a wavelength range as large as 200 nm around 1.5 µm. This wide transmission bandwidth gives a unique possibility for optics: Wavelength division switching systems.[3,42,43] An example of such a system

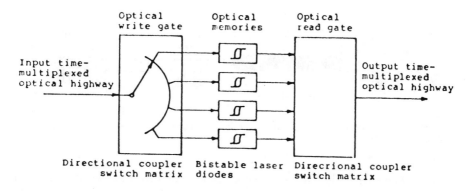

FIGURE 10.14. Experimental time division switching system using bistable semiconductor lasers (after Ref. 41).

is schematically illustrated in Fig. 10.15. In this system, each input channel supports asynchronous transfer mode signals. Each input signal is divided into two parts: the first one is used to decode the destination address of the information. The second part is injected into a wavelength converter (WC), which regenerates the input signal at a wavelength which depends on the address. All wavelength converted signals are launched into a passive coupler. At the output, a wavelength tunable filter will be used to select only one wavelength corresponding to the output channel.

In such a system, WC is one of the key components. They can be in principle realized with a hybrid optoelectronic technique, in which the input signal is

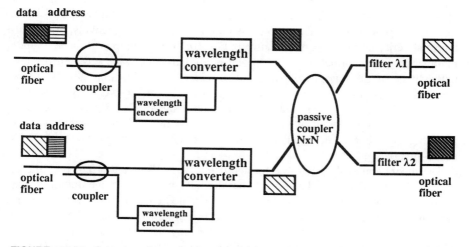

FIGURE 10.15. Schema of a wavelength division switching system using wavelength converters.

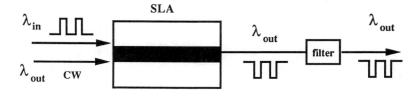

FIGURE 10.16. Principle of wavelength conversion using a gain saturated semiconductor laser amplifier.

converted into an electrical signal by a photodiode and then amplified. An electrical circuit is used to regenerate the input signal. The regenerated signal is then used to modulate the intensity or frequency of a wavelength tunable SL. This technique is now in a mature stage but is rather complicated. A very attractive solution is to realize all-optical WCs. All-optical WCs can be built with diode amplifiers, tunable laser diodes, or tunable bistable laser diodes. Physical origins involved are gain or refractive index modulation, absorption saturation, and four-wave mixing. Another important function is the filtering in the wavelength domain, which can also be realized by using tunable SL structures.

10.4.1. Semiconductor Laser Amplifier Based Wavelength Converters

10.4.1.1. Carrier Density Depletion Induced Amplitude Modulation

Amplification of an input signal results in carrier density depletion in SLAs. Optical gain in a SLA can be reduced by this carrier density depletion in the case of a high input power. This phenomenon introduces distortions to the transmitted signal for applications of SLAs as in-line amplifiers, but can be used to realize optical WCs. For this purpose, two signals, denoted as pump and probe, respectively, are simultaneously injected into a SLA, as shown in Fig. 10.16. The pump is of the amplitude modulation (AM) format while the probe is a continuous wave (cw). In the low power state of the pump, the SLA will provide the amplification for the probe signal. In the high-power state of the pump, the gain for the probe signal will be decreased due to the gain saturation by the pump signal. The degree of decrease depends to a large extent on the pump power and the injection current into the amplifier. By this way, the modulation of the pump is transferred to the probe with signal inverted.

The theoretical analysis was performed using Eqs. (10.1) and (10.6) by Valiente et al.[44] The calculated results showed that the fall time of the probe output is very short because it is dominated by the pump stimulated emission rate, while the rise time is rather long due to a long carrier recovery time, which limits the maximum operation speed of the device.[44] The factor limiting the maximum bit rate is the effective carrier lifetime, as in the case of logic gates discussed in Sec. 10.2. Optical wavelength conversion has been demonstrated at 2.5 Gbit/s by Glance[45] and at 4

Gbit/s by Durhuus et al. by using a 600 μm SLA.[46] More recently, the bit rate has been increased to 10 and 20 Gbit/s by increasing simultaneously the input pump power and the input probe power.[47,48]

It has been found that the contrast ratio of the probe output depends to a large extent on the input pump power. It takes the values of 10, 8, and 5 dB when the input power is 0, -2, and -5 dBm, respectively in the experiment reported by Glance et al.[45] We can estimate the degradation of contrast ratio (CR) which is given by for the input pump signal:

$$\mathrm{CR_{inp}} = 10 \; \log(P_{1p}/P_{0p}), \qquad (10.29)$$

where P_{0p} and P_{1p} denote the input pump power in the '0' and '1' state, respectively. At the output, the pump power at the '0' and '1' state is $G_0(\lambda_p)P_{0p}$ and $G_1(\lambda_p)P_{1p}$, with $G_0(\lambda_p)$ and $G_1(\lambda_p)$ the gain in the '0' and '1' state, respectively. The probe signal at the output is thus given by $G_0(\lambda_s)P_s$ and $G_1(\lambda_s)P_s$, respectively. Here the amplified spontaneous emission has been neglected. The sum of contrast ratios of output pump and probe is thus given by

$$\mathrm{CR_{outp}}+\mathrm{CR_{outs}} = 10 \; \log\!\left(\frac{G_1(\lambda_p)P_{1p}}{G_0(\lambda_p)P_{0p}}\right) + 10 \; \log\!\left(\frac{G_0(\lambda_s)P_s}{G_1(\lambda_s)P_s}\right) \approx \mathrm{CR_{inp}}.$$
$$(10.30)$$

In obtaining the last equality in Eq. (10.30), it has been assumed that the gain ratio for the pump and for the probe is the same. Equation (10.30) shows that the contrast ratio of the input pump is shared by the output pump and probe. In practice, it is very difficult to get zero or negative contrast ratio of the output pump signal, leading to a degradation of contrast ratio in the wavelength conversion process.

This type of WC has a wide wavelength conversion range and a high-speed response. The limiting factors of this type of WC are the necessity of an external tunable laser providing the probe cw signal and the degradation of contrast ratio. The second factor limits the number of cascaded WCs in a switching network.

10.4.1.2. Carrier Density Depletion Induced Phase Modulation

In the schema discussed above, the refractive index is also changed through carrier density depletion by the pump signal. The pumping power-dependent refractive index will result in a phase modulation of the probe signal. This phase modulation, associated with amplitude modulation, broadens the spectral linewidth of the probe signal and could degrade the system performance through dispersion in silica optical fibers. On the other hand, this phase modulation can be used to realize wavelength conversion.

The phase modulation on the probe signal can be detected by using a coherent receiver or by converting the phase modulation into amplitude modulation through an interferometric configuration as shown in Fig. 10.17(a). The probe signal at the output of the SLA should have a phase variation of π if the pumping signal is 1. The advantage of this configuration is the low pump power needed as it is only neces-

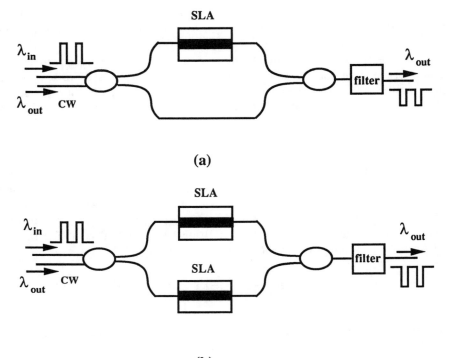

FIGURE 10.17. Wavelength conversion using the phase modulation of a semiconductor laser amplifier in an interferometer configuration, (a) asymmetric and (b) symmetric configuration.

sary to produce a phase shift of π, rather than to saturate the gain. Moreover the amplifier can be biased with a high injection current, leading to a reduced carrier lifetime and, as a consequence, to a high operation speed.

In the schema of Fig. 10.17(a), the two arms of the interferometer are not symmetric. A high contrast ratio can only be obtained in the case where two arms are symmetric, allowing the cancellation of signal in the π dephasing case. Recently symmetric interferometric configuration has been proposed,[49] consisting of a SLA in each arm, as shown in Fig. 10.17(b). Contrast ratio improvement of 2.5 dB for converted signal compared to the input signal at 2.5 Gbit/s been obtained by using this configuration.[50]

10.4.1.3. Four-Wave Mixing

Four-wave mixing (FWM) is a nonlinear phenomenon which involves optical signals of three different frequencies: pump ω_p, probe $\omega_s(= \omega_p - \Delta\omega)$ and conjugates $\omega_p + \Delta\omega$, $\omega_s - \Delta\omega$, with $\Delta\omega$ the detuning frequency. Due to nonlinear processes

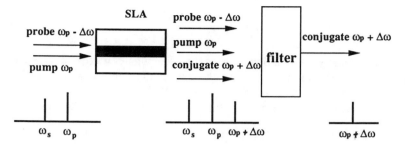

FIGURE 10.18. FWM phenomenon in a semiconductor laser amplifier.

in materials of active layer induced by the pump and the probe beams, a nonlinear susceptibility at the conjugate frequencies is created, which generates beams at those frequencies.[51-58]

FWM in a SL was first observed by Nakajima and Frey.[51] Two types of origin leading to FWM in SLAs or SLs can be distinguished: Population pulsation and intrinsic nonlinearities in semiconductors such as intraband relaxation and carrier heating.[52] The first arises from the beating between the pump and the probe, which creates a carrier density modulation at the detuning frequency. As the gain and index are related to the carrier density, temporal gain and index gratings are thus created, which lead to the generation of a modulation side band at the conjugate frequency, as shown in Fig. 10.18. The maximum detuning frequency of this type of FWM is limited to the order of $1/\tau_e$, with τ_e the carrier lifetime in the SLA, above which the carrier density can not respond to the photon density modulation.[53] This is the nearly degenerate four wave mixing (NDFWM), which will be analyzed by following a simple approach given by Großkopf et al.[53]

It is assumed that a pump signal $E_P(t) = E_P \cos(\omega_p t + \phi_p)$ and a probe signal $E_s(t) = E_s \cos(\omega_s t + \phi_s)$ are injected simultaneously into a SLA. The total photon density inside the SLA is given by

$$P = P_P + P_s + 2\sqrt{P_P P_s} \, \cos(\Delta\omega t + \Delta\phi) \tag{10.31}$$

where $\Delta\omega = \omega_p - \omega_s$ is the detuning frequency and $\Delta\phi = \phi_p - \phi_s$ is the phase difference between the pump and the probe. Thus, the total photon density contains two constant terms and a beating term. This beating creates a carrier density modulation, which can be calculated by using Eq. (10.1). In the case of a small beating term, the carrier density change due to optical amplification can be obtained by using a first order approximation from Eq. (10.1):

$$\Delta N = \Delta N_0 + \Delta N_1 \cos(\Delta\omega t + \Delta\phi'), \tag{10.32a}$$

where ΔN_0 is a static term, $\Delta\phi' = \Delta\phi - \tan^{-1}(\Delta\omega\tau_R)$, and ΔN_1 the carrier density modulation term at the detuning frequency, which is given by

$$\Delta N_1 = -\frac{v_g g_0 2\sqrt{P_P P_s}}{\sqrt{\Delta\omega^2 + (1/\tau_R)^2}} \tag{10.32b}$$

This carrier density modulation creates two conjugate waves at $\omega_p + \Delta\omega$ and $\omega_s - \Delta\omega$. A rigorous analysis requires the use of the coupled-wave formalism.[52] Analytical results exist only for traveling-wave SLAs.[52]

The pump amplitude at the output of the amplifier is related to its input $E_P(0)$ by

$$E_P(L) = E_P(0)\exp[(g-\alpha_L)L/2]\exp(-jknL)\exp(j\omega_p t). \tag{10.33}$$

Using Eqs. (10.32) and (10.33), we can also write this equation as

$$E_P(L) = E_P(0)\sqrt{G_P}$$
$$\times \exp[-jkn_0 L + j\omega_p t - \tfrac{1}{2}g_d \Delta N_1 (1 + j\alpha_H)L \cos(\Delta\omega t + \Delta\phi')]$$
$$\tag{10.34}$$

with G_P the gain for the pump signal. Equation (10.34) is only an approximation which is valid for low spatial variation of ΔN_1. It shows that the amplitude and the phase of the pump signal are simultaneously modulated. As the value of α_H is of the order of two in quantum well structures and five in bulk structures, the phase modulation is usually dominant. The component at the conjugate frequency can be evaluated by using Fourier analysis. It is a Bessel function with argument the phase modulation index β defined as $\beta = -\tfrac{1}{2}g_d \Delta N_1 \alpha_H L$. An approximate expression can be given for the output field E_c and the output power of the conjugate wave P_c at $\omega_p + \Delta\omega$ by using $J_0(\beta) \approx \beta$:

$$E_c = KE_P^2 E_s \frac{1}{\sqrt{1 + (\Delta\omega\tau_R)^2}}, \tag{10.35a}$$

$$P_c = |K|^2 P_P^2 P_s \frac{1}{1 + (\Delta\omega\tau_R)^2}, \tag{10.35b}$$

where K depends on the amplifier's parameters. One can see that the conjugate wave is a replica of the probe signal. The conversion efficiency decreases with frequency detuning between the pump and the signal. As the SLA is in the regime of linear gain, the effective carrier lifetime is the same as the carrier lifetime determined by carrier recombination rate. The maximum detuning frequency is limited by the carrier lifetime to about 2 GHz. The maximum allowed bit rate should be smaller than this maximum detuning frequency. The measured conversion efficiency as a function of the detuning frequency can be very well fitted by Eq. (10.35b).[53]

The very small wavelength conversion range and bit rate seriously limit direct applications of NDFMW in wavelength conversion. A very engineered solution is proposed by Großkopf et al.[53] consisting in injecting two pump signals simultaneously into a SLA, as shown in Fig. 10.19. Similar to the case discussed above, the

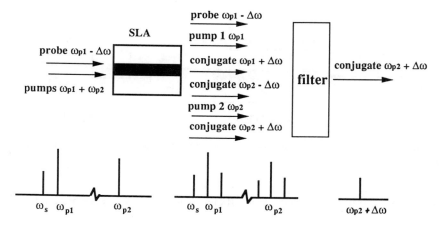

FIGURE 10.19. Wavelength conversion using double-pumping FWM in a semiconductor laser amplifier.

first pump signal and the probe create a carrier density modulation as shown in Eq. (10.34). Now a second pump $E_{P'}$ at the desired converted frequency is used. In such a configuration, the detuning frequency between the second pump and the probe signal is much higher than the SLA's bandwidth $1/\tau_e$, in order to prevent that the carrier density is modulated at this detuning frequency. However, the carrier density modulation due to the beating between the first pump and the probe generates two side bands centered at the second pump frequency. The output of the amplifier at the second pump frequency has a similar expression as Eq. (10.35). By using a narrow band filter, only the conjugate signal near the second pump frequency is selected.[53]

Using such a double-pump schema, wavelength conversion over the total gain bandwidth of a SLA can be obtained. An experiment realized by Großkopf has demonstrated wavelength conversion over 15 nm with a bit rate of 140 Mbit/s.[53] The complication is that one needs simultaneously two pump signals.

The intrinsic nonlinear phenomenon in semiconductors due to spectral-hole-burning (SHB) and carrier-heating (CB) is responsible for the highly nondegenerate four-wave mixing (HNDFWM), which could occur for detuning frequency up to about 24 nm,[54,55] as the time constant associated with spectral hole burning is about 50 fs and that associated with carrier heating is about 650 fs.[57] HNDFWM is very interesting for the study of nonlinear processes in semiconductors. For wavelength conversion application, it offers a wide wavelength conversion range and high bit rates. For example, multiterahertz frequency conversion of a picosecond pulse train and broadband wavelength conversion as large as 27 nm has been recently achieved.[56,58] The difficulty of using this type of FWM is that the generated conjugate wave is very weak due to the fact that the saturation power related to SHB and CB is very high. A second optical amplifier should be used to restore the converted

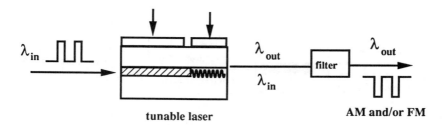

FIGURE 10.20. Wavelength conversion using a carrier density-depleted tunable semiconductor laser.

signal level. In fact, the conversion efficiency is given by the simple relation:[57]

$$10 \log \frac{P_c(L)}{P_s(0)} = 3G_s + 2P_{inp} + 20 \log \left| \sum_{m=1}^{3} \frac{c_m}{1 - j\Delta\omega\tau_m} \right| \quad (10.36)$$

with G_s the single pass gain in dB, P_{inp} the input pump power in dBm, c_m the complex coupling coefficient, and τ_m the time constant associated with interband carrier recombination, CB and SHB. For a tensile-strained MQW traveling-wave SLA, it has been found that $C_1 = 0.24 \exp(-j1.30)$, $C_2 = 0.0027 \exp(j1.30)$, and $C_3 = 0.00048 \exp(j1.53)$.[57] The effect of SHB is thus about 60 dB less than that of interband gain saturation in that SLA.

10.4.2. Tunable Semiconductor Lasers Based Wavelength Converters

10.4.2.1. Carrier Depletion Induced AM or FM

In the schema of SLA-based WCs, a SLA and a wavelength tunable source are simultaneously required. One very attractive solution is to integrate functions of tunable lasers and amplification in the same device. That is the structure of tunable single-mode semiconductor lasers, such as multielectrode DFB or DBR lasers.[1]

The principle of operation, similar to that in a saturated SLA, is shown in Fig. 10.20. The injected signal is of the AM format. In the '1' state of the injected signal, the carrier density is reduced through stimulated amplification while in the '0' state it remains the same. The carrier density variation affects simultaneously the gain and refractive index of the material in the active layer, resulting in a simultaneous variation of output power (AM-AM conversion) and of the lasing frequency (AM-FM conversion).

Theoretical analysis can be performed by using a set of rate equations including the input signal:[59]

$$\frac{dP}{dt} = (\Gamma v_g g - 1/\tau_p)P, \quad (10.37)$$

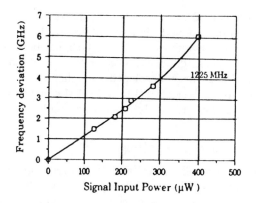

FIGURE 10.21. Frequency deviation vs input power for a DBR laser. The modulation frequency is 1225 MHz (after Ref. 50 © 1994 IEEE).

$$\frac{d\phi}{dt} = \frac{\alpha_H}{2}(\Gamma v_g g - 1/\tau_p),$$ (10.38)

$$\frac{dN}{dt} = \frac{I}{eV} - R(N) - v_g g P - v_g g_{in} P_{in},$$ (10.39)

where ϕ is the phase of the converted signal, g_{in} is the gain corresponding to the input wavelength, and P_{in} the average photon density of the input signal inside the cavity.[59] Other parameters have the same definition as in Eq. (10.1). In fact, Eqs. (10.37)–(10.39) have been largely used for the study of the laser's intrinsic response using optical modulation.[59] Small signal analysis can be performed which gives the laser's response subject to the injected optical power modulation. The traditional 3 dB modulation bandwidth can be found.[1,59] The optical modulation technique allows a bandwidth measurement free from electrical parasitics due to chip or package.

As in the case of gain-saturated SLAs, a very important input power is needed for AM-AM conversion. The required input power is of the same order of magnitude as in the case of saturated SLAs. In contrast, an AM-FM conversion is simple to achieve. Figure 10.21 gives the frequency deviation as a function of the input power of a DBR laser for a modulation frequency of 1225 MHz.[60] The conversion efficiency is about 15 GHz/mW.

In the case of AM-FM conversion, the FM signal should be converted into an AM signal by using a Mach–Zehnder or a Fabry–Perot interferometer in direct detection systems or demodulated by using a frequency discriminator in the electrical domain in coherent systems. Experiments have been realized by using a two-electrode DFB laser[61] or DBR laser.[50] The bit rate obtained is as high as 5 Gbit/s.[62] If the input signal is of the FM type, it should be converted into AM by using a frequency discriminator before being injected into the tunable laser.[63]

The advantage of this kind of WCs is that its maximum bit rate, determined by the 3 dB electrical modulation bandwidth, could be very high. Using quantum well structures, modulation bandwidth as high as 20 GHz has been obtained.[64] Furthermore, devices used in wavelength conversion are traditional tunable semiconductor lasers, which have already been developed for other applications. The limiting factors are the high input power needed for AM-AM conversion or an interferometer for AM-FM conversion and highly input wavelength-dependent performance.

10.4.2.2. FWM and Injection Locking

As in an SLA, FWM can be used to realize optical wavelength conversion in an SL. In this schema, the pump signal is provided by the tunable laser itself. The incident wave and the pump wave create a carrier density modulation at the detuning frequency, generating a conjugate wave.[51,65] Different from the case of FWM in an SLA, the FWM occurs only for detuning frequencies higher than the locking bandwidth of the laser which is given by[66]

$$\Delta \omega_L = \frac{1}{\tau_{in}} \sqrt{\frac{P_{in}}{P}} (1 + \alpha_H^2), \tag{10.40}$$

with τ_{in} the photon roundtrip time in the laser cavity. For detuning frequencies lower than the locking bandwidth, the pump laser will be locked to the injected signal. In this case, no wavelength conversion takes place. This is not really a limiting factor as the desired converted frequency is usually rather different from the input frequency.

The FWM in SLs can be used for high-speed wavelength conversion, as the detuning frequency can be as large as the 3 dB modulation bandwidth of the laser, which is much higher than that limited by the carrier lifetime in an SLA. The allowed bit rate is also increased. 5 Gbit/s wavelength conversion has been realized by using this method.[67]

Side-mode injection locking can also be used for wavelength conversion. In this case, the injected wavelength is in the vicinity of a side mode. Due to the injection, the side mode could become dominant, while the main mode power is reduced if the injection power is sufficiently high. The AM of the input signal is thus converted to the AM of the main mode of the laser. By using side-mode injection locking, wavelength conversion has been performed using Y lasers at 2.5 Gbit/s[68] and using DFB lasers at 10 Gbit/s.[69] However, such a technique presents limits related to the selected input wavelength and low conversion efficiency for lasers with high side-mode suppression ratio.

10.4.3. Tunable Bistable Laser-Based Wavelength Converters

Bistable SLs are studied as optical memories in Sec. 10.3. Single-mode BSLs incorporating a Bragg section, such as DFB or DBR, can also be used to realize optical wavelength conversion.[70–78] The principle of wavelength conversion is

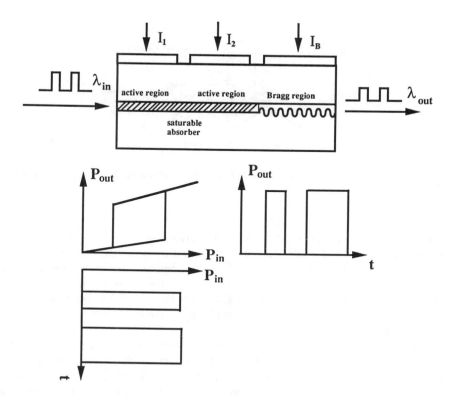

FIGURE 10.22. Principle of tunable bistable semiconductor laser as a wavelength converter.

shown in Fig. 10.22. In this application, BSLs should be biased electrically below the downward threshold. Due to the bistable behavior between the input and output powers, the input intensity modulated signal can be regenerated if the injected signal has sufficient power with photon energy greater than the energy bandgap of the material in the absorber. Thus, BSLs can regenerate the input signal but at a different wavelength, which is determined by the cavity structure of the laser.

Dynamic properties of bistable lasers have been largely discussed in Sec. 10.3. In principle, tunable single-mode BSLs should have comparable behavior as Fabry–Perot BSLs. Wavelength conversion using bistable tunable laser has been realized at 125 Mbit/s by using the DBR structure in 1991.[71] Periodic sinusoidal operation has been obtained up to 500 MHz by using a DFB structure.[72]

As the carrier lifetime in the absorber can not fall below one nanosecond, the maximum operating speed for conventional bistable lasers is limited to about 500 Mbit/s. To improve the operating speed, there exist three solutions: (i) Use of an electrical reset, (ii) ion-implanted absorber, and (iii) reverse biasing of the absorber.

The electrical reset consists in applying a negative signal to the gain or absorber section at the end of each bit, as in the case of application in optical memory. The

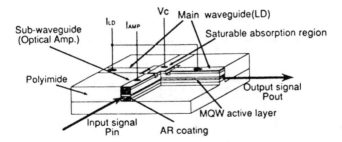

FIGURE 10.23. Side-injection bistable lasers for wavelength conversion (after Ref. 78 © 1993 IEEE).

use of electrical reset has resulted in signal conversion at a bit rate of 2.5 Gbit/s.[31] However, this method needs a clock signal and a control circuit. Recently, proton bombarded BSLs have been fabricated.[28,29] This bombardment creates defects in the absorber region and increases the carrier nonradiative recombination rate in this region. A 2.5 Gbit/s operation with BER lower than 10^{-9} has been successfully obtained.[29] A further improvement of the bit rate above 2.5 Gbit/s seems difficult to achieve by using only this solution.

The reverse biasing of the saturable absorber could make the operation speed of BSLs extremely high, as the limiting time constant in the absorber is the carrier transport time, which is much shorter than the carrier lifetime. However, the reverse biasing makes usually the bistable cycle very large, requiring a very high input power to switch-on the BSL. Small positive voltage control of the absorber was performed by using a side-injection light controlled bistable laser diode.[77,78] The structure of such a laser is shown in Fig. 10.23. Operation speed up to 2.1 GHz has been obtained. Another feature of this structure is that the input light, after amplification by the same device, is injected laterally to the absorber. Such a structure prevents the dependence of the switching behavior on the input wavelength and reduces the optical feedback by the input facet through an antireflection coating. Wavelength conversion over 100 nm at 2.1 GHz has been obtained by using this structure.

Recently, an asymmetric κ-DBR structure was proposed,[75,76] which contains Bragg reflectors with different coupling coefficients at the two sides. The input-end DBR mirror has higher coupling coefficient than that of the output-end DBR mirror. The converted light is reflected by the input-end DBR mirror and is emitted only from the output end of the device.[75,76]

The advantage of tunable BSL-based WC is that the contrast of the input signal can be improved through bistable behavior. In the 2.5 Gbit/s experiment using BSL, the measured penalty due to the WC is as low as 0.5 dB. An enhancement of sensitivity can even be obtained in some conditions. One limiting factor is that it is quite difficult to increase the operation speed beyond 2.5 Gbit/s. Another question is to know if it is possible to realize up-frequency conversion ($\omega_{out} > \omega_{in}$) in such

TABLE 10.1. Summarized characteristics of three types of wavelength converters.

Structures	Amplifiers	Tunable lasers	Bistable lasers
Mechanisms	Carrier depletion	Carrier depletion	absorption saturation
Operation speed	≤ 20 Gbit/s	≤ 20 Gbit/s	≤ 2.5 Gbit/s
Contrast ratio	degradation*	degradation	improvement possible
λ-conversion range	total bandwidth	total bandwidth	up conversion possible?
Complexity	complex	simple	simple

*Improvement possible with Mach–Zehnder configuration.

a laser. No experimental result has been published on this subject except in the case of side-injection BSLs.[77]

10.4.4. Summary on Wavelength Converters

Present performance of three types of wavelength converters is summarized in Table 10.1 in terms of operation speed, degradation of contrast ratio, wavelength conversion range, and complexity. In this table, wavelength conversion based on FWM and injection locking is not included due to the very low wavelength conversion range or low conversion efficiency.

In fact, the performance of a WC should include also polarization sensitivity, frequency chirp in the case of amplitude modulation, etc., which are not specified in this table. In view of present performance and rapid progress in this field, it is very difficult to say in general which type of WC is the most adapted for wavelength conversion applications.

10.4.5. Wavelength Tunable Filters

Another important function in WD switching systems is the wavelength selection by using a wavelength tunable filter. This can be realized with mechanically tuned Fabry–Perot etalons or gratings, etc. However, the tuning speed is very low in such a system. Tunable SLs are very good candidates to realize such a function.

For a wavelength selection application, tunable SLs are biased below threshold. Two types of tunable filters can be distinguished depending on whether the transmitted or reflected signal is used: Reflection type and transmission type. The physical origin of filtering is that the transmission and reflection coefficients depend strongly on the frequency of input signal due to facet reflection or internal grating in the laser structure.

The transmission coefficient as a function of optical frequency is shown in Fig. 10.24 for two different input powers for a DBR laser.[79] The center frequency can be tuned by changing the injection current of the Bragg section[79] or using different transmission peaks in a Fabry–Perot amplifier.[80] The bandwidth decreases when the bias current approaches the threshold current. It could be as narrow as a few hundred megahertz. For higher input powers or for bias current in the vicinity of the

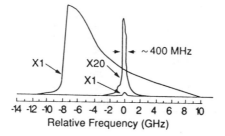

FIGURE 10.24. Transmission coefficient as a function of detuning frequency of a DBR laser for two input powers (after Ref. 79).

threshold, bistable behavior in the transmission coefficient occurs as described in Sec. 10.3.1.

Another important function of such a wavelength filter is the simultaneous detection of the input FSK signal.[80,81] In fact, the amplification coefficient depends on the instantaneous frequency of the input signal, resulting in a carrier density change of the active layer. Thus, the voltage across the diode changes also with the input frequency through the change of the quasi-Fermi levels. Simultaneous filtering and detection of FSK signals have been demonstrated for bit rates of 155 Mbit/s[79] and 1.5 Gbit/s.[80] The maximum bit rate is limited by the carrier lifetime to about 2 Gbit/s.[80]

A two-channel experiment has been realized by Hui *et al.* using a DFB laser amplifier as wavelength filter/photodetector.[81] It has been found that a minimum channel spacing of about 15 GHz is required to maintain the bit error rate below 10^{-9}.

10.4.6. Applications in Wavelength-Division Switching Systems

Many experiments involving wavelength selection or distribution have been realized in laboratories.[82-84] As an example, a 4×4 switching node is schematically shown in Fig. 10.25.[82] In this schema, two input fibers each carrying two signals at the optical frequencies f_1 and f_2 are connected to the switching module, which contains a space stage and a frequency stage. The frequency stage consists of Mach–Zehnder filters and wavelength converters. The input signals, processed by the space stage, are selected through Mach–Zehnder filters and routed by WCs to fixed output frequencies f_1 and f_2, depending on the path used. The wavelength converters used are based on FWM in DFB lasers.[67] The frequency spacing $f_2 - f_1$ was 12.5 GHz. The switching node is transparent to bit rate.

10.5. CONCLUSION

The main and the most recent progress in the study of functional SLs and SLAs has been reviewed. In summary, SLAs and SLs can be used as gating elements in SD

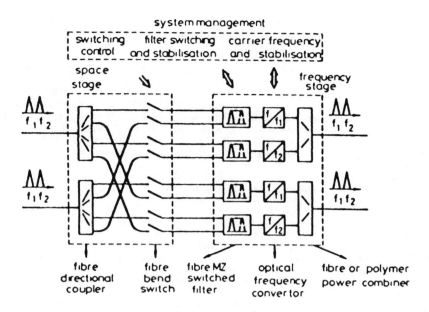

FIGURE 10.25. Architecture of a 4×4 switching node in the wavelength division (after Ref. 82).

switching systems, as optical memories in TD switching systems and as optical WCs or tunable filters in WD switching systems. These applications explore a large number of phenomena, which are undesirable for transmission but very useful for switching. Some switching functions of SLs and SLAs have already been used in laboratory demonstration systems.

The development in this field is very rapid in the last decade. Some demonstration systems have already worked in laboratories. In the future, the research trends will be towards the integration of switching components. Optical switching finds its real applications when the number of channels to be switched becomes more reasonable.

REFERENCES

1. G. P. Agrawal and N. K. Dutta, (Von Nostrand Reinhold, New York, 1986).
2. G. P. Agrawal, Chap. 8 (this book).
3. J. E. Midwinter, IEE Proc. J, **139**, pp. 1–12, 1992.
4. C. Burke, M. Fujiwara, M. Yamaguchi, H. Nishimoto, and H. Honmou, Proceedings of Topical Meeting on Photonic Switching, Salt Lake City, Utah, March 6–8, 1991.
5. J. C. Simon, J. Opt. Comm. **4**, 52 (1983).
6. J. C. Simon, J. Lightwave Technol. **5**, 1286 (1987).
7. N. A. Olsson, Proc. IEEE **80**, 375 (1992).
8. T. Saitoh and T. Mukai, IEEE J. Quantum Electron. **23**, 1010 (1987).

9. P. Doussiere, B. Mersali, A. Accard, P. Garabedian, G. Gelly, J. L. Lafragette, F. Leblond, G. Michaud, M. Monnot, and B. Fernier, Topical Meeting on Optical Amplifiers and their Applications, WE2, Snowmass Village, Colorado, 1991.
10. L. Gillner, IEE Proc. J. **139**, 331 (1992).
11. B. P. Cavanagh, I. W. Marshall, G. Sherlock, and H. Wickes, Electron. Lett. **27**, 266 (1991).
12. R. Fortenberry, A. J. Lowery, W. L. Ha, and R. S. Tucker, Electron. Lett. **27**, 1305 (1991).
13. J. D. Evankow, Jr. and R. A. Thompson, IEEE J. Sel. Areas Commun. **6**, 1087 (1988).
14. M. Eiselt, G. Grobkopf, R. Ludwig, W. Pieper, and H. G. Weber, Electron. Lett. **28**, 1438 (1992).
15. D. A. Mace, M. J. Adams, J. Singh, M. A. Fisher, and I. D. Henning, Electron. Lett. **27**, 188 (1991).
16. R. F. Kalman, L. G. Kazovsky, and J. W. Goodman, IEEE Photon. Technol. Lett. **4**, 1048 (1992).
17. R. Kishimoto, K. Yoshino, and M. Ikeda, IEEE J. Sel. Areas Commun. **6**, 1079 (1988).
18. H. Kawaguchi, Opt. Quantum Electron. **19**, S1 (1987).
19. H. Kawaguchi, IEE Proc. J, **140**, 3 (1993).
20. H. Kawaguchi, IEEE J. Quantum Electron. **23**, 1429 (1987).
21. M. J. Adams and R. Wyatt, IEE Proc. **134**, Pt.J, 35 (1987).
22. G. J. Lasher, Solid-State Electron. **7**, 707 (1964).
23. M. Ueno and R. Lang, J. Appl. Phys. **58**, 1689 (1985).
24. C. Harder, K. Y. Lau, and A. Yariv, IEEE J. Quantum Electron. **QE-18**, 1351 (1982).
25. H. F. Liu, Y. Hashimoto, and T. Kamiya, IEEE J. Quantum Electron. **QE-24**, 43 (1988).
26. H. Tsuda, T. Kurokawa, H. Uenohara, and H. Iwamura, IEEE Photon. Technol. Lett. **4**, 760 (1992).
27. U. Ohlander, P. Blixt, and O. Sahln, paper WeB16-1, Proceedings of European Conference on Optical Communication, Gothenburg, Sweden, 1989.
28. P. Landais, G.-H. Duan, E. Gaumont-Goarin, P. Garabedian, and J. Jacquet, Appl. Phys. Lett. **63**, 2615 (1993).
29. P. Landais, G.-H. Duan, C. Chabran, and J. Jacquet, Electron. Lett. **29**, 1363 (1993).
30. G.-H. Duan, P. Landais, and J. Jacquet, IEEE J. Quantum Electron., **30**, 2507 (1994).
31. T. Odagawa, T. Machida, T. Sanada, K. Wakao, and S. Yamakoshi, paper CFL4, Proceedings of Conference on Lasers and Electro-Optics, Anaheim, CA, 1990.
32. P. E. Barnsley, H. J. Wickes, G. E. Wickens, and D. M. Spirit, IEEE Photon. Technol. Lett. **3**, 942 (1991).
33. M. Mohrle, U. Feiste, J. Horer, and B. Sartorius, IEEE Photon. Technol. Lett. **4**, 976 (1992).
34. M. Jinno and T. Matsumoto, Electron. Lett. **24**, 1426 (1988).
35. M. Jinno and T. Matsumoto, Electron. Lett. **25**, 1332 (1989).
36. M. Jinno and T. Matsumoto, J. Lightwave Technol. **10**, 448 (1992).
37. G. Farrel, P. Phelan, and J. Hegarty, IEEE Photon. Technol. Lett. **5**, 571 (1993).
38. G. Farrel, P. Phelan, and J. Hegarty, Electron. Lett. **28**, 738 (1992).
39. A. Ehrhardt, D. J. As, and U. Feiste, Proceedings of European Conference on Optical Communication, Montreux, Switzerland, 1993.
40. G.-H. Duan, IEEE J. Quantum Electron. (to be published).
41. S. Suzuki, T. Terakado, K. Komatsu, K. Nagashima, A. Suzuki, and M. Kondo, J. Lightwave Technol. **LT-4**, 894 (1986).
42. C. A. Brackett, IEEE J. Sel. Areas Commun. **8**, 948 (1990).
43. J. M. Gabriagues and J. B. Jacob, Proceedings of European Conference on Optical Communication, Paris, 1991.
44. I. Valiente, J. C. Simon, and M. Le Ligne, Electron. Lett. **29**, 502 (1993).
45. B. Glance, J. M. Wiesenfeld, U. Koren, A. H. Gnauck, H. M. Presby, and A. Jourdan, Electron. Lett. **28**, 1714 (1992).
46. T. Durhuus, C. Joergensen, B. Mikkelesen, K. E. Stubjaer, B. Fernier, P. Garabedian, and E. Derouin, Proceedings of Conference on Optical Fiber Communication, San Jose, CA, 1993.
47. J. S. Perrino, J. M. Wiesenfield, and B. Glance, Electron. Lett. **30**, 256 (1994).
48. B. Mikkelesen, M. Vaa, R. J. S. Pederen, T. Durhuus, C. Joergensen, C. Braagaard, K. E. Storkfelt, K. E. Stubjaer, P. Doussire, C. Garabedian, C. Graver, E. Derouin, R. Fillion, and M. Klenk, Post-deadline paper, Proceedings of 19th European Conference on Optical Communication, Montreux, Switzerland, 1993.
49. K. Stubkjaer, B. Mikkelsen, T. Durhuus, C. G. Joergensen, C. Joergensen, T. N. Nielsen, B. Fernier, P. Doussiere, D. Leclerc, and J. Benoit, Proceedings of 19th European Conference on Optical Communication, Montreux, Switzerland, 1993.
50. T. Durhuus, C. Joergensen, B. Mikkelesen, R. J. S. Pederen, and K. E. Stubjaer, IEEE Photon. Technol. Lett. **6**, 53 (1994).

51. H. Nakajima and R. Frey, Appl. Phys. Lett. **47**, 769 (1986).
52. G. P. Agrawal, J. Opt. Soc. Am. B **5**, 147 (1988).
53. G. Großkopf, L. Kuller, R. Ludwig, R. Schnabel, and H. G. Weber, Opt. Quantum Electron. **21**, S59 (1989).
54. K. Kikuchi, M. Kakui, C-E Zah, and T-P Lee, IEEE J. Quantum Electron. **28**, 151 (1992).
55. J. H. Zhou, N. Park, J. W. Dawson, and K. J. Vahala, Appl. Phys. Lett. **62**, 2301 (1993).
56. R. Schnabel, W. Pieper, R. Ludwig, and H. G. Weber, Electron. Lett. **29**, 821 (1993).
57. J. H. Zhou, N. Park, J. W. Dawson, K. J. Vahala, M. A. Newkirk, and B. I. Miller, IEEE Photon. Technol. Lett. **6**, 50 (1994).
58. K. Vahala, J. Zhou, N. Park, M. A. Newkirk, and B. I. Miller, Post-deadline paper CPD4-1, Proceedings of the Conference on Lasers and Electro-Optics (CLEO), Anaheim, CA, 1994.
59. C. B. Su, J. Eom, C. H. Lange, C. B. Kim, R. B. Lauer, W. C. Rideout, and J. S. LaCourse, IEEE J. Quantum Electron. **28**, 118 (1992).
60. T. Durhuus, R. J. S. Pedersen, B. Mikkelsen, K. E. Stubjaer, M. Oberg, and S. Nilsson, IEEE Photon. Technol. Lett. **5**, 86 (1993).
61. P. Pottier, M. J. Chawki, R. Auffret, G. Claveau, and A. Tromeur, Electron. Lett. 2183 (1991).
62. B. Mikkelsen, T. Durhuus, R. J. Pedersen, F. Ebskamp, M. Oberg, S. Nilsson, and K. E. Stubkjaer, Proceedings of the Conference on Optical Fiber Communication, San Jose, CA, 1993.
63. K. Inoue and N. Takato, Electron. Lett. **25**, 1360 (1989).
64. P. A. Morton, T. Tanbun-Ek, R. A. Logan, P. F. Sciortino Jun, A. M. Sergent, and K. W. Wecht, Electron. Lett. **29**, 1429 (1993).
65. P. Gallion, G. Debarge, and C. Chabran, Opt. Lett. **11**, 67 (1986).
66. I. Petitbon, P. Gallion, G. Debarge, and C. Chabran, IEEE J. Quantum Electron. **24**, 148 (1988).
67. R.-P. Braun and B. Stebel, XVI International Conference on Quantum Electronics, Tokyo, 1988.
68. M. Schilling, W. Idler, D. Baums, G. Laube, K. Wunstel, and O. Hildebrand, IEEE Photon. Technol. Lett. **3**, 1054 (1991).
69. S. Gurib, A. Jourdan, J.-G. Provost, and J. Jacquet, Proceedings of the Conference on Lasers and Electro-Optics (CLEO), paper CTuT1, Anaheim, CA, 1994.
70. H. Kawaguchi, K. Oe, H. Yasaka, K. Magari, M. Fukuda, and Y. Itaya, Electron. Lett. **23**, 1088 (1987).
71. H. Rokugawa, N. Fujimoto, T. Nakagami and H. Nobuhara, Electron. Lett. **27**, 393 (1991).
72. H. Kawaguchi, K. Magari, H. Yasaka, M. Fukuda, and K. Oe, IEEE J. Quantum Electron. **QE-24**, 2153 (1988).
73. K. Kondo, M. Kuno, S. Yamakoshi, and K. Wakao, IEEE J. Quantum Electron. **28**, 1343 (1992).
74. H. Shoji, Y. Arakawa, and Y. Fujii, IEEE J. Lightwave Technol. **LT-8**, 1630 (1990).
75. K. Takahata, K. Kasaya, and H. Yasaka, Electron. Lett. **28**, 2078 (1992).
76. K. Kasaya, K. Takahata, and H. Yasaka, IEEE Photon. Technol. Lett. **5**, 321 (1993).
77. H. Tsuda, K. Nonaka, K. Hirabayashi, H. Uenohara, H. Iwamura, and T. Kurokawa, Proceedings of Conference on Optical Fiber Communication, San Jose, CA, 1993.
78. K. Nonaka, H. Tsuda, H. Uenohara, H. Iwamura, and T. Kurokawa, IEEE Photon. Technol. Lett. **5**, 139 (1993).
79. T. L. Koch, F. S. Choa, F. Heismann, and U. Koren, Electron. Lett. **25**, 890 (1989).
80. M. J. Chawki, L. Le Guiner, D. Dumay, and J. C. Keromnes, Electron. Lett. **26**, 1146 (1990).
81. R. Hui, N. Caponio, P. Gambini, M. Puleo, and E. Vezzoni, Electron. Lett. **27**, 2016 (1991).
82. R.-P. Braun, C. Caspar, H.-M. Foisel, K. Heimes, B. Strebel, N. Keil, and H. H. Yao, Electron. Lett. **29**, 912 (1993).
83. M. J. Chawki, R. Auffret, E. Le Coquil, L. Berthou, J. Le Rouzic, and L. Demeure, Electron. Lett. **28**, 147 (1992).
84. D. Chiaroni, P. Gavignet-Morin, P. A. Perrier, S. Ruggeri, S. Gauchard, D. de Bouard, J. C. Jacquinot, C. Chauzat, J. Jacquet, P. Doussire, M. Monnot, E. Grard, D. Leclerc, M. Sotom, J. M. Gabriagues, and J. Benoit, Proceedings of 19th European Conference on Optical Communication, Montreux, Switzerland, 1993.

Index